Principles and Practice of Soil Science

The Soil as a Natural Resource

Fourth Edition

ROBERT E. WHITE

Blackwell
Publishing

BLACKWELL PUBLISHING
350 Main Street, Malden, MA 02148-5020, USA
9600 Garsington Road, Oxford OX4 2DQ, UK
550 Swanston Street, Carlton, Victoria 3053, Australia

First edition published 1979
Second edition published 1987
Third edition published 1997
Fourth edition published 2006

1 2006

Library of Congress Cataloging-in-Publication Data

White, R. E. (Robert Edwin), 1937–
    Principles and practice of soil science : the soil as a natural resource / Robert E. White. – 4th ed.
        p.   cm.
    Includes bibliographical references.
    ISBN-13: 978-0-632-06455-7 (pbk. : alk. paper)
    ISBN-10: 0-632-06455-2 (pbk. : alk. paper)
    1. Soil science.   I. Title.
S591.W49   2006
631.4–dc22

A catalogue record for this title is available from the British Library.

Set in 9.5/11.5pt Sabon
by Graphicraft Limited, Hong Kong
Printed and bound in Great Britain
by TJ International Ltd, Padstow, Cornwall

The publisher's policy is to use permanent paper from mills that operate a sustainable forestry policy,
and which has been manufactured from pulp processed using acid-free and elementary chlorine-free practices.
Furthermore, the publisher ensures that the text paper and cover board used have met acceptable environmental
accreditation standards.

For further information on
Blackwell Publishing, visit our website:
www.blackwellpublishing.com

Contents

Preface to
the Fourth Edition

Dr Samuel Johnson is reputed to have said 'what is written without effort is in general read without pleasure'. This edition of *Principles and Practice of Soil Science* has certainly taken much effort to complete, so I hope it will be enjoyed and provide valuable information to as wide an audience of interested readers as possible. The people I would expect to be interested in learning more about soils are not only soil scientists and others concerned with production systems, but also the various scientists and natural historians who are concerned about Earth's ecology in its broadest sense.

At the time of the third edition (1997) I wrote about the 'new generic concept' of ecologically sustainable development (ESD) that was being promoted by international agencies and appearing with increasing frequency in government policy documents. However, through the 1990s and into the early years of the 21st century, more has been *written* about ESD than has been achieved on the ground in implementation of the policy. I have expanded on the topic of 'sustainability' in Chapter 15, drawing particularly on examples in Australia where a relatively fragile landscape continues to be put under pressure from 'development'. The largest areas affected are rural areas, especially in the better watered coastal zone and the expanding irrigation regions, and areas of urban concentration (mainly along the coasts also). In this context, the quality and quantity of water have become key issues attracting much public and political attention. In recent years in Australia, these twin issues have become enmeshed

with the question of climate change – by how much is it changing and where, and what are the possible positive and negative effects – which is directly linked to the emission of greenhouse gases from natural and human-influenced systems. Underlying these issues is soil behaviour because virtually all the precipitation that falls on land interacts with soil in some way. Hence, knowledge of the spatial distribution of different soil types and the pathways of water, with their associated physical, chemical and biological processes, in these various soil types becomes a very important component of land and water management. We need to be aware that everyone lives in a catchment and that the quality of life in that catchment depends on individual and collective human activities in that catchment. I have expounded on this subject in my 2003 G. W. Leeper Memorial Lecture 'What has soil got to do with water?', which is available on the Australian Society of Soil Science Inc. website (<wwwc>http://www.asssi.asn.au/asssi/flash/). Important tools for use in unravelling the complexity of water, energy and nutrient fluxes in catchments are models of the biophysical processes, incorporating a digital elevation model (DEM) and digital soil map, dynamically coupled with a Geographic Information System (GIS). I refer to these tools in Chapters 14 and 15.

Apart from updating and revising each chapter and adding colour photographs, I have provided sets of illustrative problems and questions at the end of each chapter, based on my experience in teaching undergraduate classes on soil resources and their management at The University of Melbourne. I have benefited from feedback from students and also from advice given by friends and colleagues, notably Dr Nick Uren and Dr Robert Edis. To all those who contributed I am most grateful, but the ultimate responsibility for any errors and omissions rests with me. I am also grateful to Debbie Seymour, Rosie Hayden and Hannah Berry at Blackwell Publishing who have been very tolerant and supportive while I was preparing this edition.

Robert E. White
Melbourne
13 December 2004

Units of Measurement and Abbreviations used in this Book

SI units

Basic unit	Abbreviation
metre	m
hectare	ha
gram	g
mole	mol
second	s
temperature	K
ampere	A
becquerel	Bq

Derived units

Unit	Abbreviation	Value
Celsius	°C	K-273
newton	N	kg.m/s
joule	J	N.m
pascal	P	N/m^2
volt	V	J/A/s
siemen	S	A/V
coulomb	C	A.s
litre	L	$m^3/1000$
tonne	t	kg.1000
bar	bar	$Pa.10^5$
Faraday's constant	F	96500 J/mol/V
Universal gas constant	R	8.3143 J/K/mol

Non-SI units used in soil science

Physical term	Unit	Abbreviation	Value
length	Angstrom	Å	10^{-10} m
concentration	moles/litre	M	mol/L
cation exchange capacity	meq/100 g	CEC	cmol charge (+)/kg
electrical conductivity	millimho/ cm	EC	dS/m

Prefixes and suffixes to units

Prefix/suffix	Abbreviation	Value
tera-	T	10^{12}
giga-	G	10^9
mega-	M	10^6
kilo-	k	10^3
deca-	da	10^1
deci-	d	10^{-1}
centi-	c	10^{-2}
milli-	m	10^{-3}
micro-	μ	10^{-6}
nano-	n	10^{-9}
pico-	p	10^{-12}

Miscellaneous symbols

( )	denotes 'activity'
[ ]	denotes 'concentration'
$\cong$	approximately equal to
$\sim$	of the order of
$<$	less than
$>$	greater than
$\leq$	less than or equal to
$\geq$	greater than or equal to
log	$\log_{10}$
ln	$\log_e$
exp	exponential of

Abbreviations

ACIAR	Australian Centre for International Agricultural Research
ACLEP	Australian Collaborative Land Evaluation Program
ADAS	Agricultural Development and Advisory Service
AEC	anion exchange capacity
AM	arbuscular mycorrhizas
AMO	ammonia mono-oxygenase
ANZECC	Australian and New Zealand Environment and Conservation Council
AR	activity ratio
ASC	Australian Soil Classification
ASRIS	Australian Soil Resources Information System
ASSSI	Australian Society of Soil Science Inc.
ATC	4-amino-1,2,4-triazole
ATP	adenosine triphosphate
AWC	available water capacity
BET	Brunauer, Emmet and Teller
BIO	microbial biomass
BMP	best management practice
BP	before present
CEC	cation exchange capacity
CFCs	chlorofluorocarbons
CPMAS	cross-polarization, magic angle spinning
CREAMS	Chemicals, Runoff and Erosion from Agricultural Management Systems
CRF	controlled-release fertilizer
CSIRO	Commonwealth Scientific and Industrial Research Organization
DAP	diammonium phosphate
DCD	dicyandiamide
DCP	dicalcium phosphate
DCPD	dicalcium phosphate dihydrate
DDL	diffuse double layer
DDT	dichlorodiphenyltrichloroethane
DEM	digital elevation model
DL	diffuse layer
DM	dry matter
DNA	desoxyribose nucleic acid
DOC	dissolved organic carbon
DPM	decomposable plant material
DTPA	diethylene triamine pentaacetic acid
E	evaporation
EC	electrical conductivity
ECEC	effective cation exchange capacity
EDDHA	ethylenediamine di (O-hydroxyphenylacetic acid)
EDTA	ethylenediamine tetraacetic acid
EMR	electromagnetic radiation
ENV	effective neutralizing value
EOC	extracted organic C
ESD	ecologically sustainable development
ESP	exchangeable sodium percentage
E_t	evapotranspiration
EU	European Union
FA	fulvic acid
FAO	Food and Agriculture Organization
FC	field capacity
FESLM	Framework for Sustainable Land Management
FTIR	Fourier Transform Infrared
FYM	farmyard manure
GIS	Geographic Information System
GLC	gas-liquid chromatography
GPS	Global Positioning System
GR	gypsum required
HA	humic acid
HAp	hydroxyapatite
HARM	hull acid rain model
HUM	humified organic matter
HYV	high-yielding variety
IBDU	isobutylidene urea
IOM	inert organic matter
IPCC	Intergovernmental Panel on Climate Change
IPM	integrated pest management
IR	infiltration rate

IS	inner sphere	Q/I	quantity/intensity
IUSS	International Union of Soil Sciences	RAW	readily available water
KE	kinetic energy	RH	relative humidity
LAI	leaf area index	RNA	ribose nucleic acid
LF	leaching fraction	RPM	resistant plant material
LR	leaching requirement	RPR	reactive phosphate rock
LRA	land resource assessment	RUSLE	Revised Universal Soil Loss Equation
LSI	Langelier saturation index		
MAFF	Ministry of Agriculture, Fisheries and Food	RWEQ	Revised Wind Erosion Equation
		SAR	sodium adsorption ratio
MAH	monocyclic aromatic hydrocarbons	SCU	sulphur-coated urea
MAP	monoammonium phosphate	SGS	Sustainable Grazing Systems
MCP	monocalcium phosphate	SI	Système International
MDB	Murray-Darling Basin	SIR	substrate-induced respiration
meq	milli-equivalent	SLM	sustainable land management
MPN	most probable number	SOM	soil organic matter
MWD	maximum potential soil water deficit	SOTER	World Soils and Terrain Database
		sp, spp.	species, singular and plural
NASIS	National Soil Information System	SRF	slow-release fertilizer
NCPISA	National Collaborative Project on Indicators for Sustainable Agriculture	SSP	single superphosphate
		ST	Soil Taxonomy
		SUNDIAL	Simulation of Nitrogen Dynamics in Arable Land
NDS	non-linear dynamic systems		
NHMRC	National Health and Medical Research Council	SWD	soil water deficit
		TCP	tricalcium phosphate
NMR	nuclear magnetic resonance	TDR	time domain reflectometer/ reflectometry
NRCS	Natural Resources Conservation Service		
		TDS	total dissolved salts
NSESD	National Strategy for Ecologically Sustainable Development	TEC	threshold electrolyte concentration
		TSP	triple superphosphate
NV	neutralizing value	UF	urea formaldehyde
OCP	octacalcium phosphate	UN	United Nations
o.d.	oven-dry	UNEP	United Nations Environment Program
OS	outer sphere		
p, pp.	page, singular and plural	USDA	United States Department of Agriculture
P	precipitation		
PAH	polycyclic aromatic hydrocarbons	USLE	Universal Soil Loss Equation
PAM	polyacrylamide	VD	vapour density
PAPR	partially acidulated phosphate rock	VP	vapour pressure
PAW	plant available water	WCED	World Commission on Environment and Development
PBC	phosphate buffering capacity		
PEG	polyethyleneglycol	WEPP	Water Erosion Prediction Project
POM	particulate organic matter	WEPS	Wind Erosion Prediction System
PR	phosphate rock	WEQ	Wind Erosion Equation
PSCU	polymer-coated sulphur-coated urea	WHO	World Health Organization
PVA	polyvinyl alcohol	WRB	World Reference Base for Soil Resources
PVAc	polyvinylacetate		
PVC	polyvinyl chloride	VFA	volatile fatty acid
PWP	permanent wilting point	XRD	X-ray diffraction
PZC	point of zero charge		

Part I

The Soil Habitat

'Soils are the surface mineral and organic formations, always more or less coloured by humus, which constantly manifest themselves as a result of the combined activity of the following agencies; living and dead organisms (plants and animals) parent material, climate and relief.'

V. V. Dokuchaev (1879), quoted by J. S. Joffe in *Pedology*

'The soil is teeming with life. It is a world of darkness, of caverns, tunnels and crevices, inhabited by a bizarre assortment of living creatures . . .'

J. A. Wallwork (1975) in *The Distribution and Diversity of Soil Fauna*

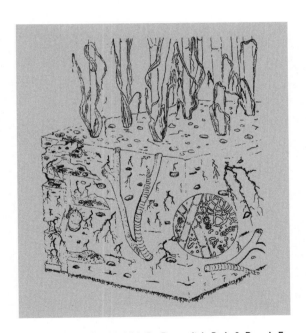

Redrawn from Reganold J. P., Papendick R. I. & Parr J. F. (1990) Sustainable agriculture. *Scientific American* **262**(6), 112–20.

Chapter 1

Introduction to the Soil

1.1 Soil in the making

With the exposure of rock to a new environment – following an outflow of lava, an uplift of sediments, recession of a water body, or the retreat of a glacier – a soil begins to form. Decomposition proceeds inexorably towards decreased free energy and increased entropy. The free energy of a closed system, such as a rock fragment, is that portion of its total energy that is available for work, other than work done in expanding its volume. Part of the energy released in a spontaneous reaction, such as rock weathering, appears as entropic energy, and the degree of disorder created in the system is measured by its entropy. For example, as the rock weathers, minerals of all kinds are converted into simpler molecules and ions, some of which are leached out by water or escape as gases.

Weathering is hastened by the appearance of primitive plants on rock surfaces. These plants – lichens, mosses and liverworts – can store radiant energy from the sun as chemical energy in the products of photosynthesis. Lichens, which are symbiotic associations of an alga and fungus, are able to 'fix' atmospheric nitrogen (N_2) and incorporate it into plant protein, and to extract elements from the weathering rock surface. On the death of each generation of these primitive plants, some of the rock elements and a variety of complex organic molecules are returned to the weathering surface where they nourish the succession of organisms gradually colonizing the embryonic soil.

A simple example is that of soil formation under the extensive deciduous forests of the cool humid areas of Europe, Asia and North America, on calcareous deposits exposed by the retreat of the Pleistocene ice cap (Table 1.1). The profile development is summarized in Fig. 1.1. The initial state is little more than a thin layer of weathered material stabilized by primitive plants. Within a century or so, as the organo-mineral material accumulates, more advanced species of sedge and grass appear, which are adapted to the harsh habitat. The developing soil is described as a Lithosol (Entisol or Rudosol*). Pioneering micro-organisms and animals feed on the dead plant remains and gradually increase in abundance and variety. The litter deposited on the surface is mixed into the soil by burrowing animals and insects, where its decomposition is hastened. The eventual appearance of larger plants – shrubs and trees – with their deeper roots, pushes the zone of rock weathering farther below the soil surface. After a few hundred more years, a Brown Forest Soil (Inceptisol or Tenosol) emerges. We shall return to the topic of soil formation, and the wide range of soils that occur in the landscape, in Chapters 5 and 9.

1.2 Concepts of soil

The soil is at the interface between the atmosphere and lithosphere (the mantle of rocks making up the Earth's crust). It also has an interface with bodies of fresh and salt water (collectively called the hydrosphere). The soil sustains the growth of many plants and animals, and so forms part of the biosphere.

* See Box 1.1 for a discussion of soil names.

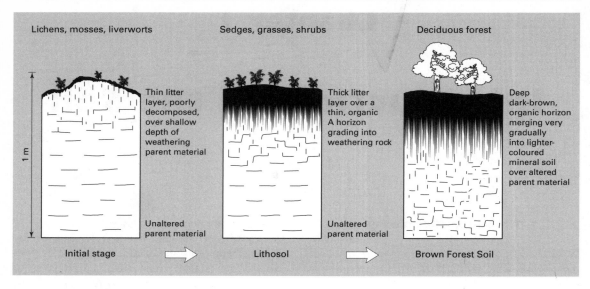

Fig. 1.1 Stages in soil formation on a calcareous parent material in a humid temperate climate.

Box 1.1 Soil variability, description and classification.

The landscape displays a remarkable range of soil types, resulting from an almost infinite variation in geology, climate, vegetation and other organisms, topography, and the time for which these factors have combined to influence soil formation (human activity is included among the effects of organisms). To bring order to such variety and to disseminate knowledge about soils, soil scientists have developed ways of classifying soils. Individual soils are described in terms of their properties, and possibly their mode of formation, and similar soils are grouped into classes that are given distinctive names. However, unlike the plant and animal kingdoms, there are no soil 'individuals' – the boundaries between different soils in the landscape are not sharp. Partly because of the difficulty in setting class limits, and because of the evolving nature of soil science, no universally accepted system of classifying (and naming) soils exists. For many years, Great Soil Group names based on the United States Department of Agriculture (USDA) Classification of Baldwin *et al.* (1938) (Section 5.3) held sway. But in the last 30 years, new classifications and a plethora of new soil names have evolved (Chapter 14). Some of these classifications (e.g. *Soil Taxonomy*, Soil Survey Staff, 1999) and the *World Reference Base for Soil Resources* (FAO, 1998) purport to be international. Others such as *The Australian Soil Classification* (Isbell, 2002) and the *Soil Classification for England and Wales* (Avery, 1980) are national in focus. This diversity of classifications creates problems for non-specialists in naming soils and understanding the meaning conveyed by a particular soil name. In this book, the more descriptive and (to many) more familiar Great Soil Group names will be used. Where possible, the approximate equivalent at the Order or Suborder level in Soil Taxonomy (ST) and the Australian Soil Classification (ASC) will be given in parentheses.

There is little merit in attempting to give a rigorous definition of soil because of the complexity of its make-up, and of the physical, chemical and biological forces that act on it. Nor is it necessary to do so, for soil means different things to different users. For example, to the geologist and engineer, the soil is little more than finely divided rock material. The hydrologist may see the soil as

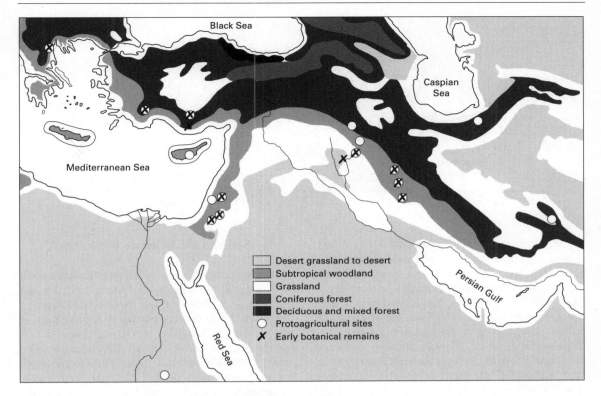

Fig. 1.2 Sites of primitive settlements in the Middle East (after Gates, 1976).

a storage reservoir affecting the water balance of a catchment, while the ecologist may be interested only in those soil properties that influence the growth and distribution of plants and animals. The farmer is naturally concerned about the many ways in which soil influences crop growth and the health of his livestock, although frequently his interest does not extend below the depth of soil disturbed by a plough (15–20 cm).

In view of this wide spectrum of potential user-interest, it is appropriate when introducing the topic of soil to readers, perhaps for the first time, to review briefly the evolution of our relationship with the soil and identify some of the past and present concepts of soil.

Soil as a medium for plant growth

Human's use of soil for food production began two or three thousand years after the close of the last Pleistocene ice age, which occurred about 11,000 years BP (before present). Neolithic people and their primitive agriculture spread outwards from settlements in the fertile crescent embracing the ancient lands of Mesopotamia, Canaan and southern Turkey (Fig. 1.2) and reached as far as China and the Americas within a few thousand years. In China, for example, the earliest records of soil survey (4000 years BP) show how soil fertility was used as a basis for levying taxes on landholders. To study the soil was a practical exercise of everyday life, and the knowledge of soil husbandry that had been acquired by Roman times was passed on by peasants and landlords, with little innovation, until the early 18th century.

From that time onwards, however, the rise in demand for agricultural products in Europe was dramatic. Conditions of comparative peace, and rising living standards as a result of the Industrial Revolution, further stimulated this demand throughout the 19th century. The period was also one of great discoveries in physics and chemistry,

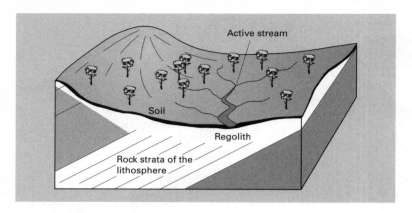

Fig. 1.3 Soil development in relation to the landscape and underlying regolith.

the implications of which sometimes burst with shattering effect on the conservative world of agriculture. In 1840, von Liebig established that plants absorbed nutrients as inorganic compounds from the soil, although he insisted that plants obtained their nitrogen (N) from the atmosphere: Lawes and Gilbert at Rothamsted subsequently demonstrated that plants (except legumes) absorbed inorganic N from the soil. In the 1850s, Way discovered the process of cation exchange in soil. During the years from 1860 to 1890, eminent bacteriologists including Pasteur, Warington and Winogradsky elucidated the role of microorganisms in the decomposition of plant residues and the conversion of ammonia to nitrate.

Over the same period, botanists such as von Sachs and Knop, by careful experiments in water culture and analysis of plant ash, identified the major elements that were essential for healthy plant growth. Agricultural chemists drew up balance sheets of the quantities of these elements taken up by crops and, by inference, the quantities that should be returned to the soil in fertilizers or animal manure to sustain growth. This approach, whereby the soil was regarded as a relatively inert medium providing water, mineral* ions and physical support for plants, has been called the 'nutrient bin' concept.

---

* The term 'mineral' is used in two contexts: first, as an adjective referring to the inorganic constituents of the soil (ions, salts and particulate matter); second, as a noun referring to specific inorganic compounds found in rocks and soil, such as quartz and feldspars (Chapter 2).

Soil and the influence of geology

The pioneering chemists who investigated a soil's ability to supply nutrients to plants tended to see the soil as a chemical and biochemical reaction medium. They little appreciated soil as part of the landscape, moulded by natural forces acting on the land surface. In the late 19th century, great contributions were made to our knowledge of soil by geologists who defined the mantle of loose, weathered material on the Earth's surface as the regolith, of which only the upper 50–150 cm, superficially enriched with organic matter, could be called soil (Fig. 1.3). Below the soil was the subsoil that was largely devoid of organic matter. However, the mineral matter of both soil and subsoil was recognized as being derived from the weathering of underlying rocks, which led to an interest in the influence of rock type on the soils formed. As the science of geology developed, the history of the Earth's rocks was subdivided into a time scale consisting of eras, periods and epochs, going back some 550 million years BP. Periods within the eras are usually associated with prominent sequences of sedimentary rocks that were deposited in the region now known as Europe. But examples of these rocks are found elsewhere, so the European time divisions have gradually been accepted worldwide (although the European divisions are not necessarily as clear-cut in all cases outside Europe). Studies of the relationship between soil and the underlying geology led to the practice of classifying soils loosely in geological terms, such as granitic (from granite), marly (derived from a mixture of limestone and clay),

Table 1.1 The geological time scale.

Era	Period	Epoch	Start time (million years BP)
Cainozoic	Quaternary	Recent	0.011
		Pleistocene	2
	Tertiary	Pliocene	5
		Miocene	23
		Oligocene	36
		Eocene	53
		Palaeocene	65
Mesozoic		Cretaceous	145
		Jurassic	205
		Triassic	250
Palaeozoic		Permian	290
		Carboniferous	360
		Devonian	405
		Silurian	436
		Ordovician	510
		Cambrian	550
Pre-Cambrian			4600

loessial (derived from wind-blown silt-size particles), glacial (from glacial deposits) and alluvial (from river deposits).

A simplified version of the geological time scale from the pre-Cambrian period to the present is shown in Table 1.1.

The influence of Russian soil science

Russia is a vast country covering many climatic zones in which, at the end of the 19th century, crop production was limited not so much by soil fertility, but by primitive methods of agriculture. Early Russian soil science was therefore concerned not with soil fertility, but with observing soils in the field and studying relationships between soil properties and the environment in which the soil had formed. From 1870 onwards, Dokuchaev and his school emphasized the distinctive features of a soil that developed gradually and distinguished it from the undifferentiated weathering rock or parent material below. This was the beginning of the science of pedology*.

---

* From the Greek word for ground or earth.

Following the Russian lead, scientists in other countries began to appreciate that factors such as climate, parent material, vegetation, topography and time interacted in many ways to produce an almost infinite variety of soil types. For any particular combination of these soil-forming factors (Chapter 5), a unique physicochemical and biological environment was established that led to the development of a distinctive soil body – the process of pedogenesis. A set of new terms was developed to describe soil features, such as:
• Soil profile – constituting a vertical face exposed by excavating the soil from the surface to the parent material;
• soil horizons – layers in the profile distinguished by their colour, hardness, texture, the occurrence of included structures, and other visible or tangible properties. The upper layer, from which materials are generally washed downwards, is described as eluvial; lower layers in which these materials accumulate are called illuvial.

In 1932, an international meeting of soil scientists adopted the notation of A and B for the eluvial and illuvial horizons, respectively, and C horizon for the parent material. The A and B horizons comprise the solum. Unweathered rock below the parent material is called bedrock R. Organic litter on the surface, not incorporated in the soil, is designated as an L layer. A typical Alfisol (ST) or Chromosol (ASC) soil profile showing a well-developed A, B and C horizon is shown in Fig. 1.4.

Soil genesis is now known to be much more complex than this early work suggested. For example, many soils are polygenetic in origin; that is, they have undergone successive phases of development due to changes in climate and other environmental factors over time. In other cases, two or more layers of different parent material are found in one soil profile. Nevertheless, the Russian approach was a considerable advance on traditional thinking, and recognition of the relationship between a soil and its environment encouraged soil scientists to survey and map the distribution of soils. The wide range of soil morphology that was revealed in turn stimulated studies of pedogenesis, an understanding of which, it was believed, would enable the copious field data on soils to be collated more systematically. Thus, Russian soil science provided the inspiration for many of the early soil classifications.

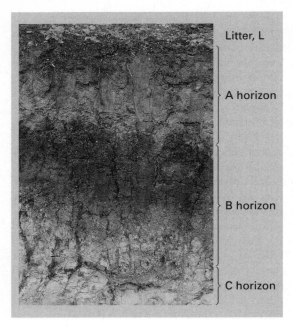

Litter, L

A horizon

B horizon

C horizon

Fig. 1.4 Profile of an Alfisol (ST) or Chromosol (ASC) showing well-developed A, B and C horizons (see Plate 1.4).

A contemporary view of soil

Between the two World Wars of the 20th century, the philosophy of the soil as a 'nutrient bin' was prevalent, particularly in the western world. More and more land was brought into cultivation, much of which was marginal for crop production because of limitations of climate, soil and topography. With the balance between crop success and failure made even more precarious than in favourable areas, the age-old problems of wind and water erosion, encroachment by weeds, and the accumulation of salts in irrigated lands became more serious. Since 1945, demand for food, fibre and forest products from an escalating world population (now > 6 billion) has led to increased use of fertilizers to improve yields, and pesticides to control pests and diseases (Chapter 12). Such practices have resulted in some accumulation of undesirable pesticide residues in soil, and in increased losses of soluble constituents such as nitrate and phosphates to surface waters and groundwater. There has also been widespread dispersal of the very stable pesticides (e.g. organo-chlorines) in the biosphere, and their accumulation to concentrations potentially toxic to some species of birds and fish.

Box 1.2 Soil as a natural body.

A soil is clearly distinguished from inert rock material by:
• The presence of plant and animal life;
• a structural organization that reflects the action of pedogenic processes;
• a capacity to respond to environmental change that might alter the balance between gains and losses in the profile, and predetermine the formation of a different soil in equilibrium with a new set of environmental conditions.
The last point indicates that soil has no fixed inheritance, because it depends on the conditions prevailing during its formation. Nor is it possible to unambiguously define the boundaries of the soil body. The soil atmosphere is continuous with air above the ground, many soil organisms live as well on the surface as within the soil, the litter layer usually merges gradually with decomposed organic matter in the soil, and likewise the boundary between soil and parent material is difficult to demarcate. We therefore speak of the soil as *a three-dimensional body that is continuously variable in time and space.*

More recently, however, scientists, producers and planners have acknowledged the need to compromise between maximizing crop production and conserving a valuable natural resource. Emphasis is now placed on maintaining the soil's natural condition by minimizing the disturbance when crops are grown, matching fertilizer additions more closely to crop demand in order to reduce losses, using legumes to fix N_2 from the air, and returning plant residues and waste materials to the soil to supply some of a crop's nutrient requirements. In short, more emphasis is being placed on the soil as a natural body (Box 1.2) and on the concept of sustainable land management (Chapter 15).

1.3 Components of the soil

We have seen that a combination of physical, chemical and biotic forces acts on organic materials and weathered rock to produce a soil with a porous fabric that retains water and gases. The mineral matter derived from weathered rock

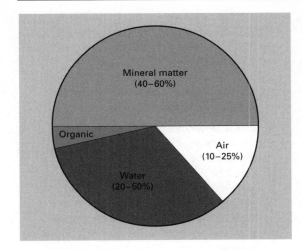

Fig. 1.5 Proportions of the main soil components by volume.

consists of particles of different size, ranging from clay (the smallest), to silt, sand, gravel, stones, and in some cases boulders (Section 2.1). The particle density ρ_p (rho p) varies according to the mineralogy (Section 2.3), but the average ρ_p is 2.65 Mg/m^3. Organic matter has a lower density of 1–1.3 Mg/m^3, depending on the extent of its decomposition. Water has a density of 1.0 Mg/m^3 at normal temperatures ($c.$ 20°C)*.

Soil water contains dissolved organic and inorganic solutes and is called the soil solution. While the soil air consists primarily of N_2 and oxygen (O_2), it usually contains higher concentrations of carbon dioxide (CO_2) than the atmosphere, and traces of other gases that are by-products of microbial metabolism. The relative proportions of the four major components – *mineral matter, organic matter, water and air* – may vary widely, but generally lie within the ranges indicated in Fig. 1.5. These components are discussed in more detail in the subsequent chapters of Part 1.

1.4 Summary

Soil forms at the interface between the atmosphere and the weathering products of the regolith. Physical and chemical weathering, erosion and redeposition, combined with the activities of

a succession of colonizing plants and animals, moulds a distinctive soil body from the milieu of rock minerals in the parent material. The process of soil formation, called pedogenesis, culminates in a remarkably variable differentiation of soil material into a series of horizons that constitute a soil profile. Soil horizons are distinguished by their visible and tangible properties such as colour, hardness, texture and structural organization. The intimate mixing of mineral and organic matter to form a porous fabric, permeated by water and air, creates a favourable habitat for a variety of plant and animal life. Soil is a fragile component of the environment. Its use for food and fibre production, and waste disposal, must be managed in a way that minimizes the off-site effects of these activities and preserves the soil for future generations. This is the basis of sustainable soil management.

References

Avery B. W. (1980) *Soil Classification for England and Wales*. Soil Survey Technical Monograph No. 14. Rothamsted Experimental Station, Harpenden.

Baldwin H., Kellogg C. W. & Thorpe J. (1938) Soil classification, in *Soils and Man*. United States Government Printing Office. Washington DC.

FAO (1998) *World Reference Base for Soil Resources*. World Resources Report No. 84. FAO, Rome.

Gates C. T. (1976) China in a world setting: agricultural response to climatic change. *Journal of the Australian Institute of Agricultural Science* **42**, 75–93.

Isbell R. F. (2002) *The Australian Soil Classification*, revised edn. Australian Soil and Land Survey Handbooks Series Volume 4. CSIRO Publishing, Melbourne.

Soil Survey Staff (1999) *Soil Taxonomy. A Basic Classification for Making and Interpreting Soil Surveys*, 2nd edn. United States Department of Agriculture Handbook No. 436. Natural Resources Conservation Service, Washington DC.

Further reading

Hillel D. (1991) *Out of the Earth: Civilization and the Life of the Soil*. The Free Press, New York.

Jenny H. (1980) *The Soil Resource – Origin and Behaviour*. Springer-Verlag, New York.

Marschner H. (1995) *Mineral Nutrition of Higher Plants*, 2nd edn. Academic Press, London.

McKenzie N. & Brown K. (2004) *Australian Soils and Landscapes: an Illustrated Compendium*. CSIRO Publishing, Melbourne.

---

* The density of water is 1.000 Mg/m^3 at 4°C and 0.998 Mg/m^3 at 20°C, which is rounded to 1.0.

Example questions and problems

1 The upper-most horizon of a soil is generally enriched with organic matter, in varying states of decomposition. Where does most of this organic matter come from?

2 (a) Give the notation for the main horizons recognized in a soil profile.
  (b) What do the terms 'eluvial' and 'illuvial' mean in the context of soil profile description?

3 What are the main external factors that cause soil variation in the landscape?

4 Soil samples were taken from the 0–10 cm depth along two transects at right angles in a pasture grazed by cattle. The samples were spaced at 5 m intervals and analysed for organic carbon (C) content. The results, in percent organic C, were as follows.

Transect 1	2.5	1.6	1.1	1.7	1.5	2.1	2.7	2.2	3.0	1.3
Transect 2	1.6	1.9	1.5	2.9	2.5	2.2	1.5	1.0	1.4	2.7

  (a) Calculate the mean organic C content for each transect, and the coefficient of variation (CV) for each set of values

$$\left(CV = \frac{\text{standard deviation}}{\text{mean}} \times 100 \right).$$

  (b) Can you suggest a reason for the spatial variation in organic C content?

5 Suppose that the volume fraction of mineral matter in a field soil is 0.5, and the organic matter fraction is 0.025.
  (a) Calculate the remaining volume fraction and say what this volume fraction is called.
  (b) (i) Calculate the weight in tonnes (t) of 1 cubic metre (1 m^3) of completely dry soil, given that the particle densities (ρ_p) of the mineral and organic fractions are 2.65 and 1.2 Mg/m^3, respectively, and (ii) calculate the weight of 5 cm^3 of dry soil (roughly 1 teaspoon).
  (c) If the depth of ploughing in this soil is 15 cm, what is the weight of dry soil (Mg) per hectare to 15 cm depth?
  (d) Suppose the 50% mineral matter (by volume) of a field soil included 10% iron oxide ($\rho_p = 5.55$ Mg/m^3) and organic matter was negligible. (i) What would be the weight of 1 m^3 of soil, and (ii) the weight of 1 ha of dry soil to 15 cm depth?

Chapter 2

The Mineral Component of the Soil

2.1 The size range

Rock fragments and mineral particles in soil vary enormously in size from boulders and stones down to sand grains and very small particles that are beyond the resolving power of an optical microscope (< 0.2 µm in diameter). Particles smaller than *c*. 1 µm are classed as colloidal. Particles that do not settle quickly when mixed with water are said to form a colloidal solution or sol; if they settle within a few hours they form a suspension. Colloidal solutions are distinguished from true solutions (dispersions of ions and molecules) by the Tyndall effect. This occurs when the path of a beam of light passing through the solution can be seen from either side at right angles to the beam, indicating a scattering of the light rays.

An arbitrary division is made by size-grading soil into material:
• That passes through a sieve with 2 mm diameter holes – the *fine earth*, and
• that retained on the sieve (> 2 mm) – *the stones or gravel*, but smaller than
• fragments > 600 mm, which are called *boulders*. The separation by sieving is carried out on air-dry soil that has been gently ground by mortar and pestle, or crushed between wooden rollers, to break up the aggregates. Air-dry soil is soil allowed to dry in air at ambient temperatures (between 20 and 40°C).

Particle-size distribution of the fine earth

The distribution of particle sizes determines the soil texture, which may be assessed subjectively in the field or more rigorously by particle-size analysis in the laboratory.

Size classes

All soils show a continuous range of particle sizes, called a frequency distribution, which is obtained by plotting the number (or mass) of particles of a given size against their actual size. When the number or mass in each size class is summed sequentially we obtain a cumulative distribution of soil particle sizes, some examples of which are given in Fig. 2.1. In practice, it is convenient to

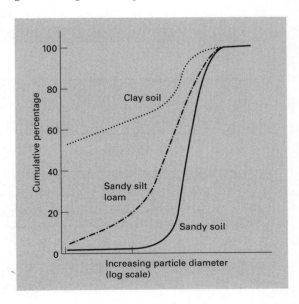

Fig. 2.1 Cumulative frequency distributions of soil particle sizes in a typical clay, sandy silt loam and sandy soil.

subdivide the continuous distribution into several class intervals that define the size limits of the *sand*, *silt* and *clay* fractions. The extent of this subdivision, and the class limits chosen, vary from country to country and even between institutions within countries. The major systems in use are those adopted by the Soil Survey Staff of the USDA, the British Standards Institution and the International Union of Soil Sciences (IUSS). These are illustrated in Fig. 2.2. All three systems set the upper limit for clay at 2 μm diameter, but differ in the upper limit chosen for silt and the way in which the sand fraction is subdivided.

Field texture

A soil surveyor assesses soil texture by moistening a sample with water until it glistens. It is then kneaded between fingers and thumb until the aggregates are broken down and the soil grains thoroughly wetted. The proportions of sand, silt and clay are estimated according to the following qualitative criteria:

• *Coarse sand* grains are large enough to grate against each other and can be detected individually by sight and feel;

• *fine sand* grains are much less obvious, but when they comprise more than about 10% of the sample they can be detected by biting the sample between the teeth;

• *silt* grains cannot be detected by feel, but their presence makes the soil feel smooth and silky and only slightly sticky;

• *clay* is characteristically sticky, although some dry clays, especially of the expanding type (Section 2.3), require much moistening and kneading before they develop their maximum stickiness.

High organic matter contents tend to reduce the stickiness of clay soils and to make sandy soils feel more silty. Finely divided calcium carbonate also gives a silt-like feeling to the soil.

Depending on the estimated proportions of sand, silt and clay, the soil is assigned to a textural class according to a triangular diagram (Fig. 2.3). The triangle in Fig. 2.3a is used by the Soil Survey of England and Wales and is based on the British Standards system of particle-size grading (Fig. 2.2); the one in Fig. 2.3b is used in Australia and is based on the International system (Fig. 2.2). The USDA system is very similar to the British Standards system. Note that in these systems there are 11 textural classes, but the Australian system has

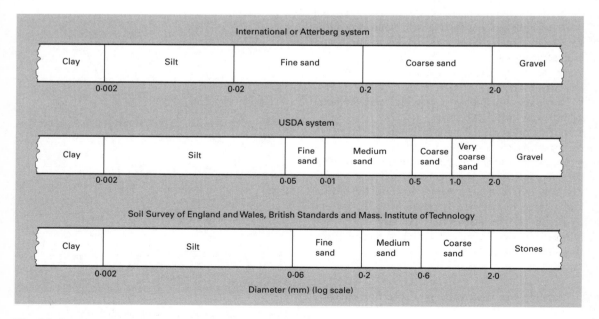

Fig. 2.2 Particle-size classes most widely adopted internationally.

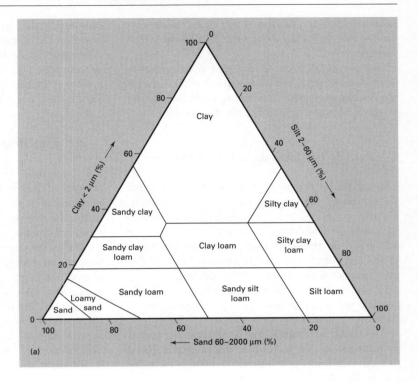

(a)

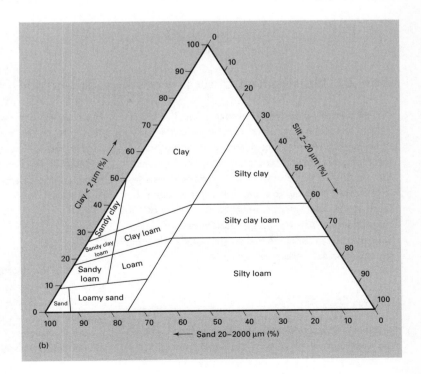

(b)

Fig. 2.3 (a) Triangular diagram of soil textural classes adopted in England and Wales (after Hodgson, 1974). (b) Triangular diagram of soil textural classes adopted in Australia (after McDonald *et al.*, 1998).

broader classes for the silty clays, silty clay loams and silty loams than the British or USDA systems.

Soil surveyors become expert at texturing after years of experience, which is gained by their checking field assessments of texture against a laboratory analysis of a soil's particle-size distribution.

Particle-size analysis in the laboratory

The success of the method relies on the complete disruption of soil aggregates and the addition of chemicals that ensure dispersion of the soil colloids in water. Full details of the methods employed are given in standard texts, for example Klute (1986) and Rayment and Higginson (1992). The sand particles are separated by sieving; silt and clay are separated using the differences in their settling velocities in suspension. The principle of the latter technique is outlined in Box 2.1.

The result of particle-size analysis is expressed as the mass of the individual fractions per 100 g of oven-dry (o.d.) soil (fine earth only). Oven-dry soil is soil dried to a constant weight at 105°C. When the coarse and fine sand fractions are combined, the soil may be represented by one point on the triangular diagrams of Fig. 2.3 (a, b). Alternatively, by stepwise addition of particle-size percentages, graphs of cumulative percentage against particle diameter of the kind shown in Fig. 2.1 are obtained.

2.2 The importance of soil texture

Soil scientists are primarily interested in the texture of the fine earth fraction. Nevertheless, in some soils the size and abundance of stones cannot be ignored because they can have a marked influence on the soil's suitability for agriculture. As the stone content increases, a soil holds less water than a stoneless soil of the same fine-earth texture, so that crops become more susceptible to drought. Conversely, such soils may be better drained and therefore warm up more quickly in spring in cool temperate regions. Large stones on the soil surface act as sinks during daytime for heat energy that is slowly released at night – this is of benefit in cool climate vineyards, such as in the Rhône Valley, France, where frost in spring and early summer can damage flowering and fruit

---

Box 2.1 Measurement of silt and clay by sedimentation.

A rigid particle falling freely through a liquid of lower density will attain a constant velocity when the force opposing movement is equal and opposite to the force of gravity acting on the particle. The frictional force acting vertically upward on a spherical particle is calculated from Stoke's law. The net gravitational force acting downwards is equal to the weight of the submerged particle. At equilibrium, these expressions can be combined to give an equation for the terminal settling velocity v, as

$$v = \frac{2}{9}\frac{g}{\eta}(\rho_p - \rho_w)r^2, \qquad (B2.1.1)$$

where g is the acceleration due to gravity, η (eta) is the coefficient of viscosity of the liquid (water),

which varies with temperature, ρ_p is the particle density, ρ_w is the density of water, and r is the particle radius.

When all the constants in this equation are collected into one term A, we derive the simple relationship

$$v = Ar^2 = \frac{h}{t}, \qquad (B2.1.2)$$

where h is the depth, measured from the liquid surface, below which all particles of radius r will have fallen in time t. To illustrate the use of Equation B2.1.2, we can calculate that all particles $> 2 \, \mu m$ in diameter settling in a suspension at 20°C will fall below a depth of 10 cm in 7.73 hours. Thus,

Box 2.1 *continued*

by sampling the mass per unit volume of the suspension at this depth after 7.73 hours, the amount of clay can be calculated. The suspension density at a particular depth can be measured in one of several ways:

• By withdrawing a sample volume of the suspension, evaporating to dryness, and weighing the mass of sediment – the pipette method;

• by using a Bouyoucos hydrometer in the suspension; or

• by calculating the loss in weight of a bulb of known volume when immersed in the suspension – the plummet balance method, illustrated in Fig. B2.1.1.

Note that constant temperature should be maintained (because of the temperature effect on the viscosity of water), and also that simplifying assumptions are made in the calculation of settling velocity by Equation B2.1.2. In particular, note that:

• Clay and silt particles are not smooth spheres, but have irregular plate-like shapes;

• the particle density varies with the mineral type (Section 2.3).

In practice, we take an average value for ρ_p of 2.65 Mg/m^3, and speak of the 'equivalent spherical diameter' of the particles being measured.

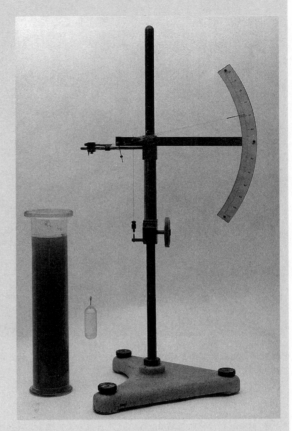

Fig. B2.1.1 A settling soil suspension and plummet balance (courtesy of J. Loveday).

set (Fig. 2.4). Stoniness also determines the ease, and to some extent the cost of cultivation, as well as the abrasive effect of the soil on tillage implements.

Texture is one of the most stable soil properties and is a useful index of several other properties that determine a soil's agricultural potential. Fine and medium-textured soils, such as clays, clay loams, silty clays and silty clay loams, are generally more desirable than coarse-textured soils because of their superior retention of nutrients and water. Conversely, where rapid infiltration and good drainage are required, as for irrigation or liquid waste disposal, sandy or coarse-textured

soils are preferred. In farming terms, clay soils are described as 'heavy' and sandy soils as 'light', which does not refer to their mass per unit volume, but to the power required to draw a plough or other implements through the soil. Because it is easy to estimate, and is routinely measured in soil surveys, texture (and more specifically clay content) has been used as a 'surrogate' variable for other soil properties that are less easily measured, such as the cation exchange capacity (Section 2.5).

Texture has a pronounced effect on soil temperature. Clays hold more water than sandy soils, and the presence of water considerably modifies the heat required to change a soil's temperature because:

Fig. 2.4 Boulders and stones covering the soil surface in a vineyard in the central Rhône Valley, France (see also Plate 2.4).

• Its specific heat capacity is 3–4 times that of the soil solids;
• considerable latent heat is either absorbed or evolved during a change in the physical state of water, for example, from ice to liquid or *vice versa*. Thus, the temperature of wet clay soils responds more slowly than that of sandy soils to changes in air temperature in spring and autumn (Section 6.6).

Texture should not be confused with tilth, of which it is said that a good farmer can recognize it with his boot, but no soil scientist can describe it. Tilth refers to the condition of the surface of ploughed soil prepared for seed sowing: how sticky it is when wet and how hard it sets when dry. The action of frost in cold climates breaks down the massive clods left on the surface of a heavy clay soil after autumn ploughing, producing a mellow 'frost tilth' of numerous small granules (Section 4.2).

2.3 Mineralogy of the sand and silt fractions

Simple crystalline structures

Sand and silt consist almost entirely of the resistant residues of primary rock minerals, although small amounts of secondary minerals (salts, oxides and hydroxides) formed by weathering also occur. The primary rock minerals are predominantly silicates, which have a crystalline structure based upon a simple unit – the silicon tetrahedron, SiO_4^{4-} (Fig. 2.5). An electrically neutral crystal is formed when cations, such as Al^{3+}, Fe^{3+}, Fe^{2+}, Ca^{2+}, Mg^{2+}, K^+ and Na^+, become covalently bonded to the O atoms in the tetrahedron and the surplus valencies of the O^{2-} ions in the SiO_4^{4-} group are satisfied. An example of this kind of structure is the primary mineral olivine, which has the composition $(Mg, Fe)_2SiO_4$.

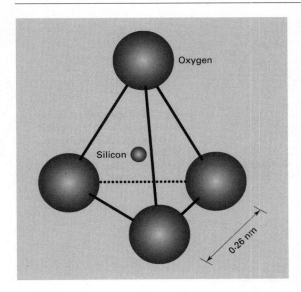

Fig. 2.5 Diagram of a Si tetrahedron (interatomic distances not to scale).

Table 2.1 Cation to oxygen radius ratios and coordination numbers for common elements in the silicate minerals. After Schulze, 1989.

Coordination no. 4		Coordination no. 6		Coordination no. 8 or greater	
Si^{4+}	0.28	Al^{3+}	0.36	Na^+	0.69
Al^{3+}	0.36	Ti^{4+}	0.49	Ca^{2+}	0.71
		Fe^{3+}	0.46	K^+	0.95
		Mg^{2+}	0.47	Ba^{2+}	0.96
		Fe^{2+}	0.53	Rb^+	1.05
		Mn^{2+}	0.57		

Coordination number

The packing of the O atoms, which are the largest of the more abundant elements in the silicates, determines the crystalline dimensions. In quartz, for example, O occupies 98.7% and Si only 1.3% of the mineral volume. The size ratio of Si to O is such that four O atoms can be packed around one Si, and larger cations such as iron (Fe) can accommodate more O atoms. The cation to oxygen radius ratio determines the coordination number of the cation (Table 2.1). Many of the common metal cations have radius ratios between 0.41 and 0.73, which means that an octahedral arrangement of six O atoms around the cation (coordination number 6) is possible. Larger alkali and alkaline earth cations, such as K^+ and Ba^{2+}, that have radius ratios > 0.73 form complexes of coordination number 8 or greater. Aluminium, which has a cation to oxygen radius ratio close to the maximum for coordination number 4 and the minimum for coordination number 6 (0.41), can exist in either fourfold (IV) or sixfold (VI) coordination.

Isomorphous substitution

Elements of the same valency and coordination number frequently substitute for one another in a silicate structure – a process called isomorphous substitution. The structure remains electrically neutral. However, when elements of the same coordination number but different valency are exchanged, there is an imbalance of charge. The most common substitutions are Mg^{2+}, Fe^{2+} or Fe^{3+} for Al^{3+} in octahedral coordination, and Al^{3+} for Si^{4+} in tetrahedral coordination. The excess negative charge is neutralized by the incorporation of additional cations, such as unhydrated K^+, Na^+, Mg^{2+} or Ca^{2+} into the crystal structure, or by structural arrangements that allow an internal compensation of charge (see the chlorites).

More complex crystalline structures

Chain structures

These are represented by the pyroxene and amphibole groups of minerals, which collectively make up the ferromagnesian minerals. In the pyroxenes, each Si tetrahedron is linked to adjacent tetrahedra by the sharing of two out of three basal O atoms to form a single extended chain (Fig. 2.6). In the amphiboles, two parallel pyroxene chains are linked by the sharing of an O atom in every alternate tetrahedron (Fig. 2.7). Cations such as Mg^{2+}, Ca^{2+}, Al^{3+} and Fe^{2+} are ionic-covalently bonded to the O^{2-} ions to neutralize the surplus negative charge.

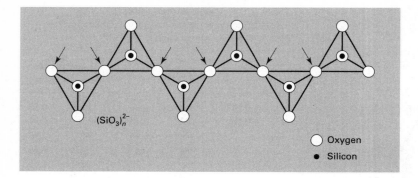

Fig. 2.6 Single chain structure of Si tetrahedra as in a pyroxene. The arrows indicate O atoms shared between adjacent tetrahedra (after Bennett and Hulbert, 1986).

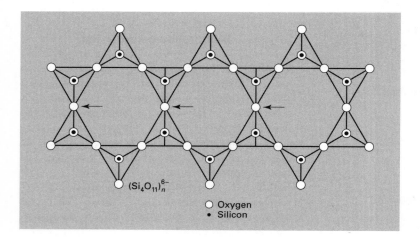

Fig. 2.7 Double chain structure of Si tetrahedra as in an amphibole. The arrows indicate O atoms shared between chains (after Bennett and Hulbert, 1986).

Sheet structures

It is easy to visualize an essentially one-dimensional chain structure being extended in two dimensions to form a sheet of Si tetrahedra linked by the sharing of all the basal O atoms. When viewed from above the sheet, the bases of the linked tetrahedra form a network of hexagonal holes (Fig. 2.8). The apical O atoms (superimposed on the Si atoms in Fig. 2.8) form ionic-covalent bonds with other metal cations by, for example, displacing OH groups from their coordination positions around a trivalent Al^{3+} ion. (Because the unhydrated proton is so small, OH occupies virtually the same space as O.) When Al octahedral units (one is shown diagrammatically in Fig. 2.9) are linked by the sharing of edge OH groups, they form an alumina sheet. Aluminium atoms normally occupy only two-thirds of the available octahedral positions – this is a dioctahedral structure, characteristic of the mineral gibbsite, $[Al_2OH_6]_n$. If Mg is present instead of Al, however, all the available octahedral positions are filled – this is a trioctahedral structure, characteristic of the mineral brucite $[Mg_3(OH)_6]_n$.

The bonding together of silica and alumina sheets through the apical O atoms of the Si tetrahedra causes the bases of the tetrahedra to twist slightly so that the cavities in the sheet become trigonal rather than hexagonal in shape. When two silica sheets sandwich one alumina sheet, the result is a 2:1 layer structure characteristic of the micas, chlorites and many soil clay minerals (Section 2.4). Two-dimensional layer crystal structures such as these are typical of the phyllosilicates, some characteristics of which are given in Box 2.2.

Additional structural complexity is introduced by isomorphous substitution. In the dioctahedral

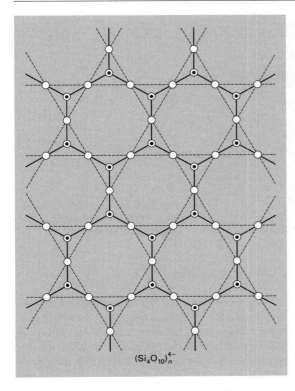

Fig. 2.8 A silica sheet in plan view showing the pattern of hexagonal holes (after Fitzpatrick, 1971).

$(Si_4O_{10})_n^{4-}$

Box 2.2 Generalized phyllosilicate structures.

Phyllosilicates are silicate minerals composed of two-dimensional tetrahedral or octahedral sheets, or covalently bonded combinations of these, stacked in regular array in the Z direction (Fig. B2.2.1). As shown in this figure, the directions of the crystal axes are X, Y and Z, and the repeat distances for atoms of the same element to occur along these axes are respectively a, b and c. The following terminology is used:

• A single *plane* of atoms (such as linked O or OH);
• a *sheet* is a combination of planes of atoms (such as a silica tetrahedral sheet);
• a *layer* is a combination of sheets (such as two silica sheets combined with one alumina sheet in mica);
• a *crystal* is made up of one or more layers;
• planes of atoms are repeated at regular intervals in multilayer crystals, which gives rise to a characteristic *d* spacing, or basal spacing, in the phyllosilicates;
• between the layers is interlayer space that may be occupied by water, organic or inorganic ions and molecules, and precipitated hydroxides;
• phyllosilicates generally have large planar surfaces and small edge faces.

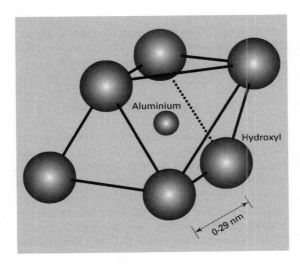

Fig. 2.9 Diagram of an Al octahedron (interatomic distances not to scale).

Aluminium

Hydroxyl

0.29 nm

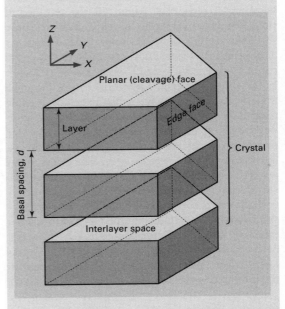

Fig. B2.2.1 General structure of a phyllosilicate crystal.

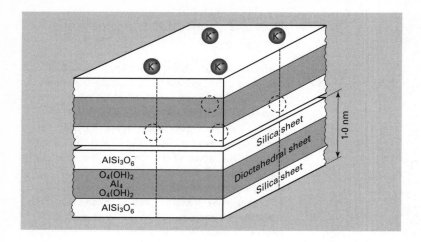

Fig. 2.10 Structure of muscovite mica – a 2 : 1 non-expanding phyllosilicate mineral.

mica muscovite, for example, one-quarter of the tetrahedral Si^{4+} is replaced by Al^{3+} resulting in a net 2 moles of negative charge per unit cell (Box 2.3). Muscovite has the structural composition:

$$[(OH)_4(Al_2Si_6)^{IV} Al_4^{VI}O_{20}]^{2-} 2K^+.$$

Note that the negative charge is neutralized by K^+ ions held in the spaces formed by the juxtaposition of the trigonal cavities of adjacent silica sheets (Fig. 2.10). Because the isomorphous substitution in muscovite occurs in the tetrahedral sheet, the negative charge is distributed over only three surface O atoms. The magnitude of the layer charge and its localization are sufficient to cause cations of relatively small ionic potential*, such as K^+, to lose their water of hydration. The unhydrated K^+ ions have a diameter comparable to that of the ditrigonal cavity formed between opposing siloxane surfaces and their presence provides very strong bonding between the layers. Such complexes, where an unhydrated ion forms an ionic-covalent bond with atoms of the crystal surface, are called inner-sphere (IS) complexes.

Biotite is a trioctahedral mica which has Al^{3+} substituted for Si^{4+} in the tetrahedral sheet and contains Fe^{2+} and Mg^{2+} in the octahedral sheet. Again, because of the localization of charge in the tetrahedral sheet, unhydrated K^+ ions are retained in the interlayer spaces to give a unit cell formula of:

$$[(OH)_4(Al_2Si_6)^{IV} (Mg, Fe)_6^{VI}O_{20}]^{2-} 2K^+.$$

The most complex of the two-dimensional structures belongs to the chlorites, which have a brucite layer sandwiched between two mica layers. In the type mineral chlorite, the negative charge of the two biotite layers is neutralized by a positive charge in the brucite, developed due to the replacement of two-thirds of the Mg^{2+} cations by Al^{3+} (Fig. 2.11). Chlorite is an example of a regular mixed-layer mineral. Predictably, the bonding between layers is strong, but the high content of Mg renders this mineral susceptible to weathering in acidic solutions. For the same reason, and also because of its ferrous (Fe^{2+}) iron content, biotite is much less stable than muscovite. Micas and chlorites weather to form vermiculites and smectites in soil (Section 2.4).

Three-dimensional structures

The most important silicates in this group are silica and the feldspars. Silica minerals consist entirely of polymerized Si tetrahedra of general composition $(SiO_2)_n$. Silica occurs as the residual mineral quartz, which is very inert, and as a secondary mineral precipitated after the hydrolysis of more complex silicates. Secondary silica initially exists as amorphous opal that dehydrates over time to form microcrystalline quartz, known as flint or chert. Silica also occurs as amorphous or microcrystalline silica of biological origin. For

---

* Determined by the charge to radius ratio. As this ratio increases, the ionic potential of the cation increases.

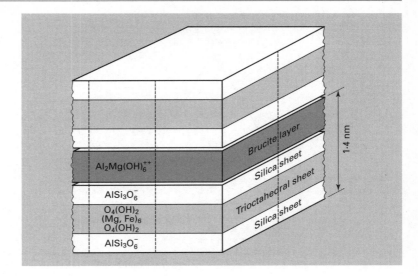

Fig. 2.11 Structure of chlorite – a 2 : 2 or mixed layer mineral.

Box 2.3 Moles of charge and equivalents.

The molar mass of an element is defined as the number of grams weight per mole (abbreviated to mol) of the element. The standard is the stable C-12 isotope of carbon. On this scale, H has a molar mass of 1 g, K a mass of 39 g, and Ca a mass of 40 g. The recommended unit of charged mass for cations, anions and charged surfaces is the mole of charge, which is equal to the molar mass divided by the ionic charge. Thus, the mass in grams of one mole of charge for the elements H, K and Ca is as follows:

- For H^+, $1/1 = 1$;
- for K^+, $39/1 = 39$;
- for Ca^{2+}, $40/2 = 20$.

For clay minerals and soils, the most appropriate unit of measurement is the centimole of charge (+) or (−) per kg (abbreviated to cmol charge/kg). For example, the cation exchange capacity (CEC) is expressed in cmol charge (+)/kg since it is measured by the moles of cation charge adsorbed (Section 2.5). In the older soil science literature, the charge on ions and soil minerals was expressed in terms of an equivalent weight, which is the atomic mass (g) divided by the valency (and identical to a mole of charge). The CEC of a mineral was expressed in milli-equivalents (meq) per 100 g, which is numerically equal to cmol charge/kg.

example, the diatom, a minute aquatic organism, has a skeleton of almost pure silica. Silica absorbed from the soil by terrestrial plants (grasses and hardwood trees in particular) forms opaline structures called phytoliths, which are returned to the soil when the plant dies.

Quartz or flint fragments of greater than colloidal size are very insoluble, and hence are abundant in the sand and silt fractions of many soils. The feldspars, on the other hand, are chemically more reactive and rarely comprise more than *c.* 10% of the sand fraction of mature soils. Of

these, the potassium feldspars are more resistant to weathering than the Ca and Na feldspars. Their structure consists of a three-dimensional framework of polymerized Si tetrahedra in which some Si^{4+} is replaced by Al^{3+}. The cations balancing the excess negative charge are all of high coordination number, such as K^+, Na^+ and Ca^{2+}, and less commonly, Ba^{2+} and Sr^{2+}. The range of composition encountered is shown in Fig. 2.12. Details of the structure, composition and chemical stability of the feldspars are given in specialist texts by Loughnan (1969) and Nahon (1991).

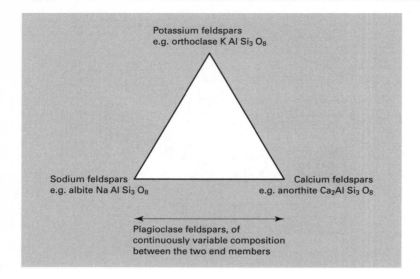

Fig. 2.12 Type minerals of the feldspar group.

2.4 Mineralogy of the clay fraction

A large assortment of minerals of varying degrees of crystallinity occurs in the clay fraction of soils. Broadly, these minerals may be divided into the crystalline clay minerals – predominantly phyllosilicates – and other minerals (oxides, hydroxides and salts). Because of their large specific surface areas and surface charges, these minerals are very important sites for physical and chemical reactions in soil (Chapters 6 and 7).

For many years the small size of clay particles prevented scientists from elucidating their mineral structure. It was thought that the clay fraction consisted of inert mineral fragments enveloped in an amorphous gel of hydrated sesquioxides ($Fe_2O_3.nH_2O$ and $Al_2O_3.nH_2O$)* and silicic acid ($Si(OH)_4$). The surface gel was amphoteric, the balance between acidity and basicity being dependent on the soil pH. Between pH 5 and 8, the surface was usually negatively charged (proton-deficient), which could account for the observed cation exchange properties of soil. During the 1930s, however, the crystalline nature of the clay minerals was established unequivocally by X-ray diffraction (XRD) (Box 2.4). Most of the

* Fe and Al oxides are collectively called sesquioxides (prefix 'sesqui' meaning one and a half), because the ratio of oxygen to metal cation is 1.5.

Box 2.4 Identification of minerals in the clay fraction.

The technique of X-ray diffraction involves directing a beam of X-rays (electromagnetic radiation of wavelength from 0.1 to 10 nm) at a clay sample (particles < 50 µm diameter). Monochromatic X-rays whose wavelength is of the same order as the spacings of atomic planes in the crystals (0.1–0.2 nm) are the most useful. The clay sample can be a powder or a suspension dried on to a glass slide, which gives a preferred orientation of the plate-like crystals. As the X-rays penetrate a crystal, a small amount of their energy is absorbed by the atoms which become 'excited' and emit radiation in all directions. Radiation from atomic planes that is in phase will form a coherent reflected beam that can be detected by X-ray sensitive film. For a beam of parallel X-rays of wavelength λ (lambda), striking a crystal at an angle θ (theta), the necessary condition for the reflected radiation from atomic planes to be in phase is

$$n \lambda = 2 d \sin \theta, \tag{B2.4.1}$$

where n is an integer and d is the characteristic spacing of the atomic planes. Equation B2.4.1 is a mathematical statement of Bragg's Law.

minerals were found to have a phyllosilicate structure similar to the micas and chlorite. The various mineral groups were identified from their characteristic d spacings (Fig. B2.2.1), as measured by XRD, and their unit cell compositions deduced from elemental analyses. Subsequent studies using scanning and transmission electron microscopes have confirmed the conclusions of the early work.

In addition, within the accessory minerals there are the weathered residues of resistant primary minerals that have been comminuted to colloidal size, and soil minerals synthesized during pedogenesis. The latter mainly comprise Al and Fe hydroxides and oxyhydroxides, which occur as discrete particles or as thin coatings on the clay minerals. The crystallinity of these minerals varies markedly depending on their mode of formation, the presence of other elements as inclusions, and their age. Some, such as the iron hydroxide ferrihydrite, were previously thought to be amorphous, but are now known to form extremely small crystals and to possess short-range order: that is, their structure is regular over distances of a few nanometres, but disordered over larger distances (tens of nanometres).

The crystalline clay minerals

Most clay minerals have a phyllosilicate structure, but a small group – the sepiolite-palygorskite series – has chain structures and another group – the allophanes – forms hollow spherical crystals. Palygorskite and sepiolite are unusual in having very high Si : Al ratios, with Mg occupying most of the octahedral positions. Sepiolite is very rare in soil and palygorskite survives only in soils of semi-arid and arid regions. They are not discussed further.

Under mild (generally physical) weathering conditions, clay minerals may be inherited as colloidal fragments of primary phyllosilicates, such as muscovite mica. Under more intense weathering, the primary minerals may be transformed to secondary clay minerals, as when soil illites, vermiculites and smectites are formed by the leaching of interlayer K from primary micas, or from the weathering of chlorites. Neoformation of clay minerals is a feature of intense weathering, or of diagenesis in sedimentary deposits (Section 5.2),

when minerals completely different from the original primary minerals are formed. When the soluble silica concentration in the weathering environment is high, 2 : 1 layer minerals such as smectites are likely to form. Leaching and removal of silica, however, can produce kaolinite and aluminium hydroxide. Increased negative charge in the crystal due to isomorphous substitution of Al^{3+} for Si^{4+} in the smectites leads to K^+ being the favoured interlayer cation, with the resultant formation of illite.

Minerals with a Si : Al mole ratio ≤ 1

Three groups of clay minerals – *imogolite*, *allophanes*, and *kaolinites* – have Si : Al ratios ≤ 1.

Imogolite and allophane are most commonly found in young soils (< 1000 years) formed on volcanic ash and pumice (order Andisol (ST)). Imogolite has also been identified in the B horizon of podzols (order Spodosol (ST) or Podosol (ASC)). Both minerals appear amorphous by XRD, but high-resolution electron microscopy has revealed their crystalline nature, so they are properly called short-range order minerals. Because of their hollow crystal structures, they have very large specific surfaces (Table 2.2) and are highly reactive, especially towards organic anions and phosphate.

Imogolite has the structural formula:

$$(OH)_6 Al_4^{VI} O_6 Si_2^{IV} (OH)_2$$

and occurs in 'threads' 10–30 nm in diameter and several μm long. Each thread consists of several tubular crystals with inner and outer diameters $c.$ 1 and 2.5 nm, respectively. They are curved to permit a reduced number of Si tetrahedra to bond to the Al octahedral sheet, as shown in Fig. 2.13.

Allophane exists as hollow, spherical particles of diameter 3.5–5 nm. One kind of allophane, which has a Si : Al ratio of 0.5 and the structural formula:

$$(H_2O)_2, (OH)_4 Al_3^{VI} O_2 (OH)_4 (Si_2, Al)^{IV} O_3, (OH)_2, H_2O$$

contains most of its Al in sixfold coordination and has charge properties very similar to imogolite, that is, very little permanent negative charge due

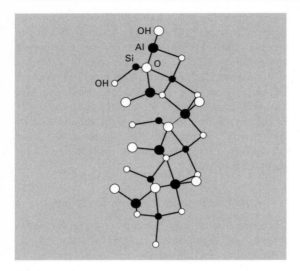

Fig. 2.13 Projection along the imogolite c axis showing the curvature produced as the Al octahedral sheet distorts to accommodate the Si tetrahedra (after Wada, 1980).

to isomorphous substitution, but variable positive and negative charge due to H^+ association or dissociation at surface OH groups (Section 7.1). At the other extreme, allophane with a Si : Al ratio of 1 and the structural formula:

$$H_2O,(OH)_2\ Al^{VI}O,(OH)_2\ H_2O(Si_2,Al)^{IV}O_3,$$
$$(OH)_2,H_2O$$

contains half its Al in the tetrahedral sheet and half in the octahedral sheet. Although a large layer charge arises because of the substitution of Al^{3+}

for Si^{4+}, it is neutralized to a variable extent, depending on the ambient pH, by the association of H^+ ions with surface OH groups. Thus, the allophanes and imogolite have pH-dependent surface charges at pH > 4.5.

Minerals of the kaolinite group have a well-defined 1 : 1 layer structure formed by the sharing of O atoms between one Si tetrahedral sheet and one dioctahedral alumina sheet. The unit cell composition is:

$$Si_2^{IV}O_5\,Al_2^{VI}OH_4$$

The common mineral of the group is kaolinite, the dominant clay mineral in many weathered tropical soils such as Oxisols (ST) or Ferrosols (ASC). Halloysite is found in weathered soils formed on volcanic ash, but is less stable than kaolinite.

Kaolinite has a d spacing fixed at 0.71 nm because of hydrogen-bonding between the H and O atoms of adjacent layers (Fig. 2.14). The layers are stacked fairly regularly in the Z direction to form crystals from 0.05 to 2 µm thick, the larger crystals occurring in relatively pure deposits of China Clay that is used for pottery. The crystals are hexagonal in plan view and usually larger than 0.2 µm in diameter, as shown in the electron micrograph of Fig. 2.15. Halloysite has the same structure as kaolinite, with the addition of two layers of water molecules between the crystal layers, which increases the d spacing of the mineral to 1 nm. The presence of this hydrogen-bonded water alters the distribution of stresses within the crystal so that the layers curve to form a tubular structure.

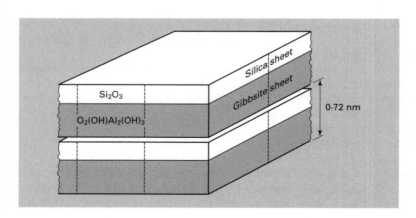

Fig. 2.14 Structure of kaolinite – a 1 : 1 phyllosilicate mineral with H-bonding between layers.

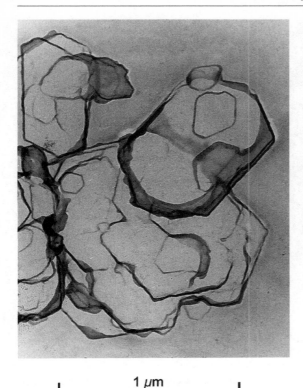

1 μm

Fig. 2.15 Electron micrograph of kaolinite crystals.

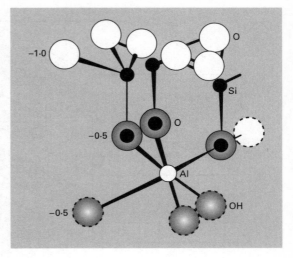

Fig. 2.16 Charges on the edge face of kaolinite at high pH (after Hendricks, 1945).

There is a small degree of substitution of Al^{3+} for Si^{4+} in kaolinite that produces < 0.005 moles of negative charge per unit cell. In addition, a variable charge can develop at the crystal edge faces (Fig. 2.16) due to the association or dissociation of H^+ ions at exposed O and OH groups. At pH > 9, this can contribute as much as 8×10^{-4} cmol charge (–) per m^2 of edge area (*cf* Table 2.3).

Minerals with a Si : Al mol ratio of 2

This category comprises the mica, vermiculite and smectite groups of clay minerals, which all have 2 : 1 phyllosilicate structures. The minerals differ mainly in the extent and location of isomorphous substitution, and hence in the type of interlayer cation that predominates (Olson *et al.*, 2000).

Within the mica group, the dioctahedral mineral illite has substitution of Al^{3+} for Si^{4+} in the tetrahedral sheets and some substitution of

Mg^{2+}, Fe^{2+} and Fe^{3+} for Al^{3+} in the octahedral sheet. An average of 1.3 moles of negative charge per unit cell of the formula:

$$[(OH)_2(Si_{3.3}Al_{0.7})^{IV}\ (Al_{1.3}Mg_{0.4}Fe_{0.1}Fe_{0.2})^{VI}\ O_{10}]^{-1.3}$$
$$1.3K^+,$$

and its location partly in the tetrahedral sheets, are sufficient to favour the formation of inner-sphere complexes with cations, primarily K^+ but sometimes also NH_4^+, which fits into the ditrigonal cavities between opposing siloxane surfaces. The *d* spacing is characteristically 0.96–1.01 nm. As illite weathers and the K^+ is gradually replaced by cations of higher ionic potential such as Ca^{2+} and Mg^{2+}, which remain partially hydrated in the interlayer regions, the mineral expands to a *d* spacing of 1.4–1.5 nm. The additional spacing is equivalent to a bimolecular layer of water molecules between the mineral layers. Consequently, the interlayer bonding is weaker than in the primary micas and the stacking of the layers to form crystals is much less regular. These minerals are sometimes called hydrous micas. The Ca^{2+} and Mg^{2+} ions are exchangeable, whereas the K^+ and NH_4^+ ions are not (Section 2.5).

Vermiculites are trioctahedral minerals with both di- and trivalent cations occupying all the available sites in the octahedral sheet. This

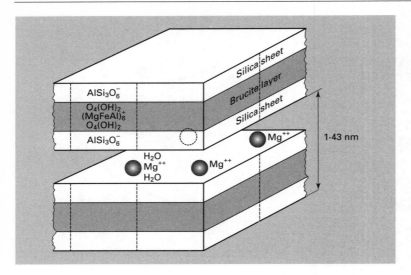

Fig. 2.17 Structure of vermiculite – *d* spacing fixed in the presence of partially hydrated Mg^{2+} ions.

arrangement results in a net positive charge that in part neutralizes the negative charge developed through substitution of Al^{3+} for Si^{4+} in the tetrahedral sheet. The net charge of the vermiculites is therefore lower than that of the soil micas at 0.6–0.9 moles of negative charge per unit cell. As a result, ions of lower ionic potential are not likely to form inner-sphere complexes and partially hydrated Ca^{2+} and Mg^{2+} are the dominant interlayer cations. The basal spacing of Mg-vermiculite, shown in Fig. 2.17, is typically 1.43 nm, but collapses to 1 nm on heating. Thus, the vermiculites show limited reversible swelling. In a moderately acid environment, hydrated Al^{3+} ions replace the Ca^{2+} and Mg^{2+} in the interlayers and 'islands' of poorly ordered $Al(OH)_3$ may form. This gives rise to an aluminous chlorite type of clay mineral.

Smectite minerals exhibit a wide range of composition with isomorphous substitution occurring in both the tetrahedral and octahedral sheets. The layer charge is lower (0.2–0.6 moles of negative charge per unit cell) than in either the soil micas or vermiculites. Two smectite minerals – nontronite and beidellite – have isomorphous substitution predominantly in the tetrahedral sheet, but the most common smectite in soil, montmorillonite, is a dioctahedral mineral with substitution in both the tetrahedral and octahedral sheets. It has the unit cell formula:

$$[(OH)_2(Si_{3.74}Al_{0.26})^{IV}(Al_{1.77}Mg_{0.23})^{VI}O_{10}]^{-0.49}$$
$$0.49M^+,$$

where M^+ is an interlayer cation. When isomorphous substitution occurs in the octahedral sheet, the excess negative charge is distributed over 10 surface O atoms and the field strength is weak. Consequently, there is no tendency for cations of low ionic potential to dehydrate and to draw adjacent mineral layers close together, as occurs in the micas. Rather, the cations neutralizing the layer charge remain hydrated and may form outer-sphere (OS) complexes with the surface, as in the case of Na- or Ca-montmorillonite. The interlayer cations are freely exchangeable with other cations in solution, and depending on the nature of the exchangeable cations, the basal spacing may vary from *c.* 1.5–4 nm, as illustrated in Fig. 2.18. Ca-montmorillonite, a form common in soil, has a basal spacing of 1.9 nm at full hydration (relative humidity > 98%), when three molecular layers of water exist between the mineral layers.

Because of the weak interlayer bonding and free movement of water and cations into and out of this region, the smectites have been called shrink-swell or expanding clays. The stacking of the layers is very irregular and the average crystal size is therefore much smaller than in the micas and kaolinites. The planar interlayer surfaces provide a large internal area that augments the external

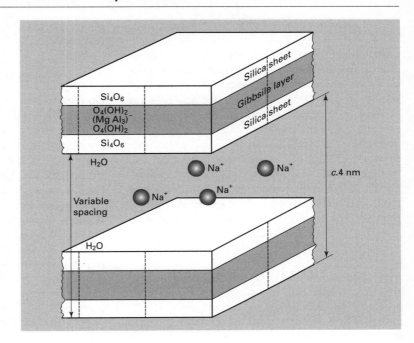

Fig. 2.18 Structure of montmorillonite – *d* spacing variable depending on the dominant exchangeable cations.

Table 2.2 Specific surface areas according to mineral type and particle size. After Fripiat, 1965; Dixon and Weed, 1989.

Mineral or size class	Specific surface (m²/g)	Method of measurement
Coarse sand	0.01	Low temperature N_2 adsorption (BET)
Fine sand	0.1	BET
Silt	1.0	BET
Kaolinites	5–40	BET
Hydrous micas (illites)	100–200	BET
Vermiculites and mixed layer minerals	300–500	Ethylene glycol adsorption
Montmorillonite (Na-saturated)	750	Ethylene glycol adsorption
Iron and aluminium oxides and oxyhydroxides	100–300	BET
Allophanes	1000	Includes internal and external surfaces

BET, Brunauer, Emmet and Teller.

planar face area, so that the total surface area of smectite clay is very large indeed (Table 2.2).

Mixed-layer or interstratified minerals

Regular and irregular interstratification of clay minerals produces mixed-layer minerals. The chlorites, for example, are formed by the regular interstratification of Al-substituted brucite sheets and biotite mica. The *d* spacing is a consistent 1.4 nm. Secondary chlorites can form in soil when

polymerized hydroxy-Al sheets (Section 7.2) build up in the interlayers of weathering vermiculites or smectites. With the improvement of XRD techniques, irregular interstratification has been found to occur commonly in soil clay minerals. Mixed layers of the 2 : 1 minerals are most common, especially intergrades of illite and smectite, and chlorite and smectite. However, interstratification of montmorillonite and kaolinite also occurs in acid environments, and may be a potential index for the intensity of soil weathering.

Other minerals of the clay fraction

Other minerals of the clay fraction are predominantly free oxides and hydroxides – compounds in which a single cation species is coordinated with O and/or OH, which include various forms of silica, iron oxides, aluminium oxides, manganese and titanium oxides. Calcite ($CaCO_3$) and dolomite ($(Ca,Mg)CO_3$) can occur in the clay fraction of soils formed on chalk and limestone, and in high pH soils of arid regions (where gypsum $CaSO_4.2H_2O$ may also occur). Accumulations of these carbonate minerals as hard layers within the soil profile are called calcrete or dolocrete.

Silica

The various forms of SiO_2 are described in Section 2.3. Silicic acid in solution has a marked tendency to polymerize as the pH rises. Thus, colloidal SiO_2 gel can occur as a cementing agent in the lower B and C horizons of leached soils (Section 4.3), where it may accumulate to such an extent that it forms a hardened silcrete layer. Amorphous SiO_2 also ocurs in soils formed on recent deposits of volcanic ash, as has been observed in Japan, New Zealand and Hawaii.

Manganese oxides

There is a variety of Mn oxides and hydroxides in soil because Mn can occur in oxidation states of 2, 3 and 4. A continuous range in composition from manganous oxide, MnO, to manganese dioxide, MnO_2, is possible, with several stable and metastable minerals having been identified. Although the most stable form of MnO_2 is pyrolusite, this mineral is rarely found in soil: the most common Mn oxides in soil are the birnessite group in which Mn^{2+}, Mn^{3+} and Mn^{4+} ions are bonded to O^{2-} and OH^-. Most Mn oxides are formed by microbial oxidation of Mn^{2+} in solution, followed by precipitation, or crystallization from poorly ordered colloidal sols. Small black deposits (1–2 mm diameter) are common in soils that experience alternating aerobic and anaerobic conditions, and the associated microbiological changes (Section 8.4). The surface of these minerals has a high affinity for metals ions, especially Co, Cu and Pb, and the availability of Co to plants is often controlled by the solubility of these Mn oxides.

Iron oxides and oxyhydroxides

Iron oxides and oxyhydroxides accumulate in soils that are highly weathered, especially those derived from the more basic rocks (Section 5.2). They have strong colours ranging from yellow to reddish-brown to black, even when disseminated through the soil profile. The most common mineral is the oxyhydroxide goethite (α-FeOOH), which precipitates as small acicular crystals, generally in clusters, from soil solutions containing Fe^{3+} ions (Fig. 2.19). Being a thermodynamically stable oxide of iron, it also forms by slow dissolution and re-precipitation from ferrihydrite. Ferrihydrite ($Fe_2O_3.2FeOOH.nH_2O$) is a poorly ordered Fe oxide that has a rusty red colour, and readily precipitates in drainage ditches and in the B horizon of podzols (order Spodosol (ST) or Podosol (ASC)). It is a necessary precursor of hematite (α-Fe_2O_3) formation, during which hematite crystals nucleate and grow within ferrihydrite clusters (Kampf et al., 2000). Because ferrihydrite formation is inhibited by the presence of organic ligands in solution, hematite (and hence red-coloured soils) are more common in regions where organic matter is rapidly oxidized than where it is not.

In poorly drained soils, where Fe^{2+} and Fe^{3+} ions can coexist in solution (Section 8.4), rapid oxidation of Fe^{2+} leads to the precipitation of lepidocrocite (γ-FeOOH), which through crystal rearrangement can slowly revert to goethite. However, under the influence of high soil temperatures due to forest or bush fires, and in the presence of organic matter, lepidocrocite is converted to maghemite (γ-Fe_2O_3). The latter is more common in soils of the tropics and subtropics.

With the exception of maghemite and hematite, which also form during weathering of the primary mineral magnetite (Fe_3O_4), all the iron oxides form by precipitation from solution. Inclusions of other elements can therefore occur, especially Al, which has been found to replace as much as 30% of the Fe in goethite on a mole for mole basis. Al is also found in the crystal structure of hematite and maghemite. Elements such as Si, Ti, V, Zn, Cu, Mo and P also occur as inclusions in iron oxides.

Fig. 2.19 Transmission electron micrograph showing clusters of many small acicular goethite crystals (courtesy of A. Suddhiprakarn and R. J. Gilkes).

Aluminium oxides and oxyhydroxides

The aluminium oxides have a non-distinctive, greyish white colour that is easily masked in soils, except where large concentrations occur as in bauxite ores. In acid soils, deposits of short-range order aluminium hydroxide form in the interlayers of expanding clays such as the vermiculites and smectites, and as surface coatings on clay minerals generally. This poorly ordered material slowly crystallizes to gibbsite (γ-Al(OH)$_3$), the principal aluminium hydroxide mineral in soil. At temperatures > 150°C, gibbsite undergoes dehydration to form the oxyhydroxide boehmite (γ-AlOOH), which is common in bauxite deposits, where it occurs with gibbsite, but is of uncertain occurrence in soil.

A mixture of sodium citrate and sodium dithionite (sometimes with sodium bicarbonate as well) is widely used to extract the free Fe oxides from soil. Acid ammonium oxalate extracts the poorly crystalline Fe oxides as well as some of the short-range order Al oxides, imogolite and allophane. Details of the methods are given in Rayment and Higginson (1992). The ratio of oxalate-extractable Fe to citrate-dithionite extractable Fe, which quantifies the proportion of reactive to less reactive Fe, has been used as an index of a soil's P adsorption capacity (Section 7.3).

Titanium oxides

The main Ti oxides found in soil are rutile and anatase (both TiO$_2$) and ilmenite ((Fe,Ti)O$_3$). On average, these oxides comprise 1–3% of the clay fraction but may be more abundant in weathered tropical soils. They exist mainly as residual minerals, although there is some evidence for their synthesis during soil formation. Because of their resistance to weathering these minerals have been used as 'markers' in studies of rates of soil formation, although in such studies it is important to identify a specific Ti mineral species and measure its concentration in both the soil and its parent material.

2.5 Surface area and surface charge

Specific surface area and adsorption

We can demonstrate that the smaller the size of a soil particle, imagined as a sphere, the greater the ratio of its surface area to volume. For example, when the radius $r = 1$ mm

$$\frac{\text{Surface area}}{\text{Volume}} = \frac{4\pi}{4/3} \frac{r^2}{r^3} = 3 \ (1/\text{mm}), \qquad (2.1)$$

and when $r = 0.001$ mm

$$\frac{\text{Surface area}}{\text{Volume}} = \frac{4\pi \times 10^{-6}}{4/3\pi \times 10^{-9}} = 3 \times 10^3 \text{ (1/mm)}.$$

(2.2)

The ratio of surface area to volume defines the specific surface area, which in practice can be measured as the surface area per unit mass, for a constant particle density. For reasons outlined in Box 2.5, a large specific surface area increases the adsorption of molecules and ions on the surface. This phenomenon has been exploited to develop methods for measuring the specific surface areas of soil minerals, as described in Box 2.5. Representative values for the specific surfaces areas of sand, silt and clay-size minerals are given in Table 2.2. Note the large range in surface areas, even within the clay fraction, from as little as 5 m²/g for kaolinite to 750 m²/g for Na-montmorillonite. The latter value is about the maximum for completely dispersed clay particles 1 nm thick.

Surface charges

Organic and inorganic cations and anions are adsorbed on clay and oxide surfaces as a result of:
• The permanent negative charges of the clays created by isomorphous substitution;
• the amphoteric properties of the edge faces of kaolinite, the surfaces of imogolite and allophone, and the oxides and oxyhydroxides of Fe and Al.

The permanent charge due to isomorphous substitution can be expressed as a surface density of charge by dividing the layer charge per unit cell by the surface area per unit cell. In practice, the charge on the mineral surface is measured as the difference between the moles of charge contributed per unit mass of mineral by the cations and anions adsorbed from an electrolyte solution of known pH. The cation and anion charge adsorbed gives rise respectively to a cation exchange capacity (CEC) and anion exchange capacity (AEC) in cmol charge per kg (or meq per 100 g, Box 2.3). For the crystalline 1 : 1 and 2 : 1 phyllosilicates, permanent negative charge and hence CEC is the more important, and representative values of CEC for the main mineral groups are given in Table 2.3. Note that the CEC

Box 2.5 Mineral surfaces and adsorption.

When a boulder is reduced to small rock fragments by weathering, part of the energy applied is conserved as free energy associated with new particle surfaces. The higher the specific surface area of a substance, therefore, the greater is its surface free energy. Because free energy tends to a minimum (Second Law of Thermodynamics), the surface energy is reduced by the work done in attracting molecules to the surface. This results in the phenomenon of adsorption. The adsorption of N_2 gas at very low temperatures on thoroughly dry clay is used to measure the specific surface area of the clay (the Brunauer, Emmet and Teller or BET method), because the adsorbed N_2 forms a monomolecular layer with an area of 0.16 nm² per molecule. Where themineral has internal surfaces, as in montmorillonite, an additional measurement is required because the non-polar N_2 molecules do not penetrate into the interlayer space. However, small polar molecules such as water or ethylene glycol do penetrate so that monolayer adsorption of these compounds gives a measure of the internal and external surfaces. In the absence of an internal surface, as in kaolinite, the BET and ethylene glycol methods should give identical results. Adsorption is also influenced by electrostatic charge and atomic interactions between the surface (the adsorbent) and the adsorbing substance (the adsorbate), which govern the adsorption of cations and anions (Sections 7.2 and 7.3).

Adsorption of a gas is called physical adsorption because it normally involves only very short-range interatomic forces. However, as indicated, adsorption of cations and anions often involves other forces between surface groups and the adsorbate, and may be called chemisorption. In the case of ions such as phosphate ($H_2PO_4^-$), at higher solution concentrations, it is difficult to distinguish between true adsorption (a surface phenomenon) and chemical precipitation, where a new phase forms on the mineral surface. The term 'sorption' is often used to describe the process where both adsorption and precipitation are possible.

Table 2.3 Cation exchange capacities and surface charge densities of the clay mineral groups.

Clay mineral group	Cation exchange capacity (CEC)* (cmol charge (+)/kg)	Surface charge density (σ) (cmol charge (−)/m^2)
Kaolinites	5–25	$1–4 \times 10^{-4}$
Illites	20–40	$1–3 \times 10^{-4}$
Smectites	100–120	$1–1.5 \times 10^{-4}$
Vermiculites	150–160	3×10^{-4}

* Although the charge on the surface is negative, it is measured by the number of moles of cation charge (+) adsorbed.

is variable within and between mineral groups, and not always the value expected from the moles of unbalanced negative charge per unit cell. Of the 2 : 1 clays, for example, the CEC of the illites is often low because much of the charge is neutralized by interlayer K^+ ions that are non-exchangeable. On the other hand, the interlayer Ca^{2+} and Mg^{2+} ions of the soil vermiculites are exchangeable and the CEC is high.

The functional surface charge density σ (sigma) of the clays is obtained from the ratio CEC/specific surface area (cmol charge/m^2). As shown in Table 2.3, despite the large range in charge per unit mass (the CEC), the values of σ are reasonably similar, suggesting that the clay crystals cannot acquire an indefinite number of negative charges per unit area and remain stable. Kaolinite, which has very little isomorphous substitution, exists as large crystals and so has a much lower specific surface area than montmorillonite; on the other hand, the large kaolinite crystals have a greater edge area at which additional negative charges can develop and so augment the total surface charge density of the clay.

For minerals bearing predominantly pH-dependent charges (the sesquioxides, imogolite and allophane), both the CEC and AEC can be significant, depending on the prevailing pH. Imogolite and allophane have roughly comparable positive and negative charges in the pH range 6–7. The sesquioxides are positively charged up to pH 8 so in most soils they make no contribution to the CEC. Maximum AEC values (measured at pH 3.5) range from 30 to 50 cmol charge/kg for

gibbsite and goethite, respectively. Because of their pH-dependent charge, these minerals buffer a soil against pH change over a wide range of pH (see Box 3.5).

The development of charges on the surfaces of clays and oxides is discussed more fully in Chapter 7.

2.6 Summary

There is a continuous distribution of particle sizes in soil from boulders and stones down to clay particles < 2 µm in equivalent diameter. Soil that passes through a 2 mm sieve, the fine earth, is divided into sand, silt and clay, the relative proportions of which determine the soil texture. Texture is a property that changes only slowly with time, and is an important determinant of the soil's response to water – its stickiness, mouldability and permeability; also its capacity to retain cations and its response to temperature change.

Adsorption of water and solutes by clay particles (< 2 µm size) depends not only on their large specific surface area, but also on the type of minerals present. Broadly, these are divided into crystalline clay minerals (phyllosilicates) and other minerals that include various Fe and Al oxides and oxyhydroxides (the sesquioxides), Mn and Ti oxides, silica and calcite. The basic crystal structure of the phyllosilicates consists of sheets of Si in tetrahedral coordination with O, $[SiO_2]_n$, and sheets of Al in octahedral co-ordination with OH, $[Al_2(OH)_6]_n$. Crystal layers are formed by the sharing of O atoms between contiguous silica and alumina sheets giving rise to minerals with Si/Al mole ratios ≤ 1 (imogolite and the allophanes), the 1 : 1 layer minerals (kaolinites), and the 2 : 1 layer minerals (illites, vermiculites and smectites). Isomorphous substitution ($Al^{3+} \rightarrow Si^{4+}$ and Fe^{2+}, $Mg^{2+} \rightarrow Al^{3+}$) in the crystal results in an overall net layer charge ranging from < 0.005 to 2 moles of negative charge per unit cell for kaolinite to muscovite, respectively. The formation of inner-sphere complexes between the surface and unhydrated K^+ ions in the interlayer spaces of the micas effectively neutralizes the layer charge and permits large crystals to form. On weathering, K^+ is replaced by cations of higher ionic potential such as Ca^{2+} and Mg^{2+} and layers of water molecules intrude into the interlayer spaces, so that the *d* spacing of the mineral increases. Hydrated Ca^{2+} and Mg^{2+} ions, which form outer-sphere complexes with the surfaces, are freely exchangeable with other cations in the soil solution.

The edge faces of clay crystals, especially kaolinite and the surfaces of Fe and Al oxides (e.g. goethite and gibbsite) bear variable charges that depend on the association or dissociation of H^+ ions at exposed O and OH groups. The surfaces are positively charged at low pH and negatively charged at high pH. The cation and anion charges adsorbed (in moles of charge per unit mass) measure the CEC and AEC, respectively, of the mineral. The CEC ranges from 3 to 160 cmol charge/kg for the clay minerals and AEC from 30 to 50 cmol charge/kg for the sesquioxides. Imogolite and allophane have roughly comparable CEC and AEC at pH 6–7.

Clays in which the interlayer surfaces are freely accessible to water and solutes are called expanding clays. The total surface area is then the sum of the internal and external areas, reaching 750 m^2/g for a fully dispersed Na-montmorillonite. Imogolite and allophane also have very large specific surface areas (*c.* 1000 m^2/g) because of the internal and external surfaces associated with their respective tubular and spherical crystal structures. However, when crystal layers are strongly bonded to one another, as in kaolinite, the surface area comprises only the external surface and may vary from 5 to 25 m^2/g, depending on the crystal size.

References

Bennett R. H. & Hulbert M. H. (1986) *Clay Microstructure.* Reidel, Boston.

Dixon J. B. & Weed S. B. (1989) (Eds) *Minerals in Soil Environments*, 2nd edn. Soil Science Society of America Book Series No. 1. Soil Science Society of America, Madison, Wisconsin.

Fitzpatrick E. A. (1971) *Pedology. A Systematic Approach to Soil Science.* Oliver & Boyd, Edinburgh.

Fripiat J. J. (1965) Surface chemistry and soil science, in *Experimental Pedology* (Eds E. G. Hallsworth & D. V. Crawford). Butterworth, London.

Hendricks S. B. (1945) Base exchange of crystalline silicates. *Industrial and Engineering Chemistry* 37, 625–30.

Hodgson J. M. (1974) (Ed.) *Soil Survey Field Handbook.* Soil Survey of England and Wales, Technical Monograph No. 5, Harpenden.

Kampf N., Scheinost A. C. & Schulze D. G. (2000) Oxide minerals, in *Handbook of Soil Science* (Ed. M. E. Sumner), CRC Press, Boca Raton, Florida, pp. F125–F168.

Klute A. (1986) (Ed.) *Methods of Soil Analysis. Part 1. Physical and Mineralogical Methods*, 2nd edn. Agronomy Monograph No. 9. American Society of Agronomy/Soil Science Society of America, Madison, Wisconsin.

Loughnan F. C. (1969) *Chemical Weathering of the Silicate Minerals.* American Elsevier, New York.

McDonald R. C., Isbell R. F., Speight J. G., Walker J. & Hopkins M. S. (1998) *Australian Soil and Land Survey Field Handbook*, 2nd edn, reprinted. Australian Collaborative Land Evaluation Program, Canberra.

Nahon D. B. (1991) *Introduction to the Petrology of Soils and Chemical Weathering.* Wiley & Sons, New York.

Olson C. G., Thompson M. L. & Wilson M. A. (2000) Phyllosilicates, in *Handbook of Soil Science* (Ed. M. E. Sumner). CRC Press, Boca Raton, Florida, pp. F77–F123.

Rayment G. E. & Higginson F. R. (1992) *Australian Laboratory Handbook of Soil and Water Chemical Methods.* Australian Soil and Land Survey Handbook. Inkata Press, Melbourne.

Schulze D. G. (1989) An introduction to soil mineralogy, in *Minerals in Soil Environments,* 2nd edn. (Eds J. B. Dixon & S. B. Weed). Soil Science Society of America Book Series No. 1. Soil Science Society of America, Madison, Wisconsin, pp. 1–34.

Wada K. (1980) Mineralogical characteristics of Andisols, in *Soils with Variable Charge.* New Zealand Society of Soil Science, Lower Hutt, New Zealand.

Further reading

Dixon J. B. & Schultze D. G. (2002) (Eds) *Soil Mineralogy with Environmental Applications.* Soil Science Society of America Book Series No. 7. Soil Science Society of America, Madison, Wisconsin.

Mitchell J. K. (1993) *Fundamentals of Soil Behavior*, 2nd edn. Wiley & Sons, New York.

Sposito G. (1989) *The Chemistry of Soils.* Oxford University Press, New York.

Velde B. (1992) *Introduction to Clay Minerals: Chemistry, Origins, Uses and Environmental Significance.* Chapman & Hall, London.

Example questions and problems

1 (a) Explain the difference between a plane of atoms and a crystal layer in the phyllosilicates.

   (b) (i) What is the *d* spacing (in nanometres) for kaolinite, and (ii) do kaolinite crystals swell in water (give a reason for your answer)?

   (c) Does kaolinite have a large CEC compared with montmorillonite (give a reason for your answer)?

2 (a) What is the origin of negative charge in (i) an allophane with a Si : Al mole ratio of 0.5 and (ii) an allophane with a Si : Al ratio equal to 1?

   (b) What is the name given to charges on crystal surfaces that depend on the association or dissociation of H^+ ions?

3 Air-dry soil equivalent to 10 g o.d. soil is completely dispersed and made into a suspension of 500 mL volume. The suspension is stirred and allowed to settle at a temperature of 20°C. The amount of silt and clay in the suspension is measured by sampling at a depth of 10 cm below the surface at particular times after the suspension was last stirred.

   (a) Calculate the appropriate times of sampling for (i) silt plus clay, and (ii) clay alone (use the International particle-size classification). Assume that the constant *A* in the settling equation B2.1.2 is 3.594×10^6 (1/m s) and the particle density is 2650 kg/m^3.

   (b) The suspension samples are taken with a 20 mL pipette, then dried and weighed. Suppose that the amounts of sediment in the first and second samples are 0.24 and 0.167 g, respectively (corrected for the amount of dissolved salts in the samples). Calculate (i) the soil's clay and (ii) silt content (% o.d. soil).

4 The silt fraction of a soil is separated from the clay and sand fractions by sedimentation and sieving. Assume the silt-size particles are approximately spherical in shape. Calculate the specific surface area for the particles at the upper and lower boundaries of the silt class size (according to the International particle-size classification). Express the specific surface area in m^2/g, assuming the particle density is 2.65 Mg/m^3 (note – the area of a sphere is $4\pi r^2$ and the volume is $4/3\pi r^3$, where *r* is the radius).

5 The clay mineral illite has formed in a soil by weathering of the primary mineral biotite over a long period of time. The content of K^+ ions in the biotite crystalline structure is reduced from 233 to 133 cmol charge/kg. Calculate:

   (a) The resultant CEC per kg of illite, and

   (b) How many moles of exchangeable Ca^{2+} ions per kg would be required to neutralize this CEC?

   (c) If the specific surface area of this illite is 200 m^2/g, calculate the surface charge density (cmol charge/m^2).

6 The clay fraction of a highly weathered Ferrosol, which has a pH of 5.5, contains 70% kaolinite, 10% smectite, 15% gibbsite and 5% goethite.

   (a) Calculate the net CEC per kg of clay, given that the CEC of kaolinite and smectite is 5 and 100 cmol charge (+)/kg, respectively, and the AEC of gibbsite and goethite at pH 5.5 is 20 and 40 cmol charge (−)/kg, respectively.

   (b) If the clay content of the soil is 50%, what is the net CEC of the soil?

   (c) Suppose that the soil is limed to raise the pH to 6.5, state with reasons what you would expect to happen to the net CEC of the soil? (Hint – check the behaviour of variable surface charges with change in pH in Section 7.1).

Chapter 3

Soil Organisms and Organic Matter

3.1 Origin of soil organic matter

The carbon cycle

The organic matter of the soil arises from the debris of green plants, animal residues and excreta deposited on the surface and mixed to a variable extent with the mineral component. The dead organic matter is colonized by a variety of soil organisms, most importantly micro-organisms, which derive energy for growth from the oxidative decomposition of complex organic molecules (the substrate). The combination of living and dead organic matter, irrespective of its source or stage of decomposition (but excluding the living parts of plants above ground), is called soil organic matter (SOM).

During decomposition, essential elements are converted from organic combination to simple inorganic forms, a process called mineralization. For example, organically combined N, P and S appear as NH_4^+, $H_2PO_4^-$ and SO_4^{2-}, ions, and about half the C is released as CO_2. Mineralization, especially the release of CO_2, is vital for the growth of succeeding generations of green plants. The remainder of the substrate C used by the micro-organisms is incorporated into their cell substance or microbial biomass, together with a variable proportion of other essential elements such as N, P and S. This incorporation makes these elements unavailable for plant growth until the organisms die and decay, so the process is called immobilization. The residues of the organisms, together with the more recalcitrant parts of

the original substrate, accumulate in the soil. The various interlocking processes of synthesis and decomposition by which C is circulated through soil, plants, animals and air – collectively the biosphere – comprise the carbon cycle (Fig. 3.1).

For the past 200 years or so, the release of CO_2 from the combustion of fossil fuels, respiration by organisms, land clearing and burning has exceeded the sequestration of C in living and dead organisms on land and in water. This has led to a steady rise in the CO_2 concentration in the atmosphere, currently at an annual rate of 0.4%, and to what has been described as the enhanced 'greenhouse effect' (Box 3.1). Globally, C in soil organic matter is a very large sink (c. 1580 Gt or 1580×10^9 t C, with an additional 800 Gt C in inorganic carbonates), so that changes in the dynamic balance between the soil, vegetation and the atmosphere can significantly affect the net flux of CO_2 to the atmosphere. Indeed, changing land management practices to achieve greater sequestration of C in SOM is part of the overall strategy to decrease net CO_2 emissions to the atmosphere. This makes the study of SOM – its composition, accumulation and decomposition – a very important topic.

Inputs of plant and animal residues

The annual return of plant and animal residues to the soil varies greatly with the climatic region and the type of vegetation or land use. The amount of above-ground material – the annual litter fall – is

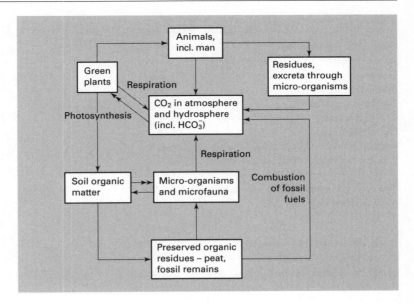

Fig. 3.1 The carbon cycle.

Box 3.1 Carbon cycling and the greenhouse effect.

The greenhouse effect is a natural phenomenon. Gases in small concentrations in the atmosphere – H_2O vapour, CO_2, methane (CH_4), nitrous oxide (N_2O), and the chlorofluorocarbons (CFCs) – are transparent to incoming short-wave solar radiation, but absorb the long-wave radiation emitted from the Earth's surface. Absorption of this energy warms the lower atmosphere (the troposphere) and traps heat which would otherwise be radiated into space. The net effect is very beneficial for life on Earth because, in the absence of this insulating effect, the average temperature would be a chilly 30°C lower than it is currently. The process is very similar to that in a greenhouse except that in this case, the glass is transparent to incoming solar radiation and absorbs much of the long-wave radiation emitted from the bodies inside.

The increase in gas emissions, mainly of CO_2 but also CH_4 and N_2O, due to human activities since the Industrial Revolution in the 18th century, has *enhanced* the greenhouse effect. Over the past 160 years, the cumulative net release of CO_2 from agriculture is estimated to have been 264 Gt C compared to 200 Gt C from fossil fuel combustion and cement production. The release from agriculture has been associated mainly with land clearing, especially of forests, burning and cultivation. Expansion of paddy rice cultivation and increased numbers of ruminant animals have contributed to the increase in CH_4 release, and greater use of N fertilizers to N_2O release (note that CH_4 and N_2O have respectively 21 and 310 times the global warming potential of CO_2 on a unit mass basis). Large-scale land clearing and deforestation now occur mainly in the tropics, and emissions from this source are currently *c.* 1.6 Gt C annually, compared with *c.* 6.4 Gt C from fossil fuel combustion.

The net annual increase in terrestrial C (soil organic and inorganic C plus vegetation C) is *c.* zero. However, because the annual increase of *c.* 3.4 Gt in $CO_2 - C$ in the atmosphere is less than half the total annual anthropogenic emissions of *c.* 8 Gt, there must be a significant sink for C in the terrestrial or marine environment that is not accounted for. Identifying this sink is difficult because of the uncertainties in calculating global net C fluxes and estimating the size of the known C sources and sinks.

Table 3.1 Annual rate of litter return to the soil.

Land use or vegetation type	Organic C* (t/ha)
Alpine and arctic forest	0.1–0.4
Arable farming (cereals)	1–2
Temperate grassland	2–4
Coniferous forest	1.5–3
Deciduous forest	1.5–4
Tropical rainforest (Columbia)	4–5
Tropical rainforest (West Africa)	10

* For an approximate conversion to organic matter, multiply these figures by 2.5.

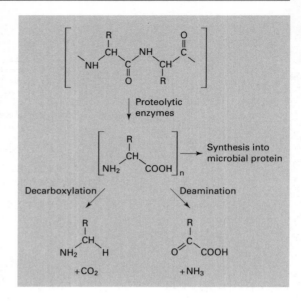

Fig. 3.2 Steps in proteolysis.

easily measured and some average values, in terms of organic C, are given in Table 3.1. To this must be added timber fall in forests (although this does not provide a readily decomposable substrate) and below-ground contributions from plant roots. Below-ground C comprises dead roots, 'sloughed off' epidermal root cells, and C compounds released from living roots into the surrounding soil – the latter two sources comprise the rhizosphere C (Section 8.1). Root contributions are difficult to measure, but experiments in which a plant shoot has been placed in an atmosphere of CO_2 labelled with the radioactive isotope of carbon (^{14}C) indicate that 5–10% of the total plant C may be released into the rhizosphere, excluding C respired as CO_2, which comes from both the roots and micro-organisms living in the rhizosphere (Tinker and Nye, 2000). Thus, the below-ground return of C (including dead roots) may range from 20 to 40% of the total photosynthate produced during a season of growth, although observations with trees suggest this figure may be as high as 60–70%. The return of C in animal excreta is obviously highly variable, but the figure for dung produced by dairy cows grazing pastures at a high stocking rate is 1.7–2.3 t C/ha/yr.

Composition of plant litter

Plant cells are made up primarily of carbohydrates, proteins and fats, plus smaller amounts of organic acids, lignin, pigments, waxes and resins. The bulk of the material is carbohydrate of which sugars (glucose, fructose and sucrose) and starch

are rapidly decomposed, while hemicellulose and cellulose are less readily decomposed. These and other decomposition reactions are catalysed by enzymes produced by specialized groups of micro-organisms, as identified in Section 3.2. Proteins are rapidly metabolized and proteolysis may begin in the senescing leaf before it has reached the soil. The ultimate products are amino-acids (Fig. 3.2), some or all of which may be used in protein synthesis by the micro-organisms. Whether or not there is amino-N surplus to the needs of the micro-organisms, so that *net* mineralization can occur, depends on the C : N ratio of the substrate and on the properties of the decomposer organisms (Section 3.3). Generally, when the substrate C : N ratio is > 20, net immobilization is likely, whereas at ratios < 20 net mineralization is favoured.

As shown in Table 3.2, the C : N ratio of plant residues and waste – fresh litter and animal excreta – returned to the soil is highly variable. But as the organic matter passes through successive cycles of decomposition in the soil, the C : N ratio gradually narrows and we find the C : N ratio of well-drained soils of pH ≈ 7 to be close to 10. Exceptions occur with poorly drained soils or those on which mor humus forms.

Nitrogen in complex heterocyclic ring compounds, such as chlorophyll, is not easily mineralized, but the chitin of insect cuticles and fungal

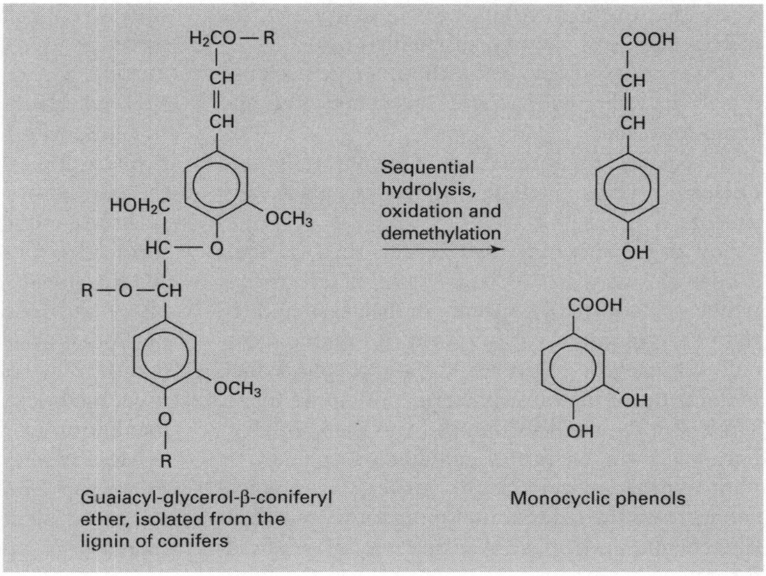

Fig. 3.3 Hydrolytic oxidation of a lignin-type compound (after Andreux, 1982).

Table 3.2 C : N ratios of freshly fallen litter, manure and soil.

	Range	Median value
Litter		
Herbaceous legumes	15–25	20
Cereal straw	40–120	80
Tropical forest species	27–32	30
Temperate hardwoods (elm, ash, lime, oak, birch)	25–44	35
Scots pine	–	91
Animal manure		
Farmyard manure	–	20
Soil		
Soils from 63 sites in the USA	7–26	12.8
Forest soils (27 sites)	11–44	21.4
Old pastures, fertilized and unfertilized	11–12	11.8
Old arable, manured and unmanured	8–10	9.4

cell walls, a glucosamine polymer, is eventually hydrolysed to glucose and NH_4^+ ions. Most plant organic acids are readily decomposed, but not the polymerized esters of long-chain hydroxy 'fatty acids' (i.e. complex phospholipids), which make up plant waxes and resins such as leaf cutin

and root suberin. These materials are particularly resistant, as is lignin, which is a complex phenolic polymer that is more stable than the carbohydrates and accumulates relative to these constituents during the decomposition of fresh residues. Lignin is a significant proportion of the dry matter of cereal straw (10–20%) and wood (20–40%). Its hydrolytic oxidation, catalysed by extracellular enzymes from various actinomycetes and fungi, produces monocyclic phenols as shown in Fig. 3.3. These degradative products may serve as precursors for the synthesis of humic macromolecules by the soil organisms (Section 3.3).

3.2 Soil organisms

The soil organisms may be grouped according to size into:
- *Macrofauna* – vertebrate animals mainly of the burrowing type, such as moles and rabbits, which live wholly or partly underground;
- *mesofauna* – small invertebrate animals representative of the phyla Arthropoda, Annelida, Nematoda and Mollusca;
- *micro-organisms* – comprising the microfauna (soil animals < 0.2 mm in length) and the microflora (bacteria, actinomycetes, fungi, algae and viruses).

According to their evolutionary development, the micro-organisms may be subdivided into:

- *Prokaryotes* (organisms without a true nucleus), which include the bacteria, actinomycetes and Cyanophyceae;
- *eukaryotes* (organisms with a membrane-bound nucleus), which include the fungi, algae and protozoa.

Soil organisms may also be classified by their mode of nutrition. Broadly, the heterotrophs, which include many species of bacteria and all the fungi, require C in the form of organic molecules for growth. However, the autotrophs, which include the remaining bacteria and most algae, synthesize their cell substance from the C of CO_2, harnessing the energy of sunlight (in the case of photosynthetic bacteria and algae), or chemical energy from the oxidation of inorganic compounds (the chemoautotrophs). Another way of subdividing the micro-organisms is on the basis of their requirement for molecular O_2; that is into:

- *Aerobes* – those requiring O_2 as the terminal acceptor of electrons in respiration;
- *facultative anaerobes* – those normally requiring O_2, but able to adapt to oxygen-free conditions by using NO_3^- and other inorganic compounds as electron acceptors in respiration;
- *obligate anaerobes* – those that grow only in the absence of O_2 because O_2 is toxic to them.

Biomass

Collectively, the mass of organisms in a given volume or mass of soil is referred to as soil biomass. Because the macro- and mesofauna can be physically separated from the soil, their mass can be measured directly and is usually expressed as kg (liveweight) per ha to a certain depth. The macro- and mesofaunal biomass ranges from 2 to 5 t/ha, with earthworms making the largest single contribution (Table 3.3). For a review of the variety of methods used to extract invertebrates from soil, in the laboratory and field, the reader should consult Edwards (1991).

However, micro-organisms are intimately mixed with the SOM and, being very small, are difficult to isolate for counting or weighing. This is particularly true of the microflora for which the individuals may be < 1 μm in size. For this group,

collectively called the soil microbial biomass, the methods used to measure numbers and/or mass include:

- Direct observations of organisms on agar plates or films, with or without dye staining;
- physiological or biochemical methods such as the extraction of adenosine triphosphate (ATP), substrate-induced respiration (SIR), and chloroform ($CHCl_3$) fumigation followed by extraction. Measurements in the second group tell nothing about the species of biomass, but give estimates of biomass size that are valuable for modelling C turnover in soil (Section 3.5). The most widely used method is $CHCl_3$ fumigation followed by incubation or extraction (Box 3.2).

Most of the micro-organisms are concentrated in the top 15–25 cm of soil because C substrates are more plentiful there. Estimates of microbial biomass C range from 500 to 2000 kg/ha to 15-cm depth, with larger values being recorded on a per hectare basis when the soil is sampled to greater depths. Alternatively, biomass C can be expressed as a percentage of total soil C, when it ranges from about 2% in arable soils to 3–4% in grassland or woodland soils.

Types of micro-organisms

Bacteria, including actinomycetes

These organisms can be as small as 1 μm in length and 0.2 μm in breadth and therefore live in water films around soil particles in all but the smallest pores. They can be motile or non-motile, coccoid (round) or rod-shaped (Fig. 3.4), and can reproduce very rapidly in the soil under favourable conditions – as little as 8–24 h for the production of two daughter organisms by division of a single parent cell. Accordingly, their number in the soil is enormous provided that living conditions, especially the food supply, are suitable. Bacterial numbers are estimated by observing the growth of colonies on special nutrient media which have been inoculated with drops of a very dilute soil suspension (the dilution plate method). If the bacteria cannot be cultured, the highest dilution that retains viable organisms is recorded and used to back-calculate the original population size (the most probable number (MPN) method). More

Box 3.2 Estimation of soil microbial biomass by $CHCl_3$ fumigation.

The method was developed for agricultural soils (generally < 5% organic C) that are not waterlogged nor very acidic (pH > 4.5). Measurements should not be made soon after large amounts of organic materials have been added to the soil. In the original method (Jenkinson and Powlson, 1976), a sample of soil (< 2 mm) was fumigated in ethanol-free $CHCl_3$ vapour to kill the organisms, the $CHCl_3$ evacuated, and the soil inoculated with fresh soil organisms before being incubated for 10 days at constant temperature. The flush of CO_2 evolved relative to a non-fumigated sample of the same soil was used to calculate the microbial biomass from the equation

$$\text{Biomass C} = \frac{(CO_2 - \text{C from fumigated soil}) - (CO_2 - \text{C from unfumigated soil})}{K_c},$$

$$(B3.2.1)$$

where K_c is the empirically derived fraction of killed biomass C that is evolved as CO_2. The appropriate K_c value depends on the composition of the microbial population, being lower for fungi than bacteria. The most commonly used values are 0.45 for incubations at 25°C (Oades and Jenkinson, 1979) and 0.41 for incubations at 22°C (Anderson and Domsch, 1978).

The $CHCl_3$ fumigation-extraction method is a more rapid variant of the fumigation-incubation technique that can also be used on very acid soils. The fumigated soil is extracted in 0.5 M K_2SO_4 for 30 minutes and the extracted organic C (EOC) is measured, relative to an unfumigated control soil. Biomass C is calculated from

$$\text{Biomass C} = \frac{(\text{EOC extracted from fumigated soil}) - (\text{EOC extracted from unfumigated soil})}{K_{ec}},$$

$$(B3.2.2)$$

where K_{ec} is an empirical factor measuring the efficiency of C extraction. K_{ec} values show a similar range to K_c, depending on the soil type, but a generally accepted value is 0.45.

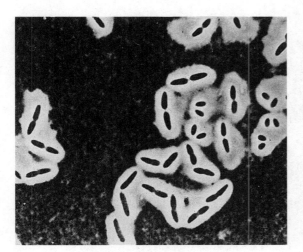

Fig. 3.4 Soil bacteria – *Azotobacter* sp. (after Hepper, 1975).

recent techniques use the extraction of bacterial desoxyribose nucleic acid (DNA) and probing for gene sequences specific to individual organisms or groups of organisms, combined with the MPN method (Angle, 2000). Estimates of $c.$ 1×10^9 organisms/g of soil, and > 20,000 species, have been obtained.

Bacteria exhibit almost limitless variety in their metabolism and ability to decompose diverse substrates. Three major subgroups are recognized:
1 *Unicellular Eubacteria* – the most numerous group including both heterotrophs and autotrophs;
2 *Branched Eubacteria* or *actinomycetes* – these are heterotrophs that form mycelial growths more delicate than the fungi (see below). They include such genera as *Streptomyces* that produce antibiotics and can degrade more recalcitrant C compounds such as lignin. Some thermophilic actinomycetes are common in high temperature (> 50°C) composts;
3 *Myxobacteria* (slime bacteria) – these unicellular organisms differ from the Eubacteria in the flexibility of their cell walls and their mode of

locomotion. Many are specialized in degrading cellulose and chitin.

The activities of some of the specialized groups of bacteria are discussed in Chapters 8 and 10.

Fungi

Some fungi, such as the yeasts, are unicellular, but the majority produce long filamentous hyphae, 1–10 μm in diameter, which may be segmented, branched or both. The network of hyphae (a mycelium) develops fruiting bodies on which the spores formed are often highly coloured. There are three main phyla of the true fungi (Thorn, 2000).

1 *Zygomycota* – these include the saprophytic Mucorales, often referred to as the 'sugar fungi' because of their ability to rapidly colonize a substrate by consuming the soluble sugars. The phylum includes the Endogonaceae, which are the endomycorrhizal fungi that form symbiotic associations with some 80% of the world's land plant species. The symbiosis is of mutual benefit, with the fungus (endophyte) deriving C compounds from the host plant, while supplying the host with mineral nutrients and in some cases improving the plant's ability to take up water (Section 10.3).

2 *Ascomycota* – largest of the fungal groups, the Ascomycota include fungi that reproduce sexually through spores and those that reproduce only asexually, the fungi imperfecti (and including the majority of yeasts). Of particular interest are the 20,000 or so species that form a stable symbiotic association with green algae or cyanobacteria (see below) to form lichens. Many species of Ascomycota are parasites or endophytes of aerial plant parts, or colonizers of animal dung.

3 *Basidiomycota* – these include the fungi that form ectomycorrhizas with trees (Section 10.3) and those that colonize woody residues because of their ability to decompose lignin, hemicellulose and cellulose – the 'white rot' and 'brown rot' fungi. Thus, they are common in forests where they attack the litter layer and wood residues, and may form spectacular colonies (Fig. 3.5). Others

Fig. 3.5 Basidomycete fungal colony on rotting wood (see also Plate 3.5).

such as *Rhizoctonia* species are serious pathogens of agricultural crops.

Although soil fungal colonies are much less numerous than bacteria ($1-4 \times 10^5$ organisms/g), because of their filamentous growth their biomass is generally larger than that of bacteria (by as much as $10:1$, but the proportion of actively growing mycelium may be less than half the total biomass). Fungal numbers are also considerably less variable than bacterial numbers because although fungi, excluding yeasts, are intolerant of anaerobic conditions, they grow better in acid soils ($pH < 5.5$) and tolerate variations in soil moisture better than bacteria. The fungi are all heterotrophic and are most abundant in the litter layer and organically-rich surface horizons of the soil, where their ability to decompose lignin confers a competitive advantage over bacteria.

Soil fungi are difficult to measure quantitatively and many, such as in the Basidiomycota, are difficult to culture because they produce few propagules from which they can be grown. Measurement methods include direct counts, with staining and sometimes bacterial inhibitors, and the extraction and isolation of fungal DNA and ribose nucleic acid (RNA). Fungal biomass is included in the microbial biomass measured by the $CHCl_3$ fumigation technique (Box 3.2). An outline of methods for identifying and enumerating the many bacterial and fungal species in soil is given in Weaver (1994).

Algae

The algae are the major group of photosynthetic organisms in soil and are therefore confined to the soil surface, although many will grow heterotrophically in the absence of light if simple organic solutes are provided. The two major subgroups are the prokaryotic Cyanophyceae (cyanobacteria), or blue-green algae, and the eukaryotic green algae. The 'blue-greens', as exemplified by the common genera *Nostoc* and *Anabaena*, are important because they can reduce atmospheric N_2 and incorporate the products into amino acids, thereby making a substantial contribution to the N status of wet soils (Section 10.2). The blue-greens prefer neutral to alkaline soils, whereas the green algae are more common in acid soils. Algae in soil are much smaller than the aquatic

or marine species and may number from 1×10^5 to 3×10^6 organisms/g.

The prolific growth of aquatic algae during past eras in the Earth's history has produced large areas of sedimentary rocks such as the English Chalk (made up of calcified algal bodies – Fig. 5.6), and deposits of diatomaceous earth (silica skeletons of diatoms).

Protozoa

The smallest of the soil animals, being $5-40\ \mu m$ in their longer dimension, the protozoa live in water films and move by means of cytoplasmic streaming (*Amoeba*), cilia, or flagellae, as in the case of *Euglena*. *Euglena* also contains chlorophyll and can live autotrophically, but nearly all of the protozoa prey upon other small organisms such as bacteria, algae, fungi and even nematodes. Protozoan biomass can be comparable to that of earthworms, so they are important in controlling bacterial and fungal numbers in the soil, and hence in nutrient cycling.

Soil enzymes

Enzymes are protein molecules, many with metal ions as cofactors, which catalyse biochemical reactions. They function inside organisms (intracellular) or outside (extracellular), where they may have been secreted by micro-organisms or plant roots, but act independently of the living organism. Extracellular enzymes avoid denaturation and degradation by being adsorbed on soil mineral or organic matter. The major types present are hydrolases, transferases, oxidases, reductases and decarboxylases. A very common and stable soil enzyme is urease, which is important in the hydrolysis of urea derived from animal urine or fertilizer (Section 10.2). Because of its biochemical function, the enzyme dehydrogenase is unlikely to occur in a free, extracellular state in soil. Thus, assay of this enzyme has been used as an indirect method of measuring soil microbial activity.

The mesofauna

Overall, the action of mesofauna in physically breaking down organic matter into smaller particles is much more important than the chemical alteration caused by their digestive processes. For this reason, the mesofauna have collectively been called reducers to distinguish them from the microflora, or decomposers, which cause chemical alteration through the action of intracellular and extracellular enzymes.

Of the mesofauna the threadlike nematodes, next to the protozoa, are the smallest, ranging from < 1 mm to a few mm in size. They are plentiful in soil and litter and may be saprophages or carnivores feeding on fungi, bacteria or other nematodes. Some species of nematodes are important pests of crop plants, living parasitically in or on the roots and causing infestations that can be fatal, especially in sandy soils. Other primitive soil animals are the molluscs (slugs and snails), many of which feed on living plants and are therefore pests, although some species feed on fungi and the faeces of other animals. Because the mollusc biomass is only 200–300 kg/ha in most soils, their contribution to the decomposition of SOM is small.

By far the most important mesofauna involved in the turnover of SOM are the arthropods and annelids.

Arthropods

This group includes the isopods (wood lice), arachnids (mainly mites), insects (winged and wingless) and myriapods (centipedes and millipedes). The more numerous are the mites and many of the insects, present as adults and larvae.

Mites and springtails (wingless Collembola) feed on plant remains and fungi in the litter, especially where thick mats build up under coniferous forests (as in mor humus, Box 3.3). Their droppings appear as characteristic pellets in the litter layer (Fig. 3.6). They may also be found deeper in the soil, living in the larger pores (0.3–5.0 mm). Predaceous adult mites and springtails feed on other mites and nematodes, while in their juvenile stage they feed on bacteria and fungi.

Many beetles and insect larvae live in the soil, some feeding saprophytically, whereas others feed on living tissues and can be serious pests of agricultural crops. Of particular value, however, are the coprophagous beetles (dung beetles) of Africa, Asia and other tropical regions that feed

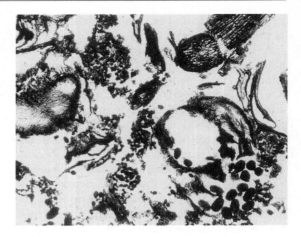

Fig. 3.6 Loose fabric of plant residues and mite droppings in mor humus (courtesy of P. Bullock).

Fig. 3.7 Dung beetle (*Sisyphus rubripes*) rolling a dung ball (after Waterhouse, 1973).

on the dung of large herbivores. There are three types of dung beetle: those that roll dung away in balls to be buried (Fig. 3.7), those that bury dung in tunnels under dung pads, and those that breed and feed in pads. Exotic dung beetles were introduced into Australia during the 1970s and 1980s to improve the rate at which cattle and sheep dung was broken down and physically mixed with soil under pastures. The goal was to decrease pasture spoilage, increase nutrient cycling and control the breeding of 'bush flies'. Although they are now widely dispersed in Australia, the coverage by the various species is patchy.

Fig. 3.8 Termite mounds in North Queensland.

The termite (a member of the Isoptera) has been called the tropical analogue of the earthworm, and indeed they are important comminuters of all forms of litter – tree trunks, branches and leaves – in the forest and especially the seasonal rainfall regions (savanna) of the tropics. Most species are surface feeders and build nests called termitaria by packing and cementing together soil particles with organic secretions and excreta. The small mounds only 30 cm or so high are the homes of *Cubitermes* species which feed under the cover of recent leaf fall (Fig. 3.8). The less frequent large mounds, some 4–6 m high, are the homes of wood-feeders and foraging termites, such as *Macrotermes* and *Odontotermes*, the latter usually colonizing the mounds of *Macrotermes* when they are abandoned. These species cultivate 'fungus gardens' within the termitaria to provide food for their larvae.

Some termite species or 'harvester ants' destroy living vegetation in African grasslands. In cool temperate regions, species of ant, such as *Formica*

fusca, are of some importance in comminuting litter under pine trees, whereas *Lasius flavus* is active in grassland soils on Chalk.

Centipedes are carnivorous whereas millipedes feed on vegetation, much of which is in the form of living roots, bulbs and tubers. Size comparisons between these and other soil invertebrates may be made by reference to the specimens in Fig. 3.9.

Annelids

These include the enchytraeid worms and the lumbricids or earthworms.

The enchytraeids are small (a few mm to 2–3 cm long) and threadlike. Their biomass is rarely > 100 kg/ha. However, the earthworms, because of their size (several cm long), numbers, and physical activity are generally more important in the consumption of litter than all the other invertebrates together. Their abundance ranges from < 100 per ha in taiga and montane forests

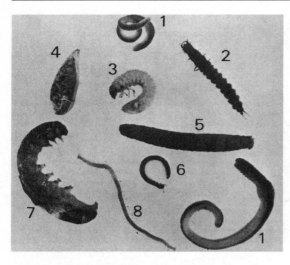

Fig. 3.9 The main litter-feeding soil animals (after Wild, 1988). 1 Earthworm, *Lumbricus* sp. (Oligochaeta); 2 Beetle larva, Carabidae (Coleoptera); 3 Chafer grub, *Phyllopertha* sp. (Coleoptera); 4 Slug, *Agriolimax* sp. (Mollusca); 5 Leatherjacket, Tipulidae (Diptera); 6 Millipede (Diplopoda); 7 Cutworm, Agrotidae (Lepidoptera); 8 Centipede (Chilopoda).

Table 3.3 Estimates of earthworm biomass in soils under different vegetation land use. Compiled from Edwards and Lofty, 1977 and Lee, 1985.

Vegetation and use	Earthworm biomass (kg liveweight/ha)
Tropical forest in Nigeria	25–100
Deciduous forests in Europe	370–980
Coniferous forest in Europe and Asia	50–350
Orchards (grassed)	640–1220
Sown pasture in New Zealand	500–3000
Arable land (cropped and fallow)	5–790

to > 20,000 per ha under productive grasslands. Numbers are also low in soils where mor humus forms (Box 3.3), and in arable soils that are frequently ploughed. An indication of the earthworm biomass under several systems of land use is given in Table 3.3.

Earthworms feed exclusively on dead organic matter, in the course of which they ingest large quantities of clay and silt-size particles. The large populations of earthworms in soils long down to grass may consume up to 90 t soil per ha

Box 3.3 Mor and mull humus.

Typically under deciduous forest species, there is a loose litter layer 2–5 cm deep under which the soil is well-aggregated and porous, dark-brown in colour and changes only gradually with depth to the lighter colour of the mineral matrix. This deep (30–50 cm) organic A horizon of C : N ratio 10–15 and rich in animals, especially earthworms, is characteristic of mull humus (Fig. B3.3.1a). Alternatively, under coniferous species, the surface litter is thick (5–20 cm) and often ramified by plant roots and a tough fungal mycelium. It is sharply differentiated from the mineral soil that is capped by a thin band of blackish humus and is usually compact, poorly drained and devoid of earthworms. This superficial organic horizon, of C : N ratio > 30, is called mor humus (Fig. B3.3.1b) and generally consists of three layers:

• *L layer* – undecomposed litter, primarily residues deposited during the previous year;

• *F layer* – the fermentation layer, partly decomposed, with abundant fungal growth and the faecal pellets of mites and springtails; and

• *H layer* – the humified layer, comprising completely altered plant residues and faecal pellets.

Mor and mull humus, and intermediate types called mor moder and mull moder, occur under a variety of vegetation types. As well as the main deciduous trees (oak, elm, ash and beech), mull usually occurs under well-drained grassland, whereas mor is found under heath vegetation (*Erica* and *Calluna* spp.) and under the major conifers (pine, spruce, larch and fir). However, the tendency towards mor or mull formation is determined not only by plant species, but also by the mineral status of the soil, particularly of Ca, N and P, so that the same species of deciduous tree may form mull on a fertile calcareous soil and mor on an infertile acid sand.

Box 3.3 *continued*

(a)   (b)

Fig. B3.3.1 (a) Uniform organic matter distribution in a soil with earthworms. (b) Litter accumulation on soil devoid of earthworms (after Edwards and Lofty, 1977).

annually. As a result, the organic and mineral matter is much more homogeneously mixed (as in mull humus, Box 3.3) when deposited in the worm faeces or casts. Epigeic species, such as *Lumbricus rubellus* and *L. castaneous*, live in the surface soil and litter layer, provided that temperature and moisture conditions are favourable, and cast on the soil surface (Fig. 3.10). The most common group is the endogeic earthworms that burrow more deeply, and deposit their casts in horizontal burrows. Included in this group are *Aporrectodea caliginosa*, *A. trapezoides* and *A. rosea*, abundant in soils of Europe, and which are the most numerous of the exotic species introduced into pasture soils in Australia. The Australian native species of Megascolecidae occur in undisturbed natural habitats, but are sparse in agricultural soils. The third group comprises the deep-burrowing or anecic species *A. longa* and *Lumbricus terrestris* that feed on

surface litter, but draw it into deep vertical burrows. When earthworms are abundant, the organic matter in litter is rapidly incorporated throughout the upper part of the soil profile, forming a mull humus profile, in contrast to the sharp boundary between the litter layer and mineral soil in the absence of earthworms (Box 3.3).

The consumption of dung and vegetable matter by earthworms can be prodigious. It has been estimated from feeding experiments that a population of 120,000 adult worms per ha is capable of consuming 25–30 t of cow dung annually. The quantities of dung and leaf litter available in the field are normally much less than this, so that the size of the earthworm population is regulated by the amount of suitable organic matter available, as well as by soil temperature, moisture and pH. For example, in hot dry weather the endogeic and anecic species burrow more deeply into the soil and aestivate. Of the common earthworms,

Fig. 3.10 Moist dark earthworm casts on the surface of a soil under pasture (see also Plate 3.10).

L. terrestris is more tolerant of low pH than the *Aporrectodea* species, and most species prefer neutral to calcareous soils. Few species of earthworms are found in soils of pH < 4.5. However, the burrowing species are effective at burying surface-applied lime and thereby increasing its effectiveness in acid soils (Section 11.3).

Grass and most herbaceous leaf litter are readily acceptable to earthworms, but there is greater variability in the palatability of litter from temperate forest species. Litter from elm, lime and birch is more palatable than pine needles, which must age considerably before being acceptable.

3.3 Changes in plant remains due to the activities of soil organisms

The first stages of decomposition

Decomposition begins with the invasion of ageing plant tissue by surface saprophytes, and proceeds in parallel with biochemical changes in the senescing tissue – the synthesis of protease enzymes, the rupture of cell membranes with consequent mixing of cellular constituents, and the auto-oxidation and polymerization of phenolic-type compounds. Decomposition accelerates, as evidenced by the rise in the rate of CO_2 production, when the plant material falls to the ground and is invaded by a host of soil organisms. The changes that occur in the plant residues lead not only to mineralization and immobilization of nutrient elements, but also to the synthesis of new compounds, less susceptible to decomposition, which collectively form a dark-brown to black, amorphous material called humus.

Colonization by soil micro-organisms

Polyphenolic compounds formed in the senescing leaf, which are not leached out by rain once it has fallen to the ground, have a considerable effect on the rate of decomposition of the leaf residue. This is well illustrated by the contrasting superficial layers and A horizons of soils under temperate forests (Fig. B3.3.1). The formation of mor humus is predisposed by high concentrations of phenols in the senescing leaf which precipitate cytoplasmic proteins onto the mesophyll cell walls, thereby making both the protein and cellulose more resistant to microbial decomposition. The process has been likened to the 'tanning' of leather, whereby plant tannins (polyphenols) are used to preserve the protein of animal skins. Where plant proteins and cellulose are not protected by tanning, these constituents and the sugars and storage carbohydrates are rapidly metabolized by soil microorganisms that colonize the litter in the sequence:

sugar fungi and non-spore-forming bacteria → spore-forming fungi and bacteria → myxobacteria → actinomycetes

As indicated in Section 3.1, whether or not net mineralization occurs during decomposition depends on whether the C : N ratio of the substrate is below or above a critical value of *c.* 20. The origin of this critical C : N ratio is shown in Box 3.4.

The role of soil animals

There is much interdependence between the activities of the microflora and small invertebrates in the decomposition of organic matter and formation of humus. Studies on the ecology of soil organisms suggest the existence of a 'food web' with several trophic (feeding) levels, and complex interactions among organisms within a trophic level and among levels. At the bottom of the web are plant litter, roots and detritus from organisms feeding at all the higher levels (including aboveground). Successively, going up the web, there are:
• Bacteria, saprophytic fungi and nematodes that feed on roots;
• protozoa, nematodes that feed on bacteria and fungi, and mites that feed on fungi;
• nematodes that feed on other nematodes and omnivorous nematodes;
• top predators – mites and collembola – that feed on subordinate organisms.

Box 3.4 Determining the critical substrate C : N ratio for net mineralization of N.

For the N requirements of the decomposers to be met, the C : N ratio of the substrate being decomposed must satisfy the condition

$$\left(\frac{C}{N}\right)\text{substrate} \leq \frac{C_a}{N_m}, \qquad \text{(B3.4.1)}$$

where C_a, the C assimilated, is equal to the sum of C_m, the C incorporated into microbial cells, and C_r, the C respired as CO_2. N_m is the microbial N. Equation B3.4.1 may be written as

$$\left(\frac{C}{N}\right)\text{substrate} \leq \frac{C_a}{C_m} \times \frac{C_m}{N_m}. \qquad \text{(B3.4.2)}$$

Note that the ratio C_m/C_a is defined as the growth yield, or microbial efficiency, of the micro-organisms. The critical C : N ratio of the substrate is therefore given by

$$\frac{\text{C : N ratio of the microbial biomass}}{\text{growth yield}}. \qquad \text{(B3.4.3)}$$

C : N ratios range from 3.3 for the bacterium *Escherichia coli* to 12.9 for the fungus *Penicillium*, with the mean C : N ratio for bacteria being *c.* 4 and for fungi *c.* 10. Thus, the critical C : N ratio will depend on the composition of the soil microflora, but taking an average microbial biomass C : N ratio of 8 and a growth yield of 0.4 gives

$$\left(\frac{C}{N}\right)\text{critical} = \frac{8}{0.4} = 20. \qquad \text{(B3.4.4)}$$

A quantitative description of energy and nutrient (N, P and S) flows through such food webs helps to explain differences in productivity and physical and biological condition among various ecosystems such as grassland, arable land or high

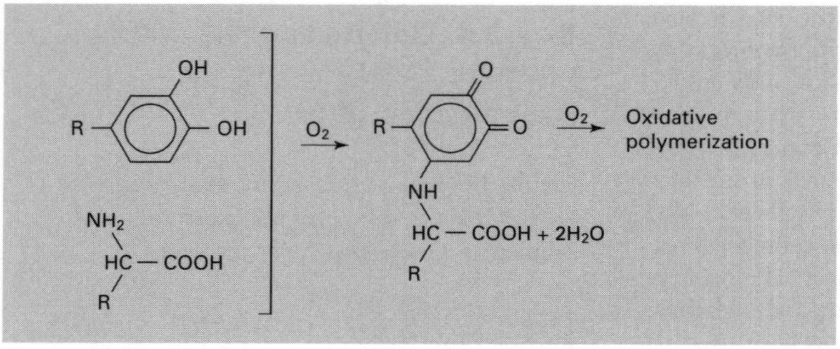

Fig. 3.11 Reaction between an amino acid and a simple phenol under oxidizing conditions (after Andreux, 1982).

input organic farms, and provides a pointer to their long-term sustainability. Although simulation models of food webs have been attempted, such is the complexity of the information required that it is currently impossible to test by direct measurement the validity of many of the input variables and model parameters. Thus, C turnover and mineralization are usually described by more pragmatic models based on substrate properties and decomposition rates, as discussed in Section 3.5.

Of the larger invertebrates, termites are unusual in that they can digest cellulose and lignin by means of their intestinal microflora, or the fungus gardens they cultivate in their termitaria. In the case of earthworms, some microbial decomposition of the litter must occur before it is readily acceptable to these animals. Once ingested, the main effect of an earthworm's digestive activity is the comminution of the plant residues, which vastly increases the surface area accessible to microbial attack. There is some chemical breakdown, usually less than 10%, which could be due to the bacteria in the earthworm gut or the effect of chitinase and cellulase enzymes secreted by the earthworms. The combined result of these processes is that earthworm casts have a higher pH, exchangeable Ca^{2+}, available P and mineral N content than the surrounding soil. Whereas the partially humified residues in the L and F layers of mor humus retain a recognizable plant structure for several years, similar residues which have passed through the earthworm's gut several times are reduced to a black, amorphous humus that is intimately associated with the mineral particles (Fig. B3.3.1a).

Biochemical changes and humus formation

The biochemical changes, which in total comprise humification, are complex because both degradative and synthetic processes occur simultaneously. Certainly plant carbohydrates and proteins are decomposed and their microbial analogues synthesized, although the latter also become substrates to be decomposed in time. Important polymerization reactions involving aromatic* compounds also occur. For example, simple o- and p-hydroxy phenols occurring naturally in plants, or produced from the degradation of lignin and polyphenolic pigments as seen in Fig. 3.3, are oxidatively polymerized to form humic precursors. The oxidation can be autocatalytic or catalysed by the polyphenol oxidases. Nitrogen is incorporated into the polymers if amino acids condense with the phenols before polymerization occurs, as shown in Fig. 3.11. Some of the quinone rings formed during polymerization may be broken and additional carboxyl groups formed, as shown in Fig. 3.12, which augment the acidity of the carboxyl groups in the non-aromatic (i.e. alkyl) side chains. Carboxyls and the remaining reactive phenolic-OH groups form salts or chelate complexes (Section 3.4) with metal ions, which increase the stability of the macromolecule. Their longevity in soil also depends on physical stabilization due to adsorption on mineral surfaces.

A typical elemental analysis of humic compounds extracted from soil is 50–58% C, 5–6% H, 3–6% N and 30–33% O. In view of the

---

* Unsaturated ring structures of C and H, of which the simplest is benzene C_6H_6.

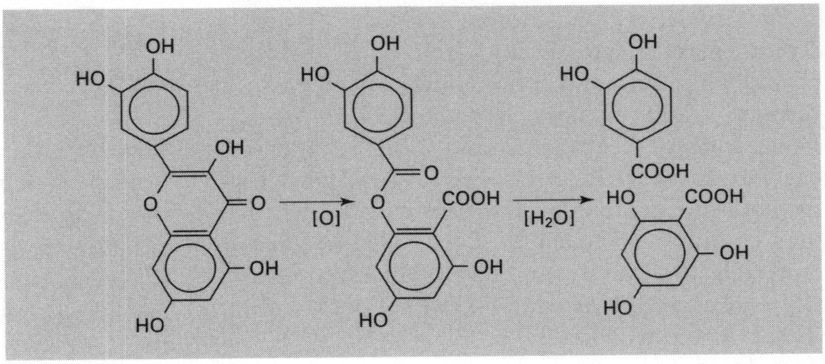

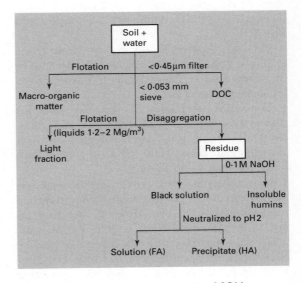

Fig. 3.12 Enzymatic decomposition of a quinone polymer, with an increase in carboxyl groups (after Andreux, 1982).

diversity of complex molecules that can be formed, it is not surprising that the identification of the chemical structure of these compounds has proved difficult. The C : H ratio is an index of aromaticity, the minimum value being 1, as for benzene. Values > 1 reflect the degree of condensation of the rings and the substitution of other elements for H in the structure. An accurate knowledge of the structure of humic compounds would help in assessing the validity of the various hypotheses of humus formation, and in explaining the humic properties described in Section 3.4.

3.4 Properties of soil organic matter

The amount of SOM

Soil organic matter is measured by combusting a soil sample in a furnace to convert C to CO_2, which is then measured using gas chromatography or by trapping in an alkali solution and titration. A correction must be made for any $CaCO_3$ present in the soil, and the result is reported as organic C. Alternatively, the SOM can be oxidized by wet digestion in a mixture of sulphuric acid and potassium dichromate, and the excess dichromate titrated with ferrous ammonium sulphate – commonly called the Walkely–Black method (Nelson and Sommers, 1996). Because oxidation of some of the more recalcitrant SOM is incomplete (including charcoal), an average multiplication factor of 1.3 is used to convert to soil organic C. Values of SOM should be reported as organic C (% or g/kg) and range from 5 to 150 g C/kg, except for peats

(Histosols (ST) or Organosols (ASC)) in which the content can be as high as 550 g C/kg.

Components of SOM

Various methods have been used to determine the components of SOM. Typically, the soil is subjected to extraction in water to obtain a dissolved organic carbon (DOC) fraction, followed by some combination of dispersion, sieving, density flotation, and chemical extraction to remove material tightly bound to mineral particles. An example of a procedure, which yields components that can be identified with some of the pools in models of C turnover (Section 3.5), is outlined in Fig. 3.13. The main non-water soluble components are:

Fig. 3.13 Scheme for the fractionation of SOM.

1 *Particulate organic matter* (POM) – mostly recent plant, animal and microbial debris, partially humified and comminuted, which passes through a 200-μm sieve, but not a 53-μm sieve. It has a density between 1 and 2 Mg/m^3 and so can be separated from heavier mineral particles by flotation in liquids of density $\leq 2\ Mg/m^3$. It may comprise up to 25% of the SOM in grassland soils, and can be used to give an initial value to the resistant plant material (RPM) pool in the ROTH C model (Box 3.7).

2 *A humified fraction* – material that passes through a 53-μm sieve, much of which adheres strongly to mineral particles, particularly clay, to form a clay–humus complex. This organic matter cannot be separated by density flotation and is incompletely dissolved in metal-chelating solvents such as acetyl-acetone. This is a composite fraction consisting of very inert organic matter, mainly charcoal in many soils, and the balance of humus corresponding to the HUM pool in the ROTH C model (see below).

3 *Humus* has traditionally been extracted with a strong alkali, such as sodium hydroxide, which acts by separating the negatively charged organic matter from clays, and by hydrolysing and depolymerizing the large organic molecules. The subdivision of humus into fulvic acid (FA) and humic acid (HA) fractions, according to its solubility in strong acids and alkalis, is shown in Fig. 3.13. Some of the problems associated with the identification of humic compounds are outlined in Box 3.5.

Although progress in elucidating the structure of humic compounds in soil is slow, because of the complexity of the task, the following points are generally accepted:

• The 'core' structure of the HA and FA fractions is composed of alkyl and aromatic groups covalently bonded to form polymers of high relative molar mass ($1000 - 1 \times 10^6$ g/mole) with much branching and folding;

• polysaccharides and proteinaceous materials occupy the voids in the folded and branched macromolecules, or are adsorbed, but they do not contribute more than *c.* 20% of the total mass. They may be covalently bonded or held by electrostatic charge effects or hydrogen bonding;

• 35–45% of the HA is aromatic, consisting mainly of single ring structures that are highly

---

Box 3.5 Characterization of soil humus.

Central to the chemical characterization of humus is the problem that, because strong solvents and fairly drastic treatment are necessary to quantitatively recover the humus from soil, it is unlikely that the compounds in the extract are the same as those originally in the soil. Further treatment with strong oxidizing agents such as $KMnO_4$ or an alkaline cupric oxide medium produces other changes in the compounds present. Nevertheless, powerful analytical techniques such as cross-polarization, magic angle spinning (CPMAS), ^{13}C (a stable isotope of C) nuclear magnetic resonance (NMR), Fourier Transform Infrared (FTIR) spectroscopy and pyrolysis-gas chromatography-mass spectrometry have been applied to the extracts of humic materials from soil. Reviews of the results of such studies are given by Schnitzer and Schulten (1995). To avoid the use of strong solvents to extract humic compounds, other scientists (Baldock and Nelson, 2000) recommend the use of ^{13}C NMR and FTIR on SOM *in situ*, and on organic matter separated on the basis of particle size and density (e.g. the 'light fraction'). Solid-state NMR spectra have shown systematic differences among the organic matter in particles of different size, changing from material of plant origin to microbial products as the particle size decreases (and degree of humification increases).

---

substituted, particularly with carboxyl (–COOH) and phenolic hydroxyl groups of varying acid strength. Alkyl chains, also containing carboxyl groups, link the aromatic groups or exist as separate side chains attached to the groups. They confer a degree of hydrophobicity on the HA molecules;

• approximately 25% of the FA is aromatic. FA molecules are smaller, more highly charged and more polar than HA. Charge repulsion causes the FA molecules to be more linear than the randomly coiled HA molecules. They do not precipitate in acid solutions and are susceptible to leaching;

Box 3.6 pK_a values and buffering.

Acids are defined as potential proton (H^+ ion) donors, and bases as potential proton acceptors. In aqueous solutions, acids dissociate H^+ ions that are accepted by H_2O. For example, the dissociation of an acid HA in water is written

$$HA + H_2O \leftrightarrow H_3O^+ + A^-. \qquad (B3.6.1)$$

The ion H_3O^+ is called a hydronium ion (normally written as H^+). The degree of dissociation of HA determines the strength of the acid, which is measured by the dissociation constant K_a, defined as

$$K_a = \frac{a_{H^+} a_{A^-}}{a_{HA} a_{H_2O}}, \qquad (B3.6.2)$$

where the terms on the right-hand-side are activities (Box 7.1). The activity of water (a_{H_2O}) is 1, and in dilute solutions, the activities of the other species can be replaced by concentrations. Thus, Equation B3.6.2 is normally written as

$$K_a = \frac{[H^+][A^-]}{[HA]}, \qquad (B3.6.3)$$

where the terms in square brackets are concentrations. If converted to the negative log form (cf pH), it is written

$$pK_a = pH - \log_{10} \frac{[A^-]}{[HA]}. \qquad (B3.6.4)$$

The smaller the value of pK_a, the stronger the acid. Most soil acids (associated with organic matter and soil minerals) are weak acids. When Equation B3.6.4 is rearranged as

$$pH = pK_a - \log_{10} \frac{[HA]}{[A^-]}, \qquad (B3.6.5)$$

we see that when there are equal concentrations of undissociated acid HA and base A^-, the $pH = pK_a$ (the half-neutralization point). At this point, the addition of small amounts of acid or base causes least change in pH. This is therefore the point of maximum buffering of the solution. The buffering capacity of a soil or solution is its resistance to pH change when acid or base is added. SOM has a buffering capacity of 10–40 cmol charge per kg SOM per pH unit.

• the type and steric arrangement of functional groups (–COOH, phenolic-OH and carbonyl C = O) facilitates the complexation of metal cations (see below);
• the non-extractable *humins* are thought to be HA-type compounds that are strongly adsorbed, or precipitated on mineral surfaces as metal salts or chelates.

Other properties of humus

Cation exchange capacity (CEC)

Humification produces an organic colloid of high specific surface area and high CEC. Of the several functional groups containing O, those which dissociate H^+ ions – the carboxylic and phenolic groups – are the most important. The former have pK_a values between 3 and 5 while the latter only begin to dissociate protons at pH > 7. A few basic $-NH_3^+$ groups may exist at pH < 3. Thus, the CEC of SOM is entirely pH-dependent (Section 7.2) and buffered over a wide range of H^+ ion concentration: the CEC ranges between 60 and 300 cmol charge/kg dry matter. Organic matter can make a substantial contribution to the CEC of the whole soil, and hence to the retention of exchangeable cations, especially in soils of low clay content. An explanation of the terms pK_a and buffering is given in Box 3.6.

Chelation

Humic compounds can form coordination complexes with metal cations by displacement of some

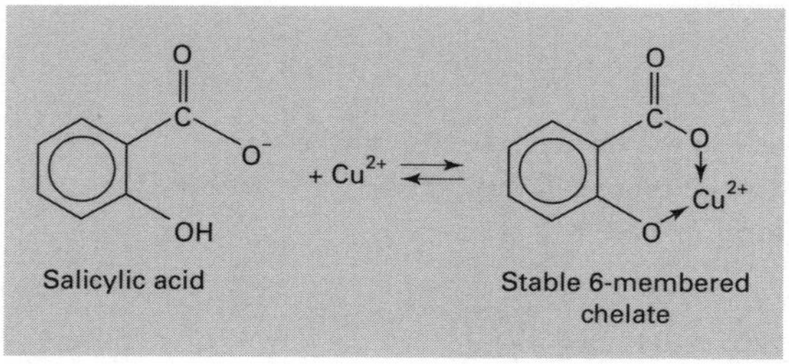

Fig. 3.14 Organo-metal chelate formation.

of the water molecules from the cation's hydration shell. The resultant inner-sphere complex (Section 2.3) acquires additional stability through the 'pincer effect' of the coordinating groups, giving rise to a chelate compound (Fig. 3.14). Chelates formed with polyvalent cations such as Fe^{3+} and Al^{3+} are the most stable, but within the divalent cation group, ion radius (unhydrated) and electronegativity determine the stability of the complex: for example, the stability constants of heavy metal cations fall in the order Cu > Ni > Pb > Co > Zn > Mn. Metal complexes formed with the HA fraction are largely immobile, but the more ephemeral Fe^{2+}–polyphenolic complexes are soluble and their downward movement contributes to profile differentiation in podzols (Section 9.2).

Organic matter and soil physical properties

The presence of organic matter is of great importance in the formation and stabilization of soil structure. Polymers of the FA and HA fractions are adsorbed on mineral surfaces by a variety of mechanisms, primarily involving the functional groups – carboxyls, carbonyl (C = O), alcoholic-OH, phenolic-OH, amino (= NH) and amine (–NH_2). These reactions are discussed in more detail in Sections 4.4 and 7.5. Uncharged non-polar groups on the humic polymers, and large uncharged polymers such as the polysaccharide and polyuronide gums synthesized by many bacteria, can be adsorbed on mineral surfaces by H-bonding or van der Waals' forces (Table 4.2). They also function as bonding agents between mineral particles.

Humified organic matter has an affinity for water and can absorb and hold water up to several times its dry weight.

3.5 Factors affecting the rate of organic matter decomposition

Turnover

Gains of organic matter through litter decomposition and root death are offset by losses through decomposition and leaching of DOC. This is called turnover, which is defined as the flux of C through the organic C in a unit volume of soil. Given a constant environment, a soil will eventually attain a steady-state equilibrium in which there is no measurable change in the organic matter content with time. The concept of turnover may be expressed mathematically by the equation

$$\frac{dC}{dt} = A - kC, \tag{3.1}$$

where dC/dt describes the rate of change in the soil organic C content (t/ha) with time, and A is the annual addition of residues, assumed to be approximately constant. The equation

$$\frac{dC}{dt} = -kC \tag{3.2}$$

describes the decomposition rate of the organic C if all the C decomposes at the same rate according to simple first-order kinetics, with k being the rate coefficient (1/year). The integral form of Equation 3.2

$$C(t) = C(0)\exp(-kt) \tag{3.3}$$

can be used to calculate how much C remains after decomposition for time t.

At steady-state equilibrium, dC/dt in Equation 3.1 is zero, and therefore

$$A = kC. \tag{3.4}$$

The turnover time is then given by

$$\frac{C}{A} = \frac{1}{k}. \tag{3.5}$$

Annual inputs of C for old arable soils at Rothamsted in England have been estimated at

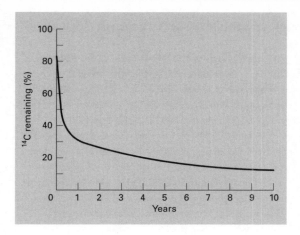

Fig. 3.15 Simplified time-course for the decomposition of ^{14}C-labelled C materials in the field (after Jenkinson, 1981).

5–6% of the soil organic C, giving turnover times of 16–20 years. Theoretically, this is the time taken for all the soil organic C to be replaced, and in a sense is inversely related to the 'biological activity' of the soil. This picture is grossly over-simplified because the age of C in these soils as measured by radioactive carbon (^{14}C) dating is 1,290–3,700 years, depending on depth. It is obvious that there must be a great range of decomposability of C compounds in soil, with some of the C being much younger than 16–20 years and the bulk of it very much older.

This conclusion is confirmed by measurements of the rate of disappearance of ^{14}C from field soils to which ^{14}C-labelled residues have been added. For various residues (ryegrass shoots and roots, green maize, wheat straw), in a temperate climate, a surprisingly constant fraction (*c.* two-thirds) of the material decomposes and disappears in the first year (Fig. 3.15). In straw, the most resistant component is lignin, but in the case of glucose and other sugars, the residue is mainly microbial metabolites produced during decomposition. After 5 years, approximately one-fifth of the added ^{14}C remains and its decomposition rate then still exceeds that of the unlabelled native soil C. Thus, although simple models such as Equation 3.1 can describe the C balance of soil–plant systems in steady state (attained only after many

years), more complex models are needed to fit data in the years following a change in the addition of residues to soil, or when changes in soil management are made. Such models divide the SOM into a number of 'pools' of different decomposability, represented by different k values. The models generally identify at least three modules corresponding to these pools:

• Litter and other inputs of organic matter (manure, root exudates and dead roots);
• microbial biomass;
• one or more components of humified organic matter.

Within these modules, further subdivisions may be made based on the age and chemical composition of the residues, protection of biomass and its functional activity, and chemical or physical stabilization of SOM by soil minerals, giving rise to a wide range of k values. Decomposition rates are moderated by factors for temperature and moisture effects, and sometimes for the effect of clay (see below). The predatory effect of soil fauna on bacteria and fungi is subsumed into the k value for biomass decay. Most models postulate a component of very old, stable organic matter that should remain constant with time. In Australian soils, this component has been identified as inert charcoal (presumed to be a relic from frequent bushfires), and comprises 5–40% of the SOM.

For the reasons outlined in Section 3.1, there is much interest in SOM models in the context of modelling global climate change, or the sustainability of current land uses (Chapter 15). Two of the more popular models are the CENTURY model (Parton *et al.*, 1988) and ROTH C (Coleman and Jenkinson, 1996). The two models are broadly similar in structure. ROTH C evolved from the concepts set out by Jenkinson and Rayner (1977) and is briefly described in Box 3.7. It has been extensively tested on long-term field data, is modest in its input requirements, and parsimonious in its parameters. Whereas the CENTURY model simulates C, N, P and S dynamics for several soil–plant systems (and includes a module for plant production and litter return), ROTH C deals only with soil and environmental processes affecting C turnover (although it has been linked to the dynamics of soil N turnover – see Box 11.4).

Box 3.7 ROTH C (ROTH C-26.3), a model to simulate SOM turnover.

SOM is divided into four active components and an amount of inert organic matter (IOM), usually < 20% of the total, which does not turnover (age 50,000 years). The other components are decomposable (DPM) and resistant (RPM) plant material, microbial biomass (BIO), and humified organic matter (HUM). Given the C input is known and RPM is equated to the particulate organic matter (POM), DPM can be calculated. The HUM pool size can be obtained by fractionation (Fig. 3.13) and the IOM equated to charcoal. Each active component decomposes by first-order kinetics according to the rate constants:

DPM $k = 10$ (1/year)
RPM $k = 0.3$ (1/year)
BIO $k = 0.66$ (1/year)
HUM $k = 0.02$ ((1/year).

These k values were set by tuning the model to long-term field data at Rothamsted. If the

radiocarbon age of the soil is known, the IOM fraction can be adjusted so that the calculated average age of C equals the measured age. The components decay in a 'cascade' effect, as illustrated in Fig. B3.7.1, producing CO_2, BIO and HUM.

The ratio DPM/RPM in the input organic matter is adjusted for different vegetative cover and whether any pre-decomposition has occurred (as in manure or compost). The ratio CO_2 : (BIO + HUM) decreases as the clay content of the soil increases to about 25%, reflecting the protection afforded organic matter when it is adsorbed on clays or occluded inside aggregates. The decay rates are modified for the effects of temperature, soil water deficit and 'plant retainment' – that is, the effect of plants in slowing the rate of organic matter decomposition relative to the rate in fallow soil. The model has been tested on long-term field data in the UK, Africa and Australia and is continually undergoing revision as new information becomes available.

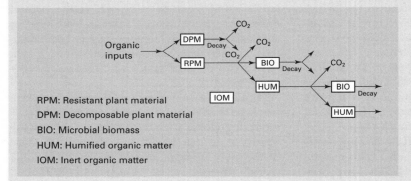

RPM: Resistant plant material
DPM: Decomposable plant material
BIO: Microbial biomass
HUM: Humified organic matter
IOM: Inert organic matter

Fig. B3.7.1 Structure of the Rothamsted Carbon Model.

Substrate availability

Priming action

It was once thought that the stimulus to the resident soil micro-organisms provided by the addition of fresh residues was sufficient to accelerate the decomposition of soil humus. This effect was called priming action. Priming action, which may be positive or negative, can be demonstrated by following the decomposition of ^{14}C-labelled residues, as illustrated in Fig. 3.16. The conclusion from this kind of experiment is that priming action generally has a negligible effect on the overall rate of

decomposition. The addition of fresh residues does stimulate microbial activity, but the new biomass has little effect on the rate of humus decomposition. The ROTH C model of SOM turnover ignores any priming effect.

Soil properties and environmental conditions

Soil organic matter content is affected by moisture, O_2 supply, pH and temperature. The first two factors tend to counteract one another because when a soil is wet, deficiency of O_2 may restrict decomposition, whereas when a soil is dry, moisture but not O_2 is limiting. pH has little

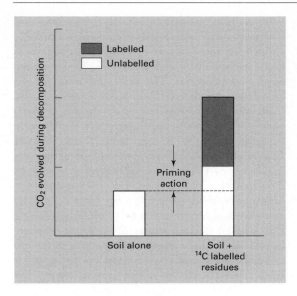

Fig. 3.16 Priming action measured by the use of ^{14}C-labelled residues (after Jenkinson, 1966).

pores, is then exposed to micro-organisms and its decomposition accelerated. Higher surface soil temperatures are also attained when the protection of the vegetative canopy and litter layer is lost.

The effect of clay content in retarding SOM decomposition is moderated according to temperature. The part played by clay minerals and particularly sesquioxides in protecting SOM under tropical conditions, where the potential rate of decomposition is fast, is much greater than in temperate regions. Also in soils derived from volcanic ash, the considerable accumulation of organic matter is attributed to the stabilization of HA and FA fractions through their adsorption on the Al sheets of imogolite and allophane.

Activity of the soil biomass

Numbers of organisms

The effect of environmental and soil factors on the size and species diversity of the soil microbial population has been referred to briefly. Soil animals are also affected by types of litter, prevailing conditions of temperature, moisture and pH, and management practices, such as cultivation and the use of agricultural chemicals.

Physiological activity

Environmental and management factors affect the physiological activity of soil organisms. Striking effects occur, for example, when the soil is partly sterilized with toxic chemicals, such as chloroform or toluene. On restoration of favourable conditions, the rapid multiplication of the few surviving organisms, feeding on the bodies of the killed organisms, produces a flush of decomposition as evidenced by a surge in CO_2 production. Similar flushes of decomposition occur in soils subjected to extremes of wetting and drying (Section 10.2), and in soils of temperate regions on passing from a frozen to thawed state. As well as changes in the availability of microbial substrates in these situations, there can be changes in other C fractions that lead to an increase in the labile soil C pool, and hence greater biomass activity.

The dynamics of soil biomass change in response to variations in soil and environmental factors are summarized graphically in Fig. 3.17.

effect, except below 4 when the decomposition rate slows, as in the case of mor humus and many upland peats (Section 9.5). On the other hand, temperature has a marked effect, not only on plant growth and hence litter return, but also on litter decomposition through an effect on microbial respiration rate (Section 8.1). In a comparison of Nigerian (mean temperature 26.1°C) and English (mean temperature 8.9°C) soils, Jenkinson and Ayanaba (1977) found that the rate of disappearance of organic residues in both cases could be represented by the curve in Fig. 3.15, provided that 10 years in southern England were equated with 2.5 years in Nigeria.

The adsorption of protein by montmorillonite, especially in the interlayer regions, protects the protein from microbial attack. The protective effect of clays and sesquioxides in slowing the decomposition of organic compounds is related primarily to the specific surface area of these materials (Baldock and Nelson, 2000). Further, organic matter held in relatively stable pores < 3 μm in diameter inside aggregates is less accessible to microbial attack. If the aggregates are large, lack of O_2 at the centre may retard the rate of organic matter decomposition (see Box 8.1). Cultivation of the soil tends to break down aggregates so that organic matter inside, and in small

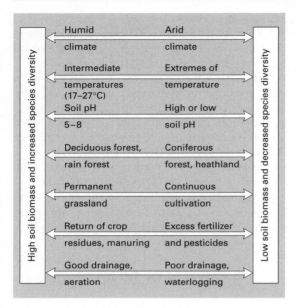

Fig. 3.17 Dynamics of soil biomass changes.

3.6 Summary

The input of C to soil in the form of plant litter, dead roots, animal remains and excreta, all of which are consumed by a heterogeneous population of soil organisms, is an important part of the global C cycle. As the organic residues decompose, releasing a proportion of the C as CO_2 to the atmosphere, SOM accumulates and serves as a repository of C and other essential elements for successive generations of organisms. Carbon dioxide is the major gas contributing to the enhanced 'greenhouse gas' effect, which is leading to global warming. At any one time, a small proportion of the C resides in the microbial biomass. This is called immobilization. Nutrient elements such as N are released – a process of mineralization – as organic matter decomposes. Whether or not there is net immobilization or mineralization depends on the nature of the decomposing substrate and the species composition of the soil organisms. Generally, C : N ratios > 20 favour an increase in microbial biomass and net immobilization of N, whereas C : N ratios < 20 favour net mineralization.

The living organisms or biomass vary in size from the macrofauna, vertebrate animals of the burrowing type, and the mesofauna, invertebrates such as mites, springtails, insects, earthworms and nematodes, to the micro-organisms – broadly subdivided into the prokaryotes and eukaryotes and comprising bacteria, actinomycetes, fungi, algae and protozoa. The most important of the mesofauna are the earthworms in temperate soils, especially under grassland, and termites in many soils of the tropics and subtropics. Earthworm biomass normally ranges from 5 to 3000 kg (liveweight) per ha, whereas the microbial biomass makes up 2–4% of the total organic C in soil, or 0.5–2 t C/ha.

Annual rates of litter fall range from 0.1 t C/ha in alpine and Arctic forests to 10 t C/ha in tropical rain forests. Carbon substrates from rhizosphere deposition and the death of roots also make important contributions to the annual C input, which is *c.* 20–40% of the above-ground input. The interaction of micro-organisms and mesofauna in decomposing litter leads not only to the release of mineral nutrients, but also to the synthesis of complex new organic compounds that are more resistant to attack. This is called humification. The chemical structure of humic compounds has proved difficult to elucidate, but analyses made of FA and HA fractions indicate that macromolecules consisting of a 'core' of substituted aromatic and alkyl C groups, with a high density of –COOH and phenolic-OH groups, are the main constituents. Polysaccharides, mostly of microbial origin, and proteinaceous material make up about 20% of the humus polymers. Soil humus has a high CEC and is active in chelating metal cations such as Cu, Fe, Al, Mn, Co and Zn. Humic compounds are also important in the stabilization of soil structure (Chapter 4).

The organic matter content of a soil reflects the balance between gains of C from plant residues and animal excreta and the loss of C through decomposition by soil organisms. The decomposition of SOM components can be described by first-order kinetics. For systems in steady state, the ratio of the soil organic C content to the annual rate of C input defines the turnover time. Carbon turnover following changes in C inputs or rates of loss can be modelled with models such as CENTURY and ROTH C, which have a range of C pools of different size and decomposability. Experiments with ^{14}C-labelled residues show that under temperate conditions, about one-third of the residue C remains after 1 year and one-fifth after 5 years. In the humid tropics, decomposition is approximately four times faster than in temperate soils.

Low pH and dry conditions retard decomposition, as does a high polyphenolic content in the senescing plant tissue, which typically leads to the formation of mor humus. However, on neutral to alkaline soils under herbaceous and deciduous forest species low in polyphenols, earthworms flourish and mull humus forms. Alternate soil wetting and drying and freeze/thaw lead to flushes of decomposition.

References

Anderson J. P. E. & Domsch K. H. (1978) Mineralization of bacteria and fungi in chloroform-fumigated soils. *Soil Biology and Biochemistry* 10, 207–23.

Andreux F. (1982) Genesis and properties of humic molecules, in *Constituents and Properties of Soils* (Eds M. Bonneau & B. Souchier). Academic Press, London, pp. 109–39.

Angle J. S. (2000) Bacteria, in *Handbook of Soil Science* (Ed. M. E. Sumner). CRC Press, Boca Raton, Florida, C14–C22.

Baldock J. A. & Nelson P. N. (2000) Soil organic matter in, *Handbook of Soil Science* (Ed. M. E. Sumner). CRC Press, Boca Raton, Florida, B25–B84.

Coleman K. & Jenkinson D. S. (1996) Roth C-26.3 – A model for the turnover of carbon in soil, in *Evaluation of Soil Organic Matter Models* (Eds D. S. Powlson, P. Smith & J. U. Smith) NATO ASI Series 1: Global Environmental Change, Volume 38. Springer, Berlin, pp. 237–46.

Edwards C. A. (1991) The assessment of populations of soil-inhabiting invertebrates. *Agriculture, Ecosystems and Environment* 34, 145–76.

Edwards C. A. & Lofty J. R. (1977) *Biology of Earthworms*, 2nd edn. Chapman and Hall, London.

Hepper C. M. (1975) Extracellular polysaccharides of soil bacteria, in *Soil Microbiology* (Ed. N. Walker). Butterworth, London.

Jenkinson D. S. (1966) The priming action, in *The Use of Isotopes in Soil Organic Matter Studies*. Proceedings of FAO/IAEA Technical Meeting, Brunswick, Volkenrode. Pergamon Press, Oxford, pp. 199–208.

Jenkinson D. S. (1981) The fate of plant and animal residues in soil, in *The Chemistry of Soil Processes* (Eds D. J. Greenland & M. H. B. Hayes). Wiley, Chichester, pp. 505–61.

Jenkinson D. S. & Ayanaba A. (1977) Decomposition of carbon-14 labelled plant material under tropical conditions. *Soil Science Society of America Journal* 41, 912–15.

Jenkinson D. S. & Powlson D. S. (1976) The effects of biocidal treatments on metabolism in soil. 1. Fumigation with chloroform. *Soil Biology and Biochemistry* 8, 167–77.

Jenkinson D. S. & Rayner J. H. (1977) The turnover of soil organic matter in some of the Rothamsted classical experiments. *Soil Science* 123, 298–305.

Lee K. E. (1985) *Earthworms: their Ecology and Relationships with Soils and Land Use*. Academic Press, Sydney.

Nelson D. W. & Sommers L. E. (1996) Total carbon, organic carbon and organic matter in, *Methods of Soil Analysis. Part 3. Chemical Methods*. (Ed. D. L. Sparks) Soil Science Society of America Book Series No. 5, pp. 961–1010.

Oades J. M. & Jenkinson D. S. (1979) Adenosine triphosphate content of the soil microbial biomass. *Soil Biology and Biochemistry* 11, 201–4.

Parton W. J., Stewart J. W. B. & Cole C. V. (1988) Dynamics of C, N, P, and S in grassland soils: a model. *Biogeochemistry* 5, 109–31.

Schnitzer M. & Schulten H.-R. (1995) Analysis of organic matter in soil extracts and whole soils by pyrolysis-mass spectroscopy. *Advances in Agronomy* 55, 168–217.

Thorn R. G. (2000) Soil fungi, in *Handbook of Soil Science* (Ed. M. E. Sumner). CRC Press, Boca Raton, Florida, C22–C37.

Tinker P. B. & Nye P. H. (2000) *Solute Movement in the Rhizosphere*. Blackwell Publishing, Oxford.

Waterhouse D. F. (1973) Pest management in Australia. *Nature New Biology* 246, 269–71.

Weaver R. W. (Ed.) (1994) *Methods of Soil Analysis. Part 2. Microbiological and Biochemical Properties*. Soil Science Society of America Book Series No. 5, Madison, Wisconsin.

Wild A. (Ed.) (1988) *Russell's Soil Conditions and Plant Growth*, 11th edn. Longman Scientific & Technical, Harlow.

Further reading

Edwards C. A. & Bohlen P. J. (1996) *Biology and Ecology of Earthworms*, 3rd edn. Chapman & Hall, London.

Killham K. (1994) *Soil Ecology*. Cambridge University Press, Cambridge.

Paul E. A. & Clark F. E. (1996) *Soil Microbiology and Biochemistry*, 2nd edn. Academic Press, San Diego.

Schnitzer M. (2000) A lifetime perspective on the chemistry of soil organic matter. *Advances in Agronomy* 68, 3–58.

Shaffer M. J., Liwang Ma & Hansen S. (Eds) (2001) *Modeling Carbon and Nitrogen Dynamics for Soil Management*. Lewis Publishers, Boca Raton, Florida.

Example questions and problems

1 (a) Why is the organic matter content of a soil usually greatest in the surface horizon?
  (b) What is meant by the turnover of soil organic matter (SOM)?
  (c) If the annual input of C to a soil is 1.5 t/ha, and the total soil C content is 40,000 kg/ha, what is the average turnover time for the soil C?

2 (a) What is the generic name for organisms that cannot live in the presence of O_2?
  (b) What is the generic description for fungi that (i) live on dead organic matter, (ii) live by feeding on living tissues, and (iii) live in symbiotic association with plant roots?
  (c) Which group of invertebrate organisms is most important for comminuting organic matter in (i) soils of moist temperate regions, and (ii) soils of the tropics?

3 The CEC of organic matter measured at pH 5 is 80 cmol charge/kg, and is entirely due to carboxyl groups that are fully dissociated at pH 5. However, there are an additional 55 cmol charge/kg of phenolic groups that have an average pK_a value of 7. Calculate the CEC of the organic matter at pH 8. (Refer to Box 3.6).

4 A soil sample containing the equivalent of 20 g o.d. soil was fumigated in $CHCl_3$ for 24 h, before being flushed free of $CHCl_3$, and incubated at 25°C for 10 days. The CO_2 released was trapped in 25 mL of 0.02 M NaOH, and the unneutralized NaOH at the end of incubation titrated with 0.02 M HCl. A control incubation was carried out with a sample of unfumigated soil. The results of the titration were as follows:

Treatment	mL 0.02 M HCl required
Control sample	22.5
Fumigated sample	2.5

Calculate the microbial biomass C (in mg/kg soil), assuming a K_c value of 0.45. Note – the atomic mass of C is 12 g. (Refer to Box 3.2).

5 Wheat straw added to soil contains 45% C and 0.5% N. Soil micro-organisms that feed on this straw and decompose it have a C : N ratio of 6 : 1. In decomposing the straw, 60% of the straw C is assimilated by the micro-organisms and the remaining 40% is respired as CO_2. Calculate
  (a) The C : N ratio of the straw.
  (b) The fraction of the N required for microbial growth that can be obtained from the straw.
  (c) If not all the microbial N can be obtained from the straw, from where is the balance of microbial N obtained?
  (d) What is the name given to the ratio of C assimilated to the total C substrate?

6 Straw residues (C : N ratio = 100) from a cereal crop are ploughed into soil at the rate of 1t C/ha, before a new crop is sown. From previous experimental work, it is known that the decay coefficient for residues of this type is 0.3 (1/year), and that the average decay coefficient for humified organic matter in this soil is 0.006 (1/year). Calculate the following.
  (a) How much of the straw (in t C/ha) will remain in the soil after 4 months?
     [Use the equation $C(t) = C(0)(\exp(-kt))$].
  (b) If the soil organic matter in the top 15 cm amounts to 20,000 kg C/ha, how much of this humified organic matter will also decompose during this period?
  (c) If the C : N ratio of the soil organic matter is 10, how much soil organic N might be mineralized during the 4 month period of crop growth?
  (d) Given that the C : N ratio of the heterotrophic soil micro-organisms is 6, and their growth yield is 0.5, would you expect any *net* mineralization of N in the soil in the 4 months following the incorporation of straw residues? Give an explanation for your answer.

Chapter 4

Peds and Pores

4.1 Soil structure

As seen in Chapter 3, one reason why a soil supports a diversity of plant and animal life is the abundance of energy and nutrients derived from the microbial decomposition of organic remains. An equally important reason lies in the physical protection and favourable temperature, moisture and O_2 supply afforded by the soil's structural organization. Weathering of parent material produces the primary soil particles – clay, silt, sand,

stones and gravel. Without the intervention of external forces, these particles will simply 'pack' randomly, as shown in Fig. 4.1a, to attain a state of minimum potential energy. However, vital forces associated with plants, animals and micro-organisms, and physical forces associated with wetting and drying, freezing and thawing, act to arrange soil particles into larger units of varied size called aggregates or peds (see below) (Fig. 4.1b). Work is done on the particles to create order out of disorder (Section 1.1), and soil

(a)
(b)

Fig. 4.1 (a) Random close packing of soil particles. (b) Structured arrangement of soil particles into aggregates.

structure is created. Inherently, the aggregated soil is less stable than if the particles had packed randomly with no obvious organization, but the existence of structure confers many advantages on organisms living in the soil.

Since aggregation of particles is the basis of soil structure, structure has been broadly defined in terms of:
• The arrangement of primary particles into peds;
• the size, shape and arrangement of the peds and of the voids or pore spaces that separate the particles and peds.

Soil structure should be described only from a freshly exposed soil profile or a large undisturbed core. Samples from augers (Chapter 14) are unsuitable because of the disturbance, as are those from old road cuttings and gullies that have been exposed for some time.

4.2 Levels of structural organization

Many soils have structural features that are readily observed in the field – these features relate to the soil's macrostructure. One of the earliest systems for describing and classifying soil macrostructure was that of the USDA Soil Survey Manual (Soil Survey Division Staff, 1993). This has been the basis for similar systems developed in other countries, such as that used by the Soil Survey of England and Wales (Hodgson, 1974), and the standard adopted in Australia (McDonald et al., 1998). However, much of the fine detail of aggregation and the distribution of pores – the microstructure – can be determined only in a laboratory under some form of magnification, ranging from a hand lens (× 10, also used in the field), to a light microscope (× 50), to an electron microscope (× 10,000). Some of the macroscopic features have identifiable counterparts at the microscopic level, in which case a descriptive term common to both can be used. Other features are peculiar to the microstructure and require new descriptive terms and sophisticated analytical techniques, which have been devised by specialists in the study of soil micromorphology (Section 4.3). The technique of fractal geometry is also being applied to the description of soil structure (Box 4.1).

Describing aggregation

Pedologists and soil surveyors use the term 'ped' to describe the naturally formed aggregates that are recognizable in the field because they are separated by voids (vacant space) and natural planes of weakness. They should persist through cycles of wetting and drying, as distinct from the less permanent units formed by the mechanical disturbance of digging or ploughing (clods), or by rupture of the soil mass across natural planes of weakness (fragments). The term 'aggregate' is more commonly used by agronomists to describe soil structural units, especially in the A horizon or topsoil, which may include clods and fragments as well as true peds. The terms 'ped' or 'aggregate' will be used in this book, depending on whether the context is pedological (e.g. soil profile description) or agronomic, respectively.

A concretion is formed when localized accumulations of an insoluble compound irreversibly cement or enclose soil particles. Nodules are similarly formed, but lack the symmetry and concentric internal structure of concretions.

Ped type

Four main types of ped are recognized, as illustrated in Fig. 4.2. A brief description of each type follows:
• *Spheroidal* – peds that are roughly equidimensional and bounded by curved or irregular surfaces not accommodated by adjacent ped faces. There are two subtypes – granules, which are relatively non-porous, and crumbs, which are very porous. These peds are usually evenly coloured throughout, and arise primarily from the interaction of soil organisms, roots and mineral particles. They are therefore common in the A horizons of soils under grassland and deciduous forest.
• *Blocky* – peds bounded by curved or planar surfaces that are more or less mirror images of surrounding ped faces. Those with flat faces and sharply angular vertices are called angular; those with a mixture of rounded and flat faces and more subdued vertices are called subangular. Blocky peds are features of B horizons where their natural surfaces may be identified by their smoothness and often distinctive colours. The ped

Box 4.1 Towards a quantitative description of soil structure.

The description of soil structure has traditionally been qualitative, whether it be at the microscopic or macroscopic level. The reason is because the shape, size and continuity of pores, and the surrounding aggregates, are so irregular. Attempts have been made, however, to use non-Euclidean geometry – the fractal geometry of Mandelbrot (1983) – not only to describe soil structure quantitatively, but also to model the physical, chemical and biological processes that operate over a range of scales to create soil structure (Section 4.4). Fractals are useful tools because they have no characteristic scale: the morphological detail of a fractal object is the same at all scales (the property of self-similarity). Fractal characterization of soil structure requires the fractal dimensions for:
- Irregular boundaries and surfaces;
- non-uniform mass distribution;
- fragmentation; and
- the dynamic properties of the fractal network (e.g. pore continuity and tortuosity)

to be measured. This can be done by computer-aided analysis of binary images of soil thin sections, or surfaces of undisturbed monoliths (Fig. B4.1.1). Examples of this approach are given in Anderson *et al.* (1996).

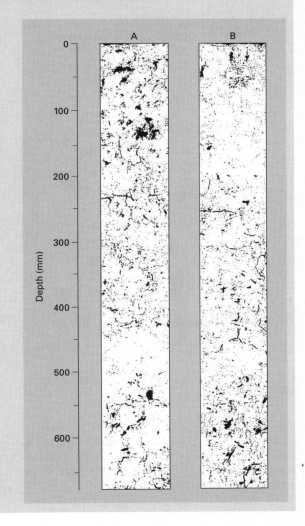

Fig. B4.1.1 Black and white images of soil monoliths of A, undisturbed soil under Eucalyptus forest, and B, same soil type but cultivated for 90 years. Black represents pore space; white represents soil solids (after McBratney *et al.*, 1992).

faces may be coated, for example, by organic matter, clay or sesquioxides (Section 4.3). The voids between peds often form paths for root penetration and water flow. The ped faces may have live or dead roots attached, or show distinct root impressions.

• *Platy* – the particles are arranged about a horizontal plane with limited vertical development; most of the ped faces are horizontal. Peds that are thicker in the middle than at the edges are called lenticular. These peds occur in the A horizon of a soil immediately above an impermeable B horizon, or in the surface of a soil compacted by machinery.

• *Prismatic* – these peds are taller than they are wide and the vertical faces end in sharp corners. They are subdivided into prisms without caps (prismatic), and prisms with rounded caps (columnar). Prismatic structure is common in the subsoil of heavy clay soils subject to frequent

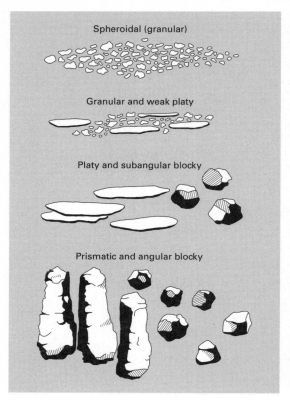

Fig. 4.2 Examples of common ped types (after Strutt, 1970).

Fig. 4.3 Prismatic structure of a clay subsoil prone to waterlogging.

waterlogging (Fig. 4.3). Columnar structure is associated particularly with the B horizon of Na-affected soils, such as a solodized solenetz or Sodosol (ASC) (Fig. 4.4).

Compound peds

All the ped types shown in Fig. 4.2 are invariably composed of smaller units that are clearly visible, and others that can only be seen under a microscope. For example, the large compound prismatic peds in Fig. 4.5 are composed of secondary subangular blocks, which in turn break down to the primary structure of angular blocky peds < 10 mm in diameter. Each of these primary peds presents an intricate internal arrangement of granules and voids which can be elucidated only by careful appraisal of the soil in thin sections (Section 4.3).

Class and grade of peds

The class of a ped is judged according to its size, and the grade according to the distinctness and durability of the visible peds. Size classes range from < 2 mm to > 500 mm, the average smallest dimension being used to determine the class (vertical for platy peds, horizontal for prismatic, and the diameter for spheroidal). The classes are described as 'fine', 'medium', 'coarse' and 'very coarse'. The reader is referred to the specialist soil survey handbooks for details.

Grade of structure may be described as follows:
• *Structureless* (or *apedal*) – no observable aggregation nor definite network of natural planes of weakness – massive if coherent, single grain if non-coherent;
• *weakly developed or weak* – poorly formed, indistinct peds that are barely observable *in situ*.

Fig. 4.4 Columnar structure in the B horizon of a solodized solentz (scale in feet) (CSIRO photograph, courtesy of G. D. Hubble).

Fig. 4.5 Compound prismatic peds showing the breakdown into secondary structure.

When disturbed, the soil breaks into a mixture of a few entire peds (< one-third), many broken peds, and much unaggregated material;

• *moderately developed or moderate* – well-formed, moderately durable, distinct peds, but not sharply distinct in undisturbed soil. When

disturbed, the soil breaks down into a mixture of many entire peds (one-third to two-thirds), some broken peds, and a little unaggregated material;

• *strongly developed or strong* – durable distinct peds in undisturbed soil that adhere weakly to one another, or become separated when the soil is disturbed. The soil consists largely of entire peds (> two-thirds), and includes a few broken peds and little or no unaggregated material.

Grades of structure reflect differences in the strength of intraped cohesion relative to interped adhesion. Strong intraped (within ped) cohesion is the basis of a strong grade of structure. The grade will depend on the moisture content at the time of sampling and the length of time the soil has been exposed to drying, which is why it is recommended that structure be described only on freshly exposed soil (Section 4.1). Prolonged drying will markedly increase the structure grade, particularly of clay soils and soils rich in iron oxides. Knowledge of how the strength of a soil's cohesion and adhesion changes with water content is important for many practical purposes, e.g. the use of soil as a foundation for roads and dam walls, its response to cultivation, and its suitability for under-drainage. These variations are described by the term consistence (Box 4.2).

4.3 Soil micromorphology

Micromorphology is concerned with the description, measurement and interpretation of soil components and pedological features at a microscopic to sub-macroscopic scale. The study of micromorphology began with Kubiena (1938), who was primarily interested in using knowledge of soil microstructure to interpret soil genesis (Chapter 9). One advantage of this approach is that some pedogenic processes may not be sufficiently advanced for their effects to be seen macroscopically. Incipient translocation down a soil profile of clay, organic matter or Fe is an example of a process that, although not obvious in the field, can be identified by the presence of microscopic-scale coatings on ped surfaces and pore walls (Fig. 4.6). From this early emphasis on soil genesis, the scope of micromorphology has widened to cover many soil physical, chemical and biological properties, especially those involved in the quantitative description of soil structure *in situ*. Some of the techniques of micromorphology are described in Box 4.3.

Box 4.2 Soil consistence.

Soil consistence depends on the grade of structure and the water content. It is measured by the resistance offered to breaking or deformation when a compressive shear force is applied. In the field, consistence is assessed on samples 20–30 mm in diameter that are subjected to a compressive shearing force between thumb and forefinger. The sample may be an individual aggregate, compound aggregate or fragment, and the water content must be specified. Usually the test is done on both air-dry and moist soil (at the field capacity, as defined in Section 4.5). The resistance of the soil to rupture or deformation is expressed on a scale ranging from loose (no force required, separate particles present) to rigid (cannot be crushed under foot by the slow application of a person's weight). Full details are given in the soil survey field handbooks cited in Section 4.1. A more quantitative relationship between consistence and soil water content, and its relevance for mole drainage, is given in Section 13.4.

0·6 mm

Fig. 4.6 Thin section through a soil pore showing clay coatings (courtesy of P. Bullock).

Box 4.3 Techniques and terminology in micromorphology.

Micromorphometric analysis is done on surfaces ranging in size from thin sections of a few cm^2 to large soil blocks (monoliths) of a few dm^2. Briefly, the method of thin sectioning involves taking an undisturbed soil sample in a Kubiena box (usual dimensions $10 \times 5 \times 3.5$ cm), removing the water in a way that minimizes soil shrinkage, and impregnating the voids with a resin under vacuum. Once the resin has cured, the soil block is sawn and polished to produce a section 25–30 μm thick which is stuck on a glass slide. The section is examined under a petrological microscope to determine the characteristic optical properties of the minerals present. Fig. 4.6 is an example of a soil thin section.

Monoliths up to $70 \times 5 \times 10$ cm can be taken from the field, usually after preliminary resin impregnation, followed by further resin impregnation, cutting and polishing in the laboratory. Such monoliths are most useful for studying the size, shape, distribution and frequency of different pore types in the void space. The polished surface is photographed or examined by computerized image-analysing equipment, capable of resolving structural differences to a few micrometres. Resolution of the pore space is aided by fluorescence under ultraviolet light from the resin or dyes incorporated in the resin. Fig. B4.1.1 is an example of a two-dimensional image showing the distribution of pore space in two soil monoliths – one from an undisturbed soil under forest (A); the other from the same soil type subjected to continuous cropping for 90 years. The effect of cultivation is seen in the loss of large pores and aggregate break-down to produce a more massive structure in the top 200 mm of soil B relative to soil A.

A more advanced technique is to quantify the size, shape and distribution of pores in undisturbed soil cores with X-ray computed tomography. The digital images are analysed with mathematical algorithms of morphology to develop a three-dimensional picture of the soil's porosity.

The terminology of micromorphology has undergone progressive development. A 'foundation' set of terms was introduced by Brewer in 1964 and revised in 1976. Fitzpatrick (1984) provided a valuable glossary of terms and their meanings, and a group commissioned by the International Society of Soil Science prepared a *Handbook for Soil Thin Section Description* (Bullock *et al.*, 1985). However, Brewer's terminology provides the basis for most current micromorphological descriptions.

Reorientation of plasma *in situ*

Brewer (1964) defined 'plasma' as colloidal particles (< 2 μm) and soluble materials not bound up in skeleton grains (sand grains), which can be moved, reorganized and concentrated during soil formation. Plasma included mineral and organic material.

During the repeated cycles of wetting and drying, and sometimes freezing and thawing, that accompany soil formation, plate-like clay particles gradually become reorganized from a randomly arranged network of small bundles into larger bundles that show a degree of preferred orientation. The bundles or plasma separations are called domains, which can be up to 5 μm long and 1–5 μm wide, depending on their mode of formation and the dominant clay mineral present. The significance of domains is discussed further in Section 4.4.

As well as the reorganization of clay particles within peds, there is often a change in the orientation of clay particles at ped faces. In dry seasons, material from upper horizons can fall into deep cracks in clay soils. On rewetting, the subsoil, which now contains more material than before, swells to a larger volume and exerts a pressure that displaces material upwards. This effect is particularly important when the clay is a smectitic type. Peds slide against each other to produce zones of preferred orientation at their faces called slickensides lying at an angle of 50–55° to the vertical (Fig. 4.7).

Plasma translocation and concentration

One of the major processes of soil formation is the translocation of colloidal and soluble material

Fig. 4.7 Intersecting slickensides in a heavy clay subsoil (CSIRO photograph, courtesy of G. D. Hubble).

from upper horizons into the subsoil. In the changed physical and chemical environment of the lower horizons, salts and oxides precipitate, clay particles flocculate and ped faces become coated with deposited materials. Brewer classified these coatings or cutans according to the nature of the surface to which they adhere and their mineralogical composition. Some of the more common examples are as follows:

• *Clay coatings* (argillans) – often different in colour from, and with a higher reflectance than the matrix of the ped; they are easily recognized in sandy and loam soils, but difficult to distinguish from slickenside surfaces in clay soils;

• *sand or silt coatings* (skeletans) – formed by the illuviation of sand or silt, or by the residual concentration of these particles after clay removal;

• *sesquioxidic coatings* (sesquans, mangans, ferrans) – formed by reduction and solution of Fe and Mn under anaerobic conditions and their subsequent oxidation and deposition in aerobic zones (Fig. 4.8). Colours range from black (mangans) through grey, red, yellow, blue and blue-green (ferrans), depending on the degree of oxidation and hydration of the Fe compounds;

• *salt coatings* (of gypsum, calcite, sodium chloride) and organic coatings (organans);

• *compound coatings*, which are mixtures of these materials, particularly of clay, sesquioxides and organic matter.

Pans. The coatings may build up to such an extent that bridges of material form between soil particles and the smallest peds, so that a cemented layer may develop in the soil. Such a layer or horizon is called a pan and is described by its thickness (if < 10 mm it is called 'thin') and its main chemical constituent. An example of a thin Fe pan is shown in Fig. 4.9. Compacted horizons that are weakly cemented by silica are called duripans. A fragipan is not cemented at all, but the size distribution of

Fig. 4.8 Manganese dioxide coatings on ped faces (courtesy of R. Brewer).

Thin Fe
pan

Fig. 4.9 A thin Fe pan developed in ferruginous sand deposits.

the soil particles is such that they pack to a very high bulk density (Section 4.5).

Voids. Cylindrical and spherical voids, usually old root channels or earthworm burrows, are called pores and those of roughly planar shape are called fissures or cracks.

Voids of indeterminate shape are called vughs. Fissures start as cracks striking vertically downwards from the surface, especially in soils containing much of the expanding-type clays (Fig. 4.10). Secondary cracks develop at right angles to the vertical fissures.

4.4 The creation and stabilization of soil structure

Formation of peds

The organization of soil material to form peds and other pedological features requires the action of physical, chemical and biotic forces. For example:

• Swelling pressures are generated through the osmotic effects of exchangeable cations adsorbed on clay surfaces;

• there are forces associated with the change in state of the soil water. Unlike any other liquid, water expands on cooling from 4° to 0°C. As the water in the largest pores freezes first, liquid water is drawn from the smaller pores thereby causing them to shrink; but the build-up of ice lenses in the large pores, and the expansion of the water on freezing, subjects these regions to intense disruptive forces;

• surface tension forces develop as water evaporates from soil pores, and the prevalence of evaporation over precipitation can lead to localized

Fig. 4.10 Close cracking pattern in the surface of a montmorillonite clay soil (scale is 15 cm long) (CSIRO photograph, courtesy of G. D. Hubble).

accumulations of salts and the formation of concretions anywhere in the soil profile;

• the mechanical impact of rain, animal treading and earth-moving machinery can reorganize soil materials, and the energy of flowing water can move them;

• burrowing animals reorganize soil materials, particularly earthworms that grind and mix organic matter with clay and silt particles in their guts, adding extra Ca in the process to form casts (Fig. 3.10), which, on drying, usually have considerable strength. The faeces of other invertebrates such as mites and collembola also form stable pellets that may exist as microaggregates, or act as nuclei for aggregate formation;

• the physical binding effect of fine roots and fungal mycelia also contributes to aggregation. The metabolic energy expended by soil microorganisms can maintain concentration gradients for the diffusion of salts, or induce a change in valency leading to the solution and subsequent re-precipitation of Mn and Fe oxides.

Interparticle forces

For a ped or aggregate to persist, the interparticle forces must be sufficient to prevent particles from separating under the continuing impact of forces responsible for reorganization. Good structure for plant growth depends on the existence of aggregates between 1 and 10 mm in diameter (the primary peds mentioned in Section 4.2) that remain stable when wetted. The response of aggregates to immersion in water is the basis of tests of aggregate stability (Box 4.4).

It is pertinent here to consider ways in which the primary constituents – sand, silt, clay and organic matter – are organized to form aggregates. Some authors distinguish between macroaggregates (> 250 μm diameter) and microaggregates

Box 4.4 Tests of aggregate stability.

A long-established test is that of measuring water-stable aggregates (Klute, 1986). This involves sieving under water an air-dry sample of soil on a vertical array of sieves of different mesh size. After a standard time, the mass of aggregates remaining on individual sieves is measured and expressed as a percentage of the total. Since one of the original aims of this test was to assess a soil's susceptibility to water erosion, several variants of the test involving simulated rain falling on soil in the field have been developed.

Another test of soil structure and aggregate stability involves placing small air-dry aggregates (0.2–0.5 g) in water of very low salt concentration (similar to rain water) (Emerson, 1991). An aggregate that quickly breaks down into subunits, which may themselves be aggregates of smaller particles, is said to slake. This indicates that the intra-aggregate forces are not strong enough to withstand the pressure of air entrapped as the aggregate wets quickly, and the pressure generated by swelling. The subunits produced by slaking may also be unstable due to internal repulsive forces, in which case individual clay particles may separate – a process of deflocculation (also called dispersion) – and the surrounding water becomes cloudy.

A scheme for the recognition of differences in aggregate stability on the basis of whether the aggregates slake or deflocculate or both is shown in Fig. B4.4.1. When there is no apparent deflocculation, a sample of dry aggregates is moistened and remoulded into small balls, approximately 5 mm in diameter. These are placed in rainwater and observed for any deflocculation (cloudiness). Possible responses are shown in Fig. B4.4.1. The main types of aggregate behaviour, based on this test, are as follows:

• *Type 1* – highly dispersible soil, very poor microaggregate stability. The soil is likely to have no consistence when wet and to set hard when dry; also likely to erode easily;
• *type 2* – partial dispersion, poor microaggregate stability; soil will probably crust and need to be protected by vegetation from erosion;
• *type 3* – complete to partial dispersion after re-moulding indicates moderate microaggregate stability. Soil structure will deteriorate under repeated cultivation, especially if cultivated when too wet;
• *type 4* – no dispersion after re-moulding indicates good microaggregate stability. The soil will not crust or erode readily.

Flocculation, deflocculation and swelling depend on reactions at surfaces, which are discussed in Chapter 7. Soil management to maintain water-stable aggregates is discussed in Section 11.4.

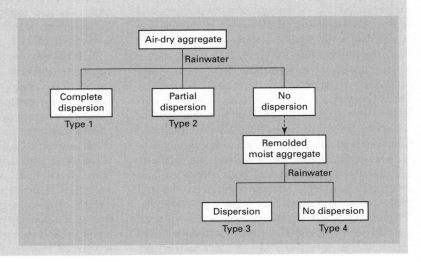

Fig. B4.4.1 Diagram of a modified Emerson test for the stability of soil aggregates (Cass, 1999). Reproduced with permission of CSIRO PUBLISHING from 'Soil Analysis: an Interpretation Manual (Eds K. I. Peverill, L. A. Sparrow & D. J. Reuter, 1999). Copyright Australasian Soil and Plant Analysis Council Inc. http://www.publish.csiro.au/pid/1998.htm.

(< 250 μm). They do so by taking account of the soil's response to disruptive treatments (e.g. sonic vibration or rapid wetting), and the inferred mechanism of aggregate stabilization. Several models of microaggregate organization have been proposed, all of which involve some combination of clay–clay and clay–organic matter interactions.

Microaggregates

Domains and quasi-crystals are thought to be the fundamental microstructural units in many soils. Domains consist of multilayer illite or vermiculite crystals up to 5 nm thick, stacked in roughly parallel alignment to form units up to 1 μm thick and up to 5 μm in the x–y direction (Box 2.2). Similarly, quasi-crystals consist of single layers of montmorillonite (1 nm thick) aligned, with much overlapping, to form units several nanometres thick and up to 5 μm in the x–y direction. When Ca^{2+} is the dominant exchangeable cation, the intercrystalline spacing in the domains or the interlayer (intracrystalline) spacing in the quasi-crystals does not exceed 0.9 nm, even when the soil is placed in distilled water, because of the stability of the trimolecular layer hydrate of Ca^{2+} ions between opposing clay surfaces (Section 2.4). Each unit – quasi-crystal or domain – acts as a stable entity in water, showing limited swelling, and is only likely to be disrupted if more than c. 15% of the exchangeable Ca^{2+} (moles of charge) is replaced by Na^+.

The generally accepted model of a microaggregate consists of arrangements of domains and quasi-crystals, sand and silt particles, combined in some way with organic polymers (Fig. 4.11). In the larger units (100–200 μm), the organic matter often consists of recognizable plant parts – leaf and root fragments. The encrustation of these plant fragments with clay domains and other particles offers protection against microbial attack. However, in smaller units (20–100 μm), the plant material is much degraded and only fragments of cell walls and lignin fibres remain. At < 20 μm, all the plant material has been degraded and microbial products – mainly polysaccharides and polyuronide gums and mucilages – predominate. Negatively charged humic polymers are attracted to positively charged clay edges, or complex with polyvalent cations (Fe^{3+}, Al^{3+} or Ca^{2+}) held on the negatively charged planar surfaces.

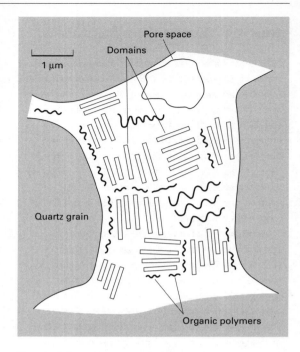

Fig. 4.11 Possible arrangements of clay domains, organic polymers and quartz grains in a microaggregate.

In highly weathered soils in which 1 : 1 clays such as kaolinite predominate, individual clay crystals can flocculate edge-to-face to form a 'cardhouse' type of microstructural unit (Section 7.4). Positively charged sesquioxide films on the planar surfaces can also act as electrostatic bridges between crystals, or form complexes with negatively charged organic polymers. A summary of these various interparticle forces of attraction is given in Table 4.1.

Flocculated clay is a prerequisite of microaggregate stability and the ability to resist disruption on wetting and mechanical disturbance. Clay–organic matter interactions are also important, as are inorganic compounds that act as interparticle cements; for example, calcium carbonate in calcareous soils, sesquioxides as films between planar clay surfaces and as discrete charged particles in many acid, highly weathered soils, and silica at depth in laterites (Section 9.4). The nature of microaggregate stabilization is such that it is relatively insensitive to changes in soil management, except when management induces:
• Marked changes in the dominant exchangeable cations (Na^+ replacing Ca^{2+}); or

Table 4.1 Summary of interparticle forces of attraction involved in aggregate formation.

ELECTROSTATIC FORCES – these involve clay–clay, clay–oxide, and clay–organic interactions and include the following:
• Positively charged edge of a clay mineral attracted to the negative cleavage face of an adjacent clay mineral.
• Positively charged sesquioxide film 'sandwiched' between two negative clay mineral surfaces.
• Positively charged groups (e.g. amino groups) of an organic molecule attracted to a negative clay mineral surface. Organic molecules, in displacing the more highly hydrated inorganic cations from the clay, modify its swelling properties.
• Clay–polyvalent cation–organic anion linkages or 'cation bridges'. If the organic anion is very large (a polyanion), it may interact with several clay particles through cation bridges, positively charged edge faces, or isolated sesquioxide films. Linear polyanions of this kind (e.g. the synthethic compound 'Krilium') are excellent flocculants of clay minerals (Section 7.5).

VAN DER WAALS' FORCES – these involve clay–organic interactions such as the following:
• Specific dipole–dipole attraction between constituent groups of an uncharged organic molecule and a clay mineral surface, as in H-bonding between a polyvinyl alcohol and an O or OH surface.
• Non-specific forces between large molecules in very close proximity, which are proportional to the number of atoms in each molecule. These forces account for the strong adsorption of polysaccharide and polyuronide gums and mucilages on soil particles. These polymers of relative molar mass 100,000–400,000 g are produced in the soil by bacteria, such as *Pseudomonas* spp., which are abundant in the rhizosphere of many grasses. When adsorption of such a large molecule results in the displacement of many water molecules, the increase in entropy of the system is large and the reaction is essentially irreversible (Section 7.5). Large, flexible polymers of this kind can make contact with several clay crystals at many points and so help to hold them together in the manner of a 'coat of paint'.

• changes in soil pH (acid → alkaline pH, which decreases the positive charges on kaolinite edge faces and sesquioxide surfaces); or
• physical disruption of aggregates that exposes 'protected' organic matter to further decomposition by soil organisms.

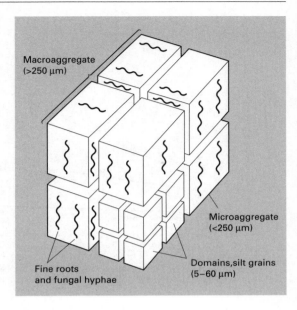

Fig. 4.12 Diagrammatic representation of a soil macroaggregate (after Greenland, 1979).

Macroaggregates

The formation of macroaggregates ($> 250\ \mu m$) depends primarily on the stabilization of micro-aggregates in larger structural units. Macro-aggregation depends more on soil management than microaggregation because its stabilization depends primarily on the action of plant roots and fungal hyphae, particularly the arbuscular mycorrhizas (Section 10.3). Roots can penetrate pores $> 10\ \mu m$ diameter and fungal hyphae pores $> 1\ \mu m$, and in so doing, they enmesh soil particles to form stable macroaggregates. A diagrammatic model of a macroaggregate is shown in Fig. 4.12. Grass swards are especially effective because the roots are so dense in the surface horizon. Further to the physical binding effect of roots, rhizosphere C deposition (Section 3.1) provides mucilage, and a general substrate for saprophytic micro-organisms producing polysaccharide gums and mucilage that bind soil particles together. These effects all depend on maintaining a high level of biotic activity – plant roots, animals (mainly earthworms) and micro-organisms – in the soil: they are most important in soils of pH 5.5–7, which derive little benefit through aggregate stabilization by either sesquioxides or $CaCO_3$.

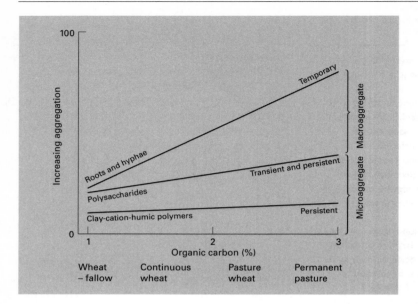

Fig. 4.13 The relative contributions of various forms of organic matter to the stabilization of aggregates in soil (after Tisdall and Oades, 1982).

Substantial inputs of organic matter and minimal soil disturbance therefore favour the stabilization of macroaggregation. The stability of macro- and microaggregates in relation to the C content of soil, and the broad structural forms of that C, are summarized in Fig. 4.13.

4.5 Soil porosity

Defining porosity

Clearly, the size, shape and arrangement of particles and peds determine the pore space or porosity of a soil. The porosity φ (phi) of a given volume of soil is defined as the ratio

$$\phi = \frac{\text{volume of pores}}{\text{total soil volume}}. \tag{4.1}$$

As can be seen from Equation 4.1, the pore volume is the product of the porosity and soil volume. Porosity depends on the water content of the soil, since the volume of pores and total volume of an initially dry soil may change differentially due to swelling as clay surfaces hydrate, or shrinkage as the soil dries. Shrinkage occurring as a montmorillonitic clay soil dries is particularly obvious, as seen by the large cracks that develop

in the surface (Fig. 4.10). The soil water content at which the porosity is measured therefore needs to be stated.

Bulk density and porosity

The porosity of a ped can be measured from the volume of a non-polar liquid, such as paraffin, that is absorbed into the dry ped under vacuum. More commonly, the porosity of a ped or larger soil volume can be measured indirectly from the bulk density and particle density. Particle density (ρ_p) was introduced in Section 1.3. Bulk density (ρ_b) is defined as the mass of oven-dry (o.d.) soil per unit volume, and depends on the densities of the constituent soil particles (clay, organic matter, etc.), and their packing and arrangement into peds. For the soil volume relevant to Equation 4.1, we may expand this equation as follows.

$$\phi = \frac{\text{total soil volume} - \text{volume of solids}}{\text{total soil volume}}$$

$$= 1 - \frac{\text{mass of o.d. soil}}{\text{total soil volume}} \times \frac{\text{volume of solids}}{\text{mass of o.d. soil}}$$

$$= 1 - \frac{\rho_b}{\rho_p}. \tag{4.2}$$

Bulk density is measured by extracting intact soil cores in the field, usually at least 6 cm in diameter and 6–10 cm deep, and oven-drying the soil before determining its mass. The particle density for mineral soils is assumed to be 2.65 Mg/m^3 (or g/cm^3) (Section 1.3). Values of ρ_b range from < 1 Mg/m^3 for soils rich in organic matter, to 1.0–1.4 for well-aggregated loamy soils, to 1.2–1.8 for sands and compacted horizons in clay soils. The use of ρ_b to calculate soil porosity, and for other calculations, is illustrated in Box 4.5.

Pore-size distribution

Total pore space indicates nothing about the actual size distribution of the pores. The distribution of pore sizes in non-swelling soils can be determined from the incremental volume of water released in response to known applied pressures, using the relationship between water potential and the radius of pores that will just retain water at that potential (Section 6.4). In soils with expanding-type clays, however, where contraction of the soil volume as well as water release occurs as pressure is applied, the use of the soil water retention relationship to determine pore-size distribution is less satisfactory. Instead, water should be removed with the minimum of shrinkage, by critical-point drying for instance, and intrusion of mercury used to determine the pore-size distribution.

Some authors have subdivided the continuous pore-size distribution of a soil into classes based on their primary functions, such as aeration, drainage or water storage, and the biotic agents associated with their formation. An example of such a classification of pore sizes is given in Table 4.2, and the significance of the pore functions is explained below and in Chapter 6.

Water-filled porosity

The pores of a soil are partly or wholly occupied by water. Water is vital to life in the soil because of its existence as a liquid at temperatures most suited to living organisms. It possesses several unique physical properties, namely the greatest specific heat capacity, latent heat of vaporization, surface tension and dielectric constant of any known liquid.

Box 4.5 Calculations involving bulk density.

Soil porosity is calculated by substituting appropriate values in Equation 4.2. For example, for a sandy soil of ρ_b 1.5 Mg/m^3 and ρ_p 2.65 Mg/m^3, we have

$$\phi = 1 - \frac{1.5}{2.65} = 0.43. \qquad \text{(B4.5.1)}$$

Porosity is sometimes expressed as a percentage, i.e. 43%.

Soil measurements made on a unit mass basis, such as soil organic C (g/kg) or 'available P' (mg/kg), can be converted to a unit volume basis by multiplying by ρ_b. Where ρ_b is not known, it is common practice to assume an 'average' value of 1.33 Mg/m^3. Thus, for example, to estimate the mass M of soil in one hectare to 0.15 m depth we have

$$M \text{ (ha-0.15 m)} = 10,000 \times 0.15 \times 1.33 = 1995 \text{ Mg,}$$
$$\text{(B4.5.2)}$$

which approximates to a soil mass of 2×10^6 kg/ha to a depth of 15 cm.

Bulk density changes should be accounted for when changes in soil elements (e.g. total C) are measured over time. For example, the Intergovernmental Panel for Climate Change (IPCC) recommends that the amount of soil C be recorded to 0.3 m depth, which corresponds to a mass of C per unit volume. However, C is measured in the laboratory as g/kg soil. A change in land use resulting in more sequestration of soil C, and a consequent increase in soil C content (g/kg), may be associated with a decrease in ρ_b and hence little apparent change in soil C to 0.3 m depth. In this example of soil C accounting, the C present in the soil mass equivalent to the mass of the original volume to 0.3 m depth should be calculated.

Table 4.2 A functional classification of pores based on size. After Greenland, 1981 and Cass et al., 1993.

Pore diameter (µm)	Biotic agent and descriptive function
5000–500	Cracks, earthworm channels, main plant roots; aeration and rapid drainage
500–30	Grass roots and small mesofauna; normally drainage and aeration
30–0.2	Fine lateral roots, fungal hyphae and root hairs; storage of 'available water'
< 0.2	Swell–shrink water in clays; residual or 'non-available' water

Soil wetness is characterized by the amount of water held in a certain mass or volume of soil; that is:

• *Gravimetric water content* (θ_g), measured by drying the soil to a constant weight at a temperature of 105°C. θ_g is then calculated from

$$\theta_g = \frac{\text{mass of water}}{\text{mass of o.d. soil}}. \tag{4.3}$$

θ_g (units of g H_2O/g o.d. soil) is often expressed as a percentage by multiplying by 100. Structural water, which is water of crystallization held in soil minerals, is not driven off at 105°C and is not measured as soil water;

• *volumetric water content* (θ_v), given by

Box 4.6 The relationship between θ_g and θ_v.

Although θ_v is a dimensionless quantity, it is important to note it has the units of m³ H_2O/m³ soil. Thus, the concentration of a solute in the soil solution (g/m³ solution) can be expressed on a soil volume basis by multiplying by θ_v.

θ_g and θ_v are related through the soil bulk density by the equation

$$\theta_v = \theta_g \times \rho_b \tag{B4.6.1}$$

because 1 Mg water occupies 1 m³ at normal temperatures. Because measurements of θ_v are more useful for most purposes than θ_g, soil water content will be expressed as a volumetric water content (symbol θ) throughout this book.

Values of θ give a direct measure of the 'equivalent depth' of water per unit area in the soil. For example, consider a soil volume of 1 m³ that has a θ value of 0.25 m³/m³. This water content is equivalent to a water depth of 0.25 m per m² of soil area in a soil volume of 1 m³. We may visualize this as a depth of 250 mm water per m depth of soil, as shown in Fig. B4.6.1. Put this way, soil water content is directly comparable with amounts of rainfall, evaporation or irrigation water, which are measured in mm. The general expression for calculating the equivalent depth d (mm) of soil water is

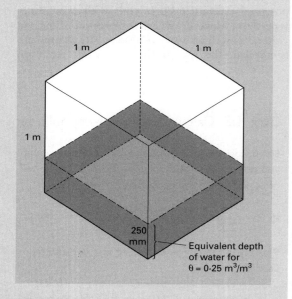

Fig. B4.6.1 Diagram of a 1 m³ soil cube filled with water to 250 mm depth (after White, 2003. From *Soils for Fine Wines* by Robert E. White, copyright © 2003 by Oxford University Press, Inc. Used by permission of Oxford University Press, Inc.).

$$d = 1000\ \theta\ z, \tag{B4.6.2}$$

where z is the soil depth in metres.

$$\theta_v = \frac{\text{volume of water}}{\text{total soil volume}}. \qquad (4.4)$$

The relationship between θ_g and θ_v is discussed in Box 4.6.

Normally, water occupies less than the total pore volume and the soil is said to be unsaturated (although during a period of prolonged rain, the soil may temporarily become saturated to the surface; also, soil may be saturated by the rise of groundwater – Section 6.1).

Air-filled porosity

The pore space that does not contain water is filled with air. The air-filled porosity ε (epsilon) is defined by

$$\varepsilon = \frac{\text{volume of soil air}}{\text{total soil volume}}, \qquad (4.5)$$

and it follows that

$$\varepsilon = \phi - \theta. \qquad (4.6)$$

Soils have been found to drain, after rain, to water contents in equilibrium with negative pressures between −5 and −10 kPa. This water content defines the field capacity (FC), which is sometimes also called the soil's drained upper limit, and is discussed further in Chapter 6. In Britain, the FC is set as the water content at −5 kPa when all pores > 60 μm diameter are drained of water (Hall *et al.*, 1977). In Australia and the USA, the FC is set as the water content at −10 kPa, which corresponds to pores of diameter > 30 μm being drained (Cass, 1999). The distinction between pores that are drained or not drained at the FC provides a simplified pore-size classification into macropores and micropores, respectively (*cf* Table 4.2). Macropores are primarily the cracks and pores between soil aggregates, which are often created by faunal and root activity (the biopores). The value of ε for a soil at FC defines the air capacity ε_a, which as Fig. 4.14 shows, may vary from 0.1 to 0.30 m³/m³ in topsoils, depending on the texture. Knowledge about the air capacity is important for:

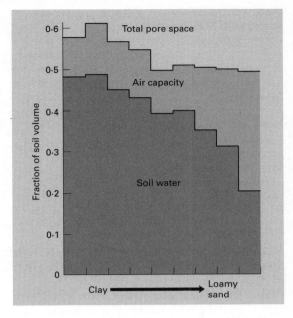

Fig. 4.14 Air and water-filled porosities at the field capacity in topsoils of varying texture (after Hall *et al.*, 1977).

- Assessing soil aeration (Chapter 8);
- soil structural quality (Section 11.4);
- designing soil drainage systems (Section 13.4).

Composition of the soil air

The proportions of the major gases in the Earth's atmosphere are normally 78% N_2, 21% O_2, 0.9% argon (Ar), and 0.0365% CO_2 by volume. According to Dalton's law of partial pressures, these concentrations correspond to partial pressures of 0.78, 0.21, 0.009 and 0.000365 of one atmosphere, which are equivalent to pressures of 79, 21.3, 0.91 and 0.037 kPa, respectively.

The composition of soil air may deviate from these proportions because of soil respiration (Section 8.1), which depletes the air of O_2 and raises the concentration of CO_2. The exchange of O_2 and CO_2 between the atmosphere and soil air establishes a dynamic equilibrium that depends on the soil's respiration rate and on the resistance to gas movement through the soil. When gases can diffuse rapidly through the macropores, the O_2 partial pressure is ~20 kPa and the CO_2 pressure

Table 4.3 The composition of the gas phase of well-aerated soils. After Russell, 1973.

Soil and land use	Usual composition (% by volume)	
	O_2	CO_2
Arable land, fallow	20.7	0.1
– unmanured	20.4	0.2
– manured	20.3	0.4
Sandy soil, manured and cropped with potatoes	20.3	0.6
Serradella crop	20.7	0.2
Pasture land	18–20	0.5–1.5

~0.2 kPa, a condition described as aerobic. The normal gas compositions for soils under a range of land uses are given in Table 4.3.

However, if the macropores begin to fill with water and the soil becomes waterlogged, undesirable chemical and biological changes ensue as the soil becomes increasingly deficient in O_2 or anaerobic: the O_2 partial pressure may drop to zero and that of CO_2 rises far above 1 kPa. In the case of soil placed as capping over refuse tips (landfill), the gas phase may consist almost entirely of CO_2 and CH_4 in the approximate proportion of 2 : 3. This is due to the proliferation of microorganisms adapted to anaerobic conditions that produce, in addition to these gases and H_2, reduced forms of N, Mn, Fe and S. These reactions are discussed more fully in Section 8.3.

4.6 Summary

Physical, chemical and biotic forces act upon the primary clay, silt and sand particles, in intimate combination with organic materials, to form an organized arrangement of peds or aggregates and intervening voids. The strength of the short-range interparticle forces within the peds must be sufficient to prevent the particles separating again under the influence of the same forces that are responsible for the reorganization. On a microscopic scale, individual crystals of montmorillonite (1 nm thick) can stack in roughly parallel alignment, with much overlapping, to form quasi-crystals that have long-range order in the x–y direction up to

5 μm. Thicker crystals of illite and vermiculite can orient in a similar way to form larger units called domains. Both kinds of unit are stable in water when Ca^{2+} is the dominant exchangeable cation. Quasi-crystals and domains are the basic units of aggregation in clay soils.

Domains, quasi-crystals, clay crystals, sesquioxides and organic polyanions are drawn together by the attraction of positively and negatively charged surfaces, reinforced by van der Waals' forces and H-bonding to form microaggregates (< 250 μm diameter). The cohesion of microaggregates to form macroaggregates (> 250 μm) is enhanced by microbial polysaccharides and polyuronide gums which act like glue, and fungal hyphae and fine plant roots which enmesh the microaggregates. Good structure for crop growth depends on the existence of water-stable aggregates between 1 and 10 mm in diameter. Aggregate stability can be assessed by the Emerson dispersion test or by wet sieving.

Naturally formed visible peds, which comprise the soil's macrostructure, have spheroidal, blocky, platy or prismatic shapes. Grade of structure relates to the strength of within-ped cohesion relative to between-ped adhesion, and is very dependent on water content. Variation in soil strength with respect to water content determines the soil's consistence. Soil also has a characteristic microstructure that is identifiable mainly under magnification (× 10 or more), and reflects quite strongly the predominant pedogenic processes of soil formation. Movement of materials in solution or suspension results in the concentration of material < 2 μm (the plasma) in pores, on ped faces, and on sand particles (skeleton grains): these coatings (cutans) are variously called clay coatings (argillans), organic coatings (organans) or iron oxide coatings (ferrans). Polished surfaces on ped faces formed under pressure are called slickensides.

The size and shape of peds and their arrangement in the soil determines the intervening volume of cylindrical and spherical pores, and planar cracks or fissures. The pore volume determines the size of the reservoir of water sustaining plant and animal life in the soil. Porosity is usually expressed as the volume fraction ϕ, which can be calculated from the bulk density ρ_b (mass of dry soil per unit volume) and the mean particle density ρ_p, that is

$$\phi = 1 - \frac{\rho_b}{\rho_p}.$$

The value of ϕ normally ranges from 0.5 to 0.6, irrespective of soil texture. Water and air occupy the pore space. Changes in soil water content are measured by changes in the volumetric water content θ_v (volume of water per unit volume of soil) or the gravimetric water content θ_g (mass of water per unit mass of soil). Since the density of water is 1 Mg/m^3, θ_g values can be converted to θ_v by multiplying by ρ_b. Within the pore space there is a continuous distribution of pores sizes. Pores from which water drains at negative pressures of -5 to -10 kPa are called macropores and exist mainly between aggregates; they include many of the pores created by burrowing earthworms and plant roots (the biopores). The water content at -5 kPa (UK) or -10 kPa pressure (Australia and USA) defines the field capacity FC. The pore volume not occupied by water defines the air-filled porosity ε (m^3/m^3) which, when the soil is at FC, is called the air capacity ε_a. In a well-aerated soil, the O_2 partial pressure in the soil air is approximately the same as in the atmosphere ($c.$ 20 kPa), and that of CO_2 lies between 0.1 and 1.5 kPa.

References

Anderson A. N., McBratney A. B. & Fitzpatrick E. A. (1996) Soil mass, surface and spectral fractal dimensions estimated from thin section photographs. *Soil Science Society of America Journal* **60**, 962–9.

Brewer R. (1964) *Fabric and Mineral Analysis of Soils.* Wiley, New York.

Brewer R. (1976) *Fabric and Mineral Analysis of Soils,* 2nd edn. Krieger. New York.

Bullock P., Fedoroff N., Jongerius A., Stoops G., Tursina T. & Babel U. (1985) *Handbook for Soil Thin Section Description.* Waine Research Publications, Wolverhampton.

Cass A. (1999) Interpretation of some physical indicators for assessing soil physical fertility, in *Soil Analysis: an Interpretation Manual* (Eds K. I. Peverill, L. A. Sparrow & D. J. Reuter). CSIRO Publications, Melbourne, pp. 95–102.

Cass A., Cockcroft B. & Tisdall J. M. (1993) New approaches to vineyard and orchard soil preparation and management, in *Vineyard Development and Redevelopment* (Ed. P. F. Hayes). Australian Society of Viticulture and Oenology, Adelaide, pp. 18–24.

Emerson W. W. (1991) Structural decline of soils, assessment and prevention. *Australian Journal of Soil Research* **29**, 905–21.

Fitzpatrick E. A. (1984) *Micromorphology of Soils.* Chapman and Hall, London.

Greenland D. J. (1979) Soil management and soil degradation. *Journal of Soil Science* **32**, 301–22.

Hall D. G. M., Reeve M. J., Thomasson A. J. & Wright V. F. (1977) *Water Retention, Porosity and Density of Field Soils.* Soil Survey of England and Wales. Technical Monograph No. 9. Harpenden, UK.

Hodgson J. M. (Ed.) (1974) *Soil Survey Field Handbook.* Soil Survey of England and Wales, Technical Monograph No. 5. Harpenden, UK.

Kubiena W. L. (1938) *Micropedology.* Collegiate Press, Ames.

Klute A. (Ed.) (1986) *Methods of Soil Analysis. Part 1. Physical and Mineralogical Methods,* 2nd edn. Agronomy Monograph No. 9. American Society of Agronomy/Soil Science Society of America, Madison, WI.

Mandelbrot B. B. (1983) *The Fractal Geometry of Nature.* Freeman, New York.

McBratney A. B., Moran C. J., Stewart J. B., Cattle S. R. & Koppi A. J. (1992) Modifications to a method of rapid assessment of soil macropore structure by image analysis. *Geoderma* **53**, 255–74.

McDonald R. C., Isbell R. F., Speight J. G., Walker J. & Hopkins M. S. (1998) *Australian Soil and Land Survey Field Handbook,* 2nd edn, reprinted. Australian Collaborative Land Evaluation Program, Canberra.

Russell E. W. (1973) *Soil Conditions and Plant Growth,* 10th edn. Longman, London.

Soil Survey Division Staff (1993) *Soil Survey Manual,* 3rd edn. United States Department of Agriculture, National Soil Survey Center. USDA-NRCS Soil Survey Division Data National STATSGO Database, Washington DC.

Strutt N. (1970) *Modern Farming and the Soil.* Report of the Agricultural Advisory Council on Soil Structure and Soil Fertility. Her Majesty's Stationery Office, London.

Tisdall J. M. & Oades J. M. (1982) Organic matter and water-stable aggregates in soils. *Journal of Soil Science* **33**, 141–63.

White R. E. (2003) *Soils for Fine Wines.* Oxford University Press, New York.

Further reading

Fitzpatrick E. A. (1993) *Soil Microscopy and Micromorphology.* Wiley & Sons, Chichester.

Golchin A., Baldock J. A. & Oades J. M. (1998) A model linking organic matter decomposition, chemistry and aggregate dynamics, in *Soil Processes and the Carbon Cycle* (Eds R. Lal, J. M. Kimble, R. F. Follett & B. A. Stewart). CRC Press, Boca Raton, FL, pp. 245–66.

Greenwood K. L. & McKenzie B. M. (2001) Grazing effects on soil physical properties and the consequences for pastures: a review. *Australian Journal of Experimental Agriculture* 41, 1231–50.

Kay B. D. & Angers D. A. (2000) Soil structure, in *Handbook of Soil Science* (Ed. M. E. Sumner). CRC Press, Boca Raton, FL, A229–A276.

Example questions and problems

1 Describe two ways in which plant roots contribute to soil aggregation.

2 (a) Describe three mechanisms by which organic polymers interact with soil mineral particles to form microaggregates.

   (b) Describe three inorganic compounds that act as interparticle cements and stabilize soil aggregates.

3 A soil is sampled to measure the bulk density of the A horizon. Four intact cores (A–D) are taken with a steel cylinder 6.5 cm internal diameter and 10 cm long. The soil cores are dried at 105°C for 48 hours and the mass of oven-dry (o.d.) soil calculated. The results were as follows (note – the soil did not shrink on drying).

Core sample	Mass of o.d. soil (g)	Soil volume (cm^3)	Bulk density	Porosity
A	380			
B	419			
C	390			
D	432			

   (a) Complete this table, assuming that the density of the soil particles is 2.65 Mg/m^3. Calculate the mean bulk density and mean porosity of the A horizon, and the standard errors of these means.

   (b) What would be the mass of soil in one hectare to a depth of 15 cm (1 ha = 10,000 m^2)?

   (c) What is the maximum fraction (by volume) of the A horizon soil that could be occupied by water?

4 A soil is sampled to measure the water content. Two samples of moist soil weighing 125.1 and 130.6 g were oven-dried and then weighed again, recording 95 and 97.2 g dry weight, respectively.

   (a) Calculate the mean gravimetric water content.

   (b) The bulk density of this soil was measured as 1.42 Mg/m^3. Estimate the volumetric water content of the soil when sampled.

5 The volumetric water content of a soil at field capacity is 0.39 m^3/m^3 in the A horizon, which is 20 cm deep, and 0.43 m^3/m^3 in the B horizon (20–80 cm deep). Calculate the equivalent depth of water (mm) in the soil profile (0–80 cm).

6 (a) Given that the soil in Q.5 has a bulk density of 1.20 and 1.35 Mg/m^3 in the A and B horizons, respectively, calculate the air capacity ε_a (%) of each horizon.

   (b) How much rain (mm) must infiltrate the soil to wet the soil profile (to 80 cm depth) from field capacity to saturation?

7 Commercial lawn turf is grown on a soil that has a deep sandy loam A horizon (40 cm deep). The turf is stripped in rolls 5 cm deep for sale to sports arenas, landscape gardeners, etc. After stripping, the soil is resown at 6 monthly intervals. If the amount of soil removed in each turf roll is 40% of the volume, and the soil's bulk density is 1.2 Mg/m^3, calculate

   (a) the weight of dry soil removed per ha with each turf roll, and

   (b) how long it will be before all the A horizon is removed and the B horizon exposed.

   (c) If the B horizon is a compacted clay (bulk density 1.65 Mg/m^3), would this be suitable for commercial turf production? Give a reason for your answer.

Part 2

Processes in the Soil Environment

'Do you think that to be an agronomist you must till the soil . . . with your own hands? No: you have to study the composition of various substances – geological strata, atmospheric phenomena, the properties of the various soils, minerals, types of water, the density of different bodies, their capillary attraction. And a hundred other things.'

From *Madame Bovary* by Gustave Flaubert (1856)

Chapter 5

Soil Formation

5.1 The soil-forming factors

The seminal book *Factors of Soil Formation* (Jenny, 1941) presented an hypothesis that drew together many of the current ideas on soil formation, the inspiration for which owed much to the earlier studies of Dokuchaev and the Russian school (Section 1.2). The hypothesis was that soil is formed as a result of the interaction of many variables, the most important of which are:

- Parent material;
- climate;
- organisms;
- relief; and
- time.

Jenny called these variables soil-forming factors that controlled the direction and speed of soil formation. He attempted to define the relationship between any soil property (*s*) and the variables climate (*cl*), organisms (*o*), relief (*r*), parent material (*p*) and time (*t*), by a function of the form

$$s = \int(cl, o, r, p, t \dots). \qquad (5.1)$$

The dots in parenthesis indicated that factors of lesser importance such as mineral accessions from the atmosphere, or fire, might need to be taken into account. A combination (*S*) of soil properties *s*, each property being determined by a specific relationship with the variables in Equation 5.1 (the *clorpt* equation), makes up an individual soil type at a particular site. Subsequently, Jenny (1980) redefined the soil-forming factors as 'state' variables and added terms on the left of Equation 5.1 to include total ecosystem properties, vegetation and animal properties, as well as soil properties.

Since 1941, there has much debate among soil scientists about ways in which Equation 5.1 could be solved to give quantitative relationships between soil properties and the state variables. A summary of possible approaches is given in Box 5.1. Note that the inclusion of 'time' as a state variable in Equation 5.1 allows for the possibility that any one of the other state variables (and their interactions) could change during the course of soil formation, which makes the outcome of Equation 5.1 even for one location very uncertain (and hence leads to a wide range of soil types). In practice, however, pedologists have simplified the application of Equation 5.1 by considering situations where a soil property *s* (or a soil type *S*) is expressed as a function of only one state variable, under conditions where the others are constant or nearly so. For example, if the controlling variable is climate, the relationship is then called a climofunction

$$S = \int(cl)_{o,r,p,t\dots} \qquad (5.2)$$

and the range of soils $(S_1 - S_n)$ formed under different climates $(1 - n)$ is called a *climosequence* (Fig. 5.1). Similarly, where other soil-forming factors are the predominant controlling variable, we have:

- Parent material   a *lithosequence*;
- organisms   a *biosequence*;
- relief   a *toposequence*; and
- time   a *chronosequence*.

Thus far, the main virtue of Jenny's attempt to define the relationship between soil properties and

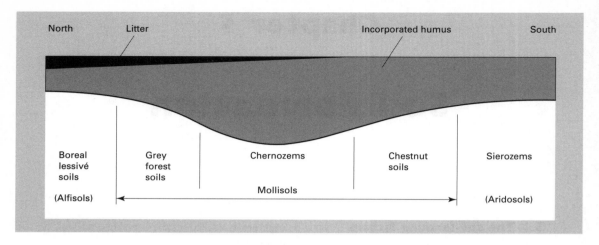

North   Litter   Incorporated humus   South

Boreal   Grey   Chernozems   Chestnut   Sierozems
lessivé   forest   soils
soils   soils

(Alfisols) ◄——————————————— Mollisols ———————————————►   (Aridosols)

Fig. 5.1 A climosequence of soils found along a north–south transect through Russia showing soil depth and degree of humus incorporation (after Duchaufour, 1982).

soil-forming factors lies not in the prediction of the exact value of *s* or *S* at a particular site. Rather its value lies in identifying trends in properties and soil types that are associated with readily observable changes in climate, parent material, etc. The concept of a causal relationship between *S* and the factors *cl*, *o*, *r*, *p* and *t* has been implicit, and at times explicit (Section 5.3), in attempts at imposing order on the natural variation of soils in the field through soil classification (Chapter 14). Furthermore, factors such as climate and organisms determine the rate at which chemical and biological reactions occur in the soil (the pedogenic processes, discussed in Section 9.2); others such as parent material and relief define the initial state for soil development, and time measures the extent to which a reaction will have proceeded.

In summary, there is a logical progression in considering soil formation from:

the environment (an expression of soil-forming factors) → physical, chemical and biological pedogenic processes → the development of characteristic soil properties.

This chapter deals briefly with the main factors affecting soil formation, and some of the outcomes expected for particular combinations of factors. The approach is essentially qualitative and intended to give the reader some insight as to why soils are so variable in the field. Subsequent chapters (6, 7 and 8) deal with important physical, chemical and biological processes (also covered in Chapters 2, 3 and 4), culminating in processes of profile development (Chapter 9). Soil properties and their effect on soil management are discussed in Chapter 10 *et seq*.

Box 5.1 Are solutions to Jenny's *clorpt* equation possible?

Given the range of possible initial values for the state variables *cl*, *o*, *r* and *p* in Equation 5.1, and the possibility that they may change independently, or interactively, over time *t*, a soil has been described as an 'unstable, chaotic, self-organizing system' (Phillips, 1998). As such, the mathematics of non-linear dynamic systems (NDS) can be applied to modelling soil formation. The properties of NDS are such that variations in initial conditions, and of perturbations in those conditions, can grow and persist over time, so that the end-point of a mature or zonal soil (Sections 5.3 and 5.6) is only one of many possible outcomes of soil formation. In support of his argument, Phillips quotes the example of short-range variability (over tens of metres) in soil caused by minor variations in texture and watertable height, which occurred in a landscape (extending over kilometres) where there were no

Box 5.1 *continued*

apparent differences in *cl*, *o*, *r* and *p* at this larger scale. We return to the question of the scale of soil variability in Chapter 14, but it is important to note here Bouma and Hoosbeek's (1996) scale hierarchy for pedological models, namely:

i + 5	Continent
i + 4	Region
i + 3	Watershed (catchment)/landscape
i + 2	Catena/landscape
i + 1	Polypedon
i	Pedon
i − 1	Soil horizon
i − 2	Soil macrostructure
i − 3	Soil microstructure
i − 4	Molecular interactions

In this hierarchy, the *i*th level (pedon) defines the soil unit: levels below *i* relate to properties and processes in the soil, and levels above *i* relate to landscape processes. Jenny's *clorpt* equation has been applied at the continental scale down to the catena/landscape scale, as evidenced by the climosequence (Section 5.3) and toposequence concepts (Section 5.5), respectively. Implicitly at these higher levels, the effects of processes occurring at lower levels and the inevitable transfer of materials between pedons are 'averaged out'. Others such as Hoosbeek and Bryant (1992) and Paton *et al.* (1995) argue that, rather than through the *clorpt* equation, soil formation should be quantitatively described through combinations of mechanistic models of pedogenic processes. This has usually been attempted on soil profiles or pedons (one-dimensional processes), but Minasny and McBratney (2001) have recently extended this approach to the landscape scale in two dimensions, with some success, although much further work needs to be done.

5.2 Parent material

Soil may form by the weathering of:
- Consolidated rock *in situ* (a residual soil); or
- unconsolidated superficial deposits which have been transported by ice, water, wind or gravity. These deposits originate from past denudation and erosion of consolidated rock.

Fig. 5.2 Stones of a desert soil surface fractured due to extreme variations in temperature (J. J. Harris Teal in Holmes, 1978).

Weathering

Generally, weathering proceeds by physical disruption of the rock structure that exposes the constituent minerals to chemical alteration. Forces of expansion and contraction induced by diurnal temperature variations cause rocks to shatter and exfoliate, as illustrated by surface features in a desert soil (Fig. 5.2). This effect is most severe when water trapped in the rock repeatedly freezes and thaws, resulting in irresistible forces of expansion followed by contraction.

Water is the dominant agent in weathering, not only because it initiates solution and hydrolysis (Section 9.2), but also because it sustains plant life on the rock surface. Lichens play a special part in weathering, because they produce chelates that trap metallic elements of the decomposing rock in organo-metallic complexes. In general,

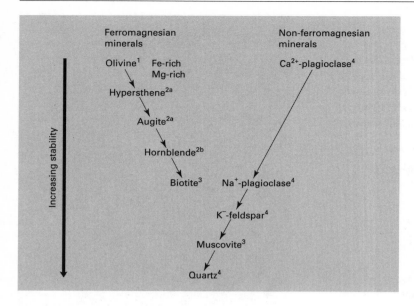

Fig. 5.3 Ranking of the common silicate minerals in order of increasing stability. Note that the superscripts indicate the number of O atoms shared between adjacent SiO_4^{4-} tetrahedra: 1 = none; 2a = 2 (single chain); 2b = 2 and 3 (double chain); 3 = 3 (sheet structure); 4 = 4 (three-dimensional structure) (after Goldich, 1938 and Birkeland, 1999).

the growth, death and decay of plants and other organisms markedly enhances the solvent action of rainwater through an increase in CO_2 concentration from respiration. Plant roots also contribute to the physical disintegration of rock.

Water carrying suspended rock fragments has a scouring action on surfaces. The suspended material can vary in size from the finest 'rock flour' formed by the grinding action of a glacier, to the gravel, pebbles and boulders moved along and constantly abraded by fast-flowing streams. Particles carried by wind also have a 'sand-blasting' effect, and the combination of wind and water to form waves produces powerful destructive forces along shorelines.

Mineral stability

The rate of weathering depends on:
• Temperature;
• rate of water percolation;
• oxidation status of the weathering zone;
• surface area of rock exposed (largely determined by fracturing and exfoliation); and
• types of mineral present.

Much attention has been paid to the relationship between the crystalline structure of a rock mineral and its susceptibility to weathering. One scheme for the order of stability, or weathering sequence, of the more common silicate minerals in coarse rock

fragments (> 2 mm diameter) is shown in Fig. 5.3. This sequence demonstrates that the resistance of a mineral to weathering increases with the degree of sharing of O atoms between adjacent Si tetrahedra in the crystal lattice (Section 2.3). The Si-O bond has the greatest energy of formation, followed by the Al-O bond, and the even weaker bonds formed between O and the basic metal cations* (Na^+, K^+, Mg^{2+}, Ca^{2+}, Sr^{2+}, Ba^{2+}). Thus, olivine weathers rapidly because the Si tetrahedra are held together only by O-metal cation bonds, whereas quartz is very resistant because it consists entirely of linked Si tetrahedra. In the chain (amphiboles and pyroxenes) and sheet (phyllosilicate) structures, the weakest points are the O-metal cation bonds. Isomorphous substitution (Section 2.3) of Al^{3+} for Si^{4+} also contributes to instability because the proportion of Al-O to Si-O bonds increases and more O-metal cation bonds are necessary. These factors account for the decrease in stability of the Ca-plagioclases compared with the Na-plagioclases (Fig. 5.3).

Weathering continues in the finely comminuted materials (< 2 mm diameter) of soils and unconsolidated sedimentary deposits, with the least stable primary minerals surviving only in the early stages, and the more resistant minerals becoming dominant

---

* Cations of elements that form strongly alkaline solutions such as NaOH and KOH when they react with water.

Table 5.1 Stages in the weathering of minerals (< 2 mm particle size) in soils. After Jackson et al., 1948.

Stage	Type mineral	Soil characteristics
Early weathering stages		
1	Gypsum	These minerals occur in the
2	Calcite	sand, silt and clay fractions of
3	Hornblende	young soils all over the world,
4	Biotite	and in soils of arid regions
5	Albite	where lack of water inhibits
		chemical weathering and
		leaching
Intermediate weathering stages		
6	Quartz	Found mainly in soils of
7	Muscovite	temperate regions, frequently
	(also illite)	on parent materials of glacial
8	Vermiculite and	or periglacial origin; generally
	mixed layer	fertile, with grass or forest as
	minerals	the natural vegetation
9	Montmorillonite	
Advanced weathering stages		
10	Kaolinite	Found in the clay fractions of
11	Gibbsite	many highly weathered soils on
12	Hematite	old land surfaces of humid and
	(also goethite)	hot inter-tropical regions;
13	Anatase	often of low fertility

as weathering advances. The stages of weathering can be identified by certain type minerals, as shown in Table 5.1, which also includes some of the broad characteristics of the associated soils.

Rock types

Consolidated rocks are of igneous, sedimentary or metamorphic origin.

Igneous rocks

These rocks are formed by the solidification of molten magma in or on the Earth's crust, and are the ultimate source of all other rocks. Material can be released explosively from the vents of volcanoes (volcanic ash and pyroclastic flows, collectively called tephra), or flow over the land surface as streams of lava that may cover many hundreds of square kilometres (extrusive formations). Other volcanic material wells up through lines of weakness in the Earth's crust to form sills and dykes (intrusive formations). Volcanic magma that does not reach the surface accumulates in magma chambers where it solidifies as a pluton or batholith. Subsequent weathering of the surrounding rock may leave the hard, intrusive igneous rock remaining as a dominant feature in the landscape, as illustrated by the massive granite dome in Fig. 5.4.

Fig. 5.4 Granite dome in Zimbabwe.

Table 5.2 Chemical composition (%) of typical granitic and basaltic rock. After Mohr and van Baren, 1954.

Element	Granite	Basalt
O	48.7	46.3
Si	33.1	24.7
Al	7.6	10.0
Fe	2.4	6.5
Mn	–	0.3
Mg	0.6	2.2
Ca	1.0	6.1
Na	2.1	3.0
K	4.4	0.8
H	0.1	0.1
Ti	0.2	0.6
P	–	0.1
Total	100.2	100.7

Fig. 5.5 Deep, uniform, coarse-textured soil formed on granite in south-central Chile. The scale is 15 cm long (see also Plate 5.5).

Igneous rocks are broadly subdivided on mineral composition into:
• *Acidic* rocks, which are relatively rich in quartz and the light-coloured Ca or K, Na-feldspars – mainly intrusive; and
• *basic* rocks, which contain little quartz but much dark-coloured ferromagnesian minerals (hornblende, micas, pyroxene) – mainly extrusive.

The chemical composition of typical granite and basalt, rocks representative of these two groupings, is given in Table 5.2. The higher the ratio of Si to (Ca + Mg + K), the more probable that some of the Si is not bound in silicates, and may remain as residual quartz as the rock weathers. Soils formed from granite therefore usually have a higher content of quartz grains (sand and silt size) than those formed from basalt. Fig. 5.5 shows a profile of a deep, coarse sandy soil derived from granite. Such soils are usually freely drained (Section 5.5).

Sedimentary rocks

These are composed of the weathering products of igneous, metamorphic and older sedimentary rocks, and are formed after deposition by wind and water. Cycles of geologic uplift, weathering, erosion and subsequent deposition of eroded materials in rivers, lakes and seas have produced thick sequences of sediments in many parts of the world. Under the weight of overlying sediments, deposits gradually consolidate and harden – the process of diagenesis – to form sedimentary rocks. Faulting, folding and tilting movements, followed by erosion, can cause the sequence of rock strata in the geological column to be revealed by their surface outcrops, as illustrated in Fig. 5.6. Clastic sedimentary rocks are composed of fragments of the more resistant minerals. The size of fragments decreases in order from conglomerates to sandstones to siltstones to mudstones. Other sedimentary rocks are formed by precipitation or flocculation from solution. The most common are limestones including chalk (Fig. 5.7), which vary in composition from pure $CaCO_3$ to mixtures of Ca and Mg carbonates (dolomite), or carbonates with detrital sand, silt and clay minerals.

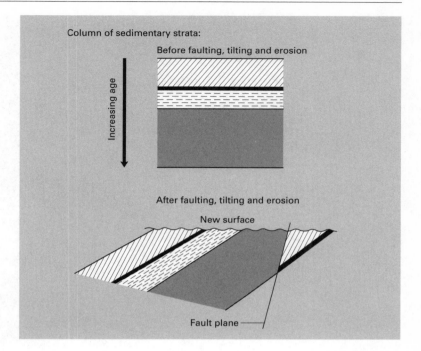

Column of sedimentary strata:

Fig. 5.6 Diagram to show how sedimentary rocks of different age and type can provide the parent material for present-day soils.

Fig. 5.7 Electron micrograph of a sample of Middle Chalk (Berkshire, England) showing microstructure of calcified algae (courtesy of S. S. Foster).

Metamorphic rocks

Igneous and sedimentary rocks that are subjected to intense heat and great pressures are transformed into metamorphic rocks. There are two broad types of metamorphism, namely:

- Contact or thermal metamorphism; and
- regional metamorphism.

Contact metamorphism occurs when volcanic magma intrudes into sedimentary rocks. At the higher temperatures around the zone of intrusion, minerals in the original rock are transformed to more resistant forms, resulting in a harder and denser rock. An example is quartzite, which is formed by the recrystallization of quartz grains in sandstone. An example of regional metamorphism arises when a soft sedimentary rock such as mudstone is progressively metamorphosed (depending on the intensity of the process) to shale, slate, schist or phyllite. Fig. 5.8 shows an example of weathered schist.

Fig. 5.8 Weathering schist exposed in Languedoc-Roussillon, southern France.

Soils on transported material

Soil layers and buried profiles

Most parts of the Earth's surface have undergone several cycles of submergence, uplift, erosion and denudation over hundreds of millions of years. During the mobile and depositional phases there is much opportunity for materials from different rock formations to be mixed, and the interpretation of soil genesis on such heterogeneous parent material is often complex. In regions of active seismicity, such as along the crustal plate boundaries of the Pacific Ocean rim, volcanic eruptions and lava flows have, over millions of years, buried existing land surfaces and provided fresh parent material for soil formation (e.g. in the North Island of New Zealand).

When two or more layers of transported material derived from different rocks are superimposed they form a lithological discontinuity. The resultant layered profile is often difficult to distinguish from the pedogenically differentiated profile of a soil formed on the same parent material. In other cases, a soil may be buried by transported material during a period of prolonged erosion/deposition of an ancient landscape. The old soil is then preserved as a fossil soil, as illustrated by the buried 'mottled zone' of an old laterite (Oxisol (ST) or Ferrosol (ASC)) in Fig. 5.9.

Water

Transport by water on land produces alluvial, terrace and footslope deposits. Lacustrine deposits are laid down in lakes, and marine deposits under the sea. During transport the rock material is sorted according to size and density and abraded, so that fluviatile (river) deposits characteristically have smooth, round pebbles. The larger fragments are moved by rolling or are lifted by the turbulence of water flowing over the irregular stream bed, and carried forward some distance by the force of the water. This 'bouncing' process is called saltation (*cf* saltation of wind-blown particles, Section 11.5). The smallest particles are carried in suspension and salts move in solution. Colloidal material may not be deposited until the stream discharges into the sea, when flocculation occurs because of the high total salt concentration. The modes of sediment transport in relation to the depth and velocity of the water are shown in Fig. 5.10.

Ice

Ice was an important agent in transporting rock materials during the 2 million years of the Pleistocene epoch (Table 1.1).

During the colder glacial phases, the ice cap advanced from polar and mountainous regions to cover a large part of the land surface, especially in the Northern Hemisphere (the ice reached a line only a few kilometres north of the present River Thames in southern England). In the Southern Hemisphere, where the Pleistocene ice sheets were much less extensive because of the greater area of sea relative to land, the influence of recent glaciation on soil formation is less pronounced. However, rock residues from much earlier glaciations do provide parent material for present-day soils in the Southern Hemisphere.

The moving Pleistocene ice ground down rock surfaces and the 'rock flour' became incorporated in the ice; rock debris also collected on the surface of valley glaciers by colluviation (see under 'Gravity'). During the warmer interglacial

Fig. 5.9 The 'mottled zone' of an ancient laterite (Oxisol (ST) or Ferrosol (ASC)) buried under more recent alluvium in southern New South Wales (see also Plate 5.9).

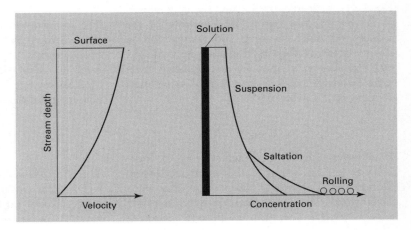

Fig. 5.10 Sediment transport in relation to depth and velocity of water (after Mitchell, 1993).

Fig. 5.11 Soil formation on glacifluvial sands and gravels of the Lleyn Peninsular, Wales.

phases, the ice retreated leaving extensive deposits of glacial drift or till, which is very heterogeneous. Streams flowing out of the glaciers produced glacifluvial deposits of sands and gravels (Fig. 5.11). Repeated freezing and thawing in periglacial regions resulted in frost-heaving of the surface and much swelling and compaction in subsurface layers (forming fragipans, Section 4.3). Such deposits may often be distinguished from water-borne materials by the vertical orientation of many of the pebbles.

Wind

Wind moves small rock fragments by the processes of rolling, saltation and aerial suspension (Section 11.5). Material swept from dry periglacial regions has formed aeolian deposits called loess in central USA, central Europe, northern China and South America. These deposits are many metres thick in places and form the parent material of highly productive soils, such as in the central Mississippi Basin, USA and in northern China (Fig. 5.12). In the Murray-Darling Basin of southern Australia, such wind-blown deposits are called parna. Erosion and transport of loess from plateaux in northern China by the Yellow River, and other large rivers, have led to the formation of fertile soils on the North China Plain. Other characteristic wind deposits are the dunes of old sea and lake shores, and in desert areas.

Gravity

Gravity produces colluvial deposits, of which the scree slopes formed by rock slides in mountainous regions are an obvious example. Less obvious, but widespread under periglacial conditions, are solifluction deposits at the foot of hillslopes. These are formed when a frozen soil surface thaws and the saturated soil mass slips over the frozen ground below, even on gentle slopes. Creep occurs when the surface mantle of soil moves by slow colluviation away from the original rock. The junction between creep layer and weathering rock is frequently delineated by a stone line. Slips and larger earth flows can occur on steep slopes on unstable parent materials, especially when forest is cleared and replaced by short-rooted crop or pasture species, as on the east coast of the North Island of New Zealand.

5.3 Climate

The importance of climate in soil formation is demonstrated by its explicit and implicit inclusion in the criteria for separating soil classes at a high level in classifications such as Soil Taxonomy (Soil Survey Staff, 1996). The key components of climate are moisture and temperature.

Moisture

The effectiveness of moisture depends on:
• The form and intensity of the precipitation;
• its seasonal variation;
• the evaporation rate from vegetation and soil;

Fig. 5.12 Terraces cut into deep loess deposits near Yan Ling, northern China.

• slope of the land; and
• the permeability of the parent material.

Various functions of meteorological variables have been proposed to measure moisture effectiveness. Most focus on the balance between precipitation and evaporation, as shown by one of the earliest and simplest indices due to Thornthwaite (1931) (Box 5.2). Another example is a moisture effectiveness index used for the British Isles, which is defined in terms of the maximum potential soil water deficit (MWD). The MWD for a year is obtained as the difference between rainfall and potential evaporation (Section 6.5) (both in mm). The value is calculated for each year in a series of years and the results averaged to give the MWD for that site (Avery, 1990).

Where precipitation generally exceeds evaporation, as in wet or humid climates, there is a net downward movement of water in the soil for most of the year, which results in the leaching of soluble materials, sometimes out of the soil entirely, and the translocation of clay particles from upper to lower horizons. In arid climates, where evaporation generally exceeds precipitation, salts are not leached out of the profile and can be carried upwards as water evaporates from the soil surface. More is said about these processes, and soil profile differentiation, in Chapter 9.

Temperature

Temperature (T) varies with latitude and altitude, and the extent of absorption and reflection of solar radiation by the atmosphere. Temperature affects the rate of mineral weathering and synthesis, and the biological processes of growth and decomposition. Reaction rates are roughly doubled for each 10°C rise in temperature, although enzyme-catalysed reactions are sensitive to high temperatures and usually attain a maximum between 30 and 35°C.

Box 5.2 Thornthwaite's climate indices.

As a measure of moisture effectiveness, Thornthwaite developed the P-E index, which he defined as the average of the sum for each year of the monthly ratio

$$\frac{\text{Precipitation, P}}{\text{Evaporation, E}} \times 10.$$

(The multiplier of 10 was introduced to eliminate fractions.) Based on the P-E index, five major categories of climate and associated vegetation types were recognized as follows:

P-E index	Category	Vegetation
128 and above	Wet	Rainforest
64–127	Humid	Forest
32–63	Sub-humid	Grassland
16–31	Semi-arid	Steppe
< 16	Arid	Desert

Since the effects of high and low temperatures on plant growth and soil formation are similar to those of abundant and deficient moisture supply, respectively, Thornthwaite also developed an index of thermal efficiency analogous to the P-E index. The temperature index T-E was calculated as the average of the sum for each year of the monthly values of

$9 \times$ mean monthly temperature (°C)/20.

This empirical expression was chosen so that the cold margin of the Tundra had a T-E index of zero and the cooler margin of the rainforest an index of 128. The T-E indices and associated categories are as below:

T-E index	Category
128 and above	Tropical
64–127	Mesothermal
32–63	Microthermal
16–31	Taiga
1–15	Tundra
0	Frost

Table 5.3 A climatic classification for the British Isles. After Avery, 1990.

Climatic regime	Mean annual accumulated temperature above 5.6°C (day-degrees)	Average maximum potential water deficit (MWD) (mm)
Sub-humid temperate	> 1375	> 125
Humid temperate	> 1375	50–125
Perhumid temperate	> 1375	< 50
Humid (oro)*boreal	675–1375	50–125
Perhumid (oro)*boreal	675–1375	< 50
Perhumid oroarctic*	< 675	< 50

* Altitude > 150 m.

One of the simplest indices of the thermal efficiency of the climate is Thornthwaite's T-E index (Box 5.2). Another is a temperature index used in Britain, which is based on the accumulated degrees Celsius for days each year when the mean daily temperature exceeds 5.6°C. The combination of the average MWD and the temperature index produces a climatic classification for the British Isles, as used by Avery (1990) (Table 5.3).

The concept of soil zonality

From Dokuchaev around 1870 and later, many pedologists in Europe and North America regarded climate as a pre-eminent factor in soil formation. The obvious relationship between climatic zones, with their associated vegetation, and the broad belts of similar soils that stretched roughly east–west across Russia inspired the zonal concept of soils, as shown in Fig. 5.1. The broad categories recognized were as follows:
• *Zonal* soils – those in which the influence of the regional climate and its associated vegetation, acting for a sufficient length of time, overrides the influence of any other factor. Zonal soils are mature soils (Section 5.6);
• *intrazonal* soils – those in which some local anomaly of relief, parent material or vegetation is strong enough to modify the primary influence of the regional climate;

• *azonal* soils – immature soils that have poorly differentiated profiles, either because of their youth or because some factor of the parent material or environment has arrested their development.

The zonal concept gave rise to systems of classification in which broad soil classes were related to climatic and vegetation zones on a continental scale, as in Russia, the USA and Australia. A key categorical level in such classifications was that of the Great Soil Groups, each of which embodied a central concept based on current understanding of soil genesis and environmental controls on genesis (Section 14.4). As these classifications evolved and higher and lower categories were developed, the definition of the Great Soil Groups became more 'factual'; that is, based on observable soil properties – their presence or absence, and limiting values. Pedogenic processes and their response to environmental factors continue to underpin many current classification systems, and the Great Soil Group names (Table 5.4), continue to have currency in agricultural pursuits where there is a long association between the produce and the soil type, as in viticulture for wine-making.

Predictably, the concept of soil zonality is less applicable in the subtropics and tropics where land surfaces are generally much older than in continental Europe, Central Asia and North America, and consequently have undergone many cycles of erosion and deposition associated with climatic change: here the soil's age (Section 5.6) and local topography (Section 5.5) are factors of major importance. The zonal concept is also less applicable in regions, such as Britain and Scandinavia, where much of the parent material of present-day soils is young – of Pleistocene or more recent origin, or derived from recent volcanic activity (e.g. New Zealand). In these cases, the nature of the parent material strongly influences many soil properties, and relief again plays an important role in soil formation.

5.4 Organisms

Soil and the organisms living on and in it comprise an ecosystem. The active components of the soil ecosystem are the plants, animals, microorganisms and humans.

Vegetation

The primary succession of plants that colonize a weathering rock surface (Section 1.1) culminates in the development of a climax community, the species composition of which depends on the climate and parent material, but which, in turn, has a profound influence on the soil that is formed. For example, in the Mid-West and Great Plains of the USA, deciduous forest seems to accelerate soil formation compared with prairie (grassland) on the same parent material and under similar climatic conditions. In parts of England such as on the North York moors and heathlands of the Bagshot Sands, soil profiles preserved under the oldest burial mounds of the Bronze Age people are typical of Brown Forest Soils (Inceptisols (ST) or Tenosols (ASC)) formed under deciduous forests at least 4000 years BP. In contrast, podzols (Spodosols (ST) or Podosols (ASC)) have formed on the surfaces of the mounds under the heathland vegetation, which has developed since early settlers removed the forest. Differences in the chemical composition of leaf leachates can partly account for this divergent pattern of soil formation over time (Section 9.2).

Fauna

The species of invertebrate animals in soil and the type of vegetation are closely related. The acidic litter of pines, spruce and larch creates an environment unfavourable to earthworms, so that litter accumulates on the soil surface where it is only slowly broken down by mites and collembola and decomposed by fungi – similarly under many native tree species in Australia (Fig. 5.13). The litter of elm and ash, and to a lesser extent oak and beech, is more readily ingested by earthworms and incorporated into the soil as faecal material (earthworm casts). Earthworms are the most important of the soil-forming fauna in temperate regions, being supported to a variable extent by small arthropods and the larger burrowing animals (rabbits and moles). Earthworms are also important in tropical soils, under moist conditions; but in general the activities of termites, ants and dung-eating beetles are of greater significance,

Table 5.4 Higher soil categories according to the Zonal concept. After Thorp and Smith, 1949.

Order	Suborder	Great Soil Groups
Zonal soils	1. Soils of the cold zone	Tundra soils
	2. Light-coloured soils of arid regions	Desert soils Red desert soils Sierozems Brown soils Reddish-brown soils
	3. Dark-coloured soils of semi-arid, sub-humid, and humid grasslands	Chestnut soils Reddish chestnut soils Chernozems Prairie soils Reddish prairie soils
	4. Soils of the forest–grassland transition	Degraded chernozems Non-calcic brown or Shantung brown soils
	5. Light-coloured podzolized soils of the forest regions	Podzols Grey wooded or Grey podzolic soils Brown podzolic soils Grey-brown podzolic soils Red-yellow podzolic soils
	6. Lateritic soils of forested mesothermal and tropical regions	Reddish-brown lateritic soils Yellowish-brown lateritic soils Laterites
Intrazonal soils	1. Halomorphic (saline and alkali) soils of imperfectly drained arid regions and littoral deposits	Solonchak or saline soils Solonetz Soloth (solod)
	2. Hydromorphic soils of marshes, swamps, seep areas and flats	Humic gley soils Alpine meadow soils Bog soils Half-bog soils Low-humic gley soils Planosols Groundwater podzols Groundwater laterites
	3. Calcimorphic soils	Brown forest soils (Braunerde) Rendzinas
Azonal soils		Lithosols Regosols (includes Dry sands) Alluvial soils

particularly in the sub-humid to semi-arid savanna of Africa and Asia (Section 3.2).

Surface-casting earthworms can build up a stone-free layer at the soil surface, as well as intimately mixing the litter with fine mineral particles they have ingested. The surface area of the organic matter that is accessible to microbial attack is then much greater. Termites are also

Fig. 5.13 Litter layer colonized by fungi under Acacia and Eucalypt trees (see also Plate 5.13).

important soil-formers in that they carry fine particles, water and dissolved salts from depths of 1 m or more to their termitaria on the surface (see Fig. 3.8). When abandoned, the termitaria gradually disintegrate due to weathering and trampling by animals to form a thin mantle on the soil surface.

Human influence

Humans influence soil formation through manipulation of the natural vegetation, and its associated animals, and the effects of agricultural practices and urban and industrial development. Domestic stock and feral animals can denude the vegetation so that the soil is exposed to a greater erosion risk. Compaction by the trampling of domestic grazing animals and the traffic of farm machinery decreases the rate of water infiltration into the soil, thereby increasing runoff and the risk of erosion, especially on slopes. Further, the course

of soil formation around many old settlements has been influenced by the deposition of organic residues, such as domestic waste, and in areas near the sea, seaweed. Shallow vineyard soils on Chalk in the Champagne region, and gravelly soils in the Medoc region of France, have been modified by inputs of organic materials over several centuries. Soils that have been built up with substantial inputs of organic materials by humans are called plaggen soils (Anthroposols (ASC)). In *Keys to Soil Taxonomy* (Soil Survey Staff, 1996), a man-made surface layer 50 cm or more thick and produced by long-continued manuring is called a plaggic epipedon (Section 9.1).

Human influences on the management of the soil resource are discussed in Part 3.

5.5 Relief

Relief has an important influence on the local climate, vegetation and drainage of a landscape.

Major topographical features are easily recognized – mountains and valleys; escarpments, ridges and gorges; hills, plateaux and floodplains. Changes in elevation affect the temperature (a decrease of approximately 0.5°C per 100 m increase in height), the amount and form of the precipitation, and the intensity of storm events. These factors interact to influence the type of vegetation. More subtle changes in local climate and vegetation are associated with the slope and aspect of valley sides and escarpments.

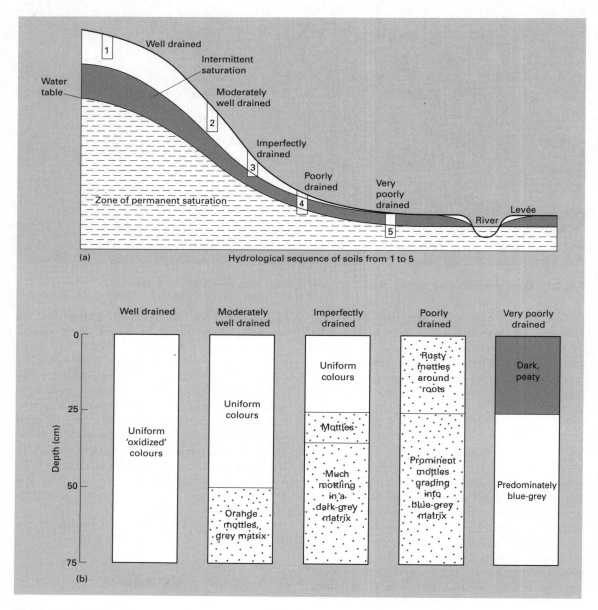

Fig. 5.14 (a) Section of slope and valley bottom showing a hydrological soil sequence.
(b) Changes in profile morphology, principally colour, with the deterioration in soil drainage (after Batey, 1971).

Slope

Angle of slope (steepness) and vegetative cover affect moisture effectiveness by governing the proportion of runoff to infiltration. Soils with impermeable B horizons, especially where they occur in an undulating landscape, can develop a perched water table (Section 6.3) that predisposes to reducing conditions, and mottling and gleying in the profile. This condition is common, for example, during wet winters on 'duplex' soils, such as red and yellow podzolics – (Alfisols/Ultisols (ST) or Chromosols/Kurosols (ASC) – on the lower slopes and foothills of the Great Dividing Range in southeastern Australia. Normally, however, in undulating landscapes on permeable materials, the soils at or near the top of a slope tend to be freely drained with a deep water table, whereas at the valley bottom the soils are poorly drained, with the water table near or at the surface (Fig. 5.14a). The succession of soils forming under different drainage conditions on relatively uniform parent material comprises a hydrological sequence, an example of which is shown in Fig. 5.14b. As the drainage deteriorates, the oxidized soil profile, with its warm orange-red colours due to ferric oxides and oxyhydroxides (Section 2.4), is transformed into the mottled and gleyed profile of a waterlogged soil (Fe in the ferrous state). Such a sequence of soils on a slope is an example of a catena (Box 5.3).

In addition to modified drainage, there is soil movement on slopes. A mid-slope site is continually receiving solids and solutes from sites immediately up-slope by wash and creep, and continually losing material to sites below. Here the form of the slope is important – whether it is smooth or uneven, convex or concave, or broken by old river terraces. Frost-heaving in cold regions or faunal activity (rabbits, moles and termites) promotes the slow creep of soil down even slight slopes, which creates a deeper profile at the footslope than on the slope above.

Small variations in elevation can also be important in flat lands, especially if the groundwater is saline and rises close to the surface. A curious pattern of microrelief of this kind, called gilgai*, is widespread in the black earths (Vertisols (ST)

* From the Aboriginal word for a small waterhole.

Box 5.3 Catenas and catenary complexes.

The importance of relief on soil formation was highlighted by Milne (1935) in central East Africa, who recognized a recurring sequence of soils forming on slopes in an undulating landscape. He introduced the term catena (Latin for a chain) to describe the suite of contiguous soils extending from hill top to valley bottom. A catena is restricted to soils of the one climatic zone, that is, the members are variations within a zonal soil. Originally, there were two types of catenas:
• One in which the parent material is the same, and soil profile differences develop as a result of variations in drainage and aeration (as in Fig. 5.14). There can also be differential sorting of eroded material and accumulation of solutes according to the slope;
• another in which the slope has been carved out of two or more superimposed parent materials. The soils formed on the upper ridges and escarpment are old and denuded, whereas those formed on the lower slopes are younger, reflecting the rejuvenation of the *in situ* parent material by materials derived from above.

In the USA, the term has been restricted to sequences of soils on uniform parent material. In this sense, the catena is synonymous with Jenny's toposequence, which defines the suite of soils formed when the influence of relief predominates over that of the other soil-forming factors. In the USA, when two or more members of a catena occur together in the landscape, they form a catenary complex that can be mapped as a catenary soil association. Other examples of catenas in the broader sense, as defined by Milne, are given in Chapter 9.

or Vertosols (ASC)) of semi-arid regions in Australia (Fig. 5.15). The soils of the mound ('puff') and hollow ('shelf') are quite different, with $CaCO_3$ and gypsum ($CaSO_4$) occurring much closer to the surface in the yellow-brown crumbly clay of the puff than in the grey, compact cracking clay of the shelf. Comparable features are

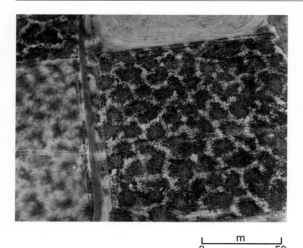

m
0   50

Fig. 5.15 Aerial view of gilgai microrelief (CSIRO photograph, courtesy of G. D. Hubble).

found in the black clay soils of Texas, the Free State of South Africa, and in the Black Cotton soils of the Deccan, India. In the prairies of western Canada, glacial drift covers much of the landscape and forms a characteristic knob and kettle (depression) or ridge and swale topography. This markedly influences the local drainage and hence the type of soil formed, which ranges from being freely drained (even droughty) on the knobs to poorly drained (possibly saline) in the depressions.

5.6 Time

Time acts on soil formation in two ways:
• First, the value of a soil-forming factor may change with time (e.g. climatic change, new parent material, etc.);
• second, the extent of a pedogenic reaction depends on the time for which it has operated. Monogenetic soils are those that have formed under one set of factor values for a certain period of time; soils that have formed under more than one set of factor values at different times are called polygenetic. The latter are much more common than the former.

We know that the world climate has changed over geological time: the most recent large changes were associated with the alternating glacial and interglacial periods of the Pleistocene. Major

climatic changes were accompanied by the rising and falling of sea level, erosion and deposition and consequent isostatic processes in the Earth's crust, all of which produced radical changes in the distribution of parent material and vegetation, and the shape of the landscape. On a much shorter timescale, covering a few thousand years, changes can occur in the biotic factor of soil formation, as evidenced by the primary succession of species on weathering rock; and also in the relief factor through changes in slope form and the distribution of groundwater.

Rate of soil formation

Knowledge of a soil's age, together with information about any change in the factor values, enables rates of soil formation in different environments to be estimated. One criterion of a soil's age is its profile morphology, in particular, the number of horizons, their degree of differentiation and their thickness. The use of profile criteria is feasible for a soil derived from uniform parent material. Frequently, however, the past history of climatic, geological and geomorphological change is so complex that an accurate assessment of a soil's age by the use of these criteria is impossible.

The organic matter in individual horizons can be dated from the natural radioactivity of the C (the $^{14}C : ^{12}C$ ratio). Where the radio-C ages down a profile are not too dissimilar, an average age for the whole soil may be calculated. The relation between soil development and time can also be determined in special cases where the time that has elapsed since the exposure of parent material is known, and the subsequent influence of the other soil-forming factors has been reasonably steady. Such studies suggest that the rate of soil development is extremely variable, ranging from very rapid (several cm in 100 years or so) on volcanic ash in the tropics to very slow (1 cm per 5000 years) on Chalk weathering in a cool temperate environment. Theoretically, the rate of soil formation sets the upper limit for an acceptable rate of soil loss by erosion (Section 11.5): in many cases, however, soil formation is so slow relative to erosion under agriculture that the soil profile becomes thinner with time.

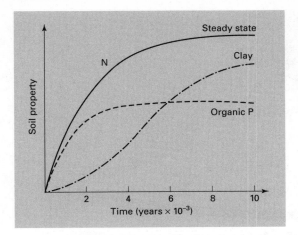

Fig. 5.16 The approach to steady state for selected soil properties under temperate conditions (after Jenny, 1980).

The concept of steady state

When the rate of change of a soil property is negligibly small, the soil is said to be in steady state with respect to that property. For example, for a soil forming on moderate slopes, the rate of natural erosion may just counteract the rate of rock weathering so that the soil, although immature, appears to be in steady state. In many cases, however, because pedogenic processes do not all operate at the same rate, individual soil properties do not all attain steady state at the same time. Soil formation is not linear with time, but tends to follow an exponential-type 'law of diminishing returns' (Section 11.2). This trend has been demonstrated for the accumulation of C, N and organic P in soils of temperate regions (Fig. 5.16). The time course of clay accumulation, on the other hand, is probably more sigmoidal in nature.

Because colonization by plants and animals of weathering parent material is an integral part of soil formation, it is more appropriate to consider steady state in the context of a soil and its climax vegetation (Section 5.4), that is, the soil–plant ecosystem. Studies of such ecosystems that have developed on parent material exposed by the last retreat of the Pleistocene ice (c. 11,000 years BP) suggest that a stable combination of soil and vegetation (in the absence of human intervention)

can be achieved in 1000–10,000 years. The system is stable in the sense that any changes occurring are immeasurably small within the time period since scientific observation began (c. 150 years). The soil component of such systems, the zonal soil, is sometimes referred to as mature.

By contrast, in many subtropical and tropical regions where the land surface has been exposed for much longer (10^5–10^7 years), the concept of a soil–plant ecosystem attaining steady state is more problematic. The effect of time in the older landscapes is confounded by changes in other factors – climate, parent material and relief. Although the rate of mineral weathering is faster than in temperate regions, and leaching in humid climates potentially more severe, weathering occurs to much greater depths and the deep horizons are less influenced by surface organic matter. Crystalline clay minerals break down mainly to sesquioxides, with the release of silica and basic cations (Section 9.2). This trend in soil formation with time on volcanic ash weathering in a wet tropical climate is illustrated in Fig. 5.17. Soil depth and vegetation biomass attain maxima in the 'virile' stage, only to decline in the 'senile' and final 'lateritic' stages. Climate exerts less effect than parent material and relief on the ultimate result of soil formation under such conditions.

5.7 Summary

Jenny (1941) sought to quantify the effects of *climate*, *organisms*, *relief*, *parent material* and *time* on soil formation by writing the equation

Soil property, $s = \int (cl, o, r, p, t \ldots)$.

Less significant factors such as fire and atmospheric inputs are represented by the dots. Attempts have been made to solve this equation by examining the effect of one factor, such as climate, on s (or a combination of soil properties, S) when the other factors are constant or nearly so. Such conditions give rise to suites of soils called climosequences, biosequences, toposequences, lithosequences and chronosequences, respectively.

The successful application of the *clorpt* concept to model soil formation quantitatively and

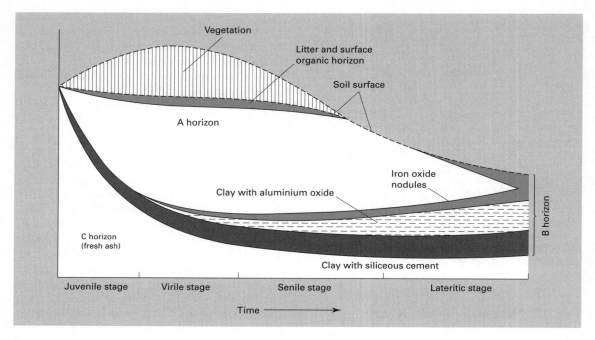

Fig. 5.17 Soil formation with time on volcanic ash under a wet tropical climate (after Mohr and van Baren, 1954).

predict the type of soil to be formed under particular conditions is a 'work in progress'. However, recognition of the role of these factors, and their interactions, has helped scientists to understand why certain types of soil occur where they do. Thus, the concept has underpinned many of the attempts to impose order, through soil classification, on the wide-ranging natural distribution of soils in the field. For example, the perceived dominant effect of climate and its associated vegetation on soil formation in continental regions, such as in Russia, the USA and Australia, gave rise to zonal systems of soil classification, in which the Great Soil Group is the most durable hierarchical category. Originally developed around a central pedogenic concept, Great Soil Groups have evolved in later classifications into classes with specified ranges of soil property values. Great Soil Group names continue to be used in soil science communication.

Where only one set of factor values has operated during soil formation, the resultant soil is said to be monogenetic. Polygenetic soils are much more common, and arise when the value of one or more of the factors changes significantly during soil formation.

Parent material affects soil formation primarily through the type of rock minerals, their rate of weathering and the weathering products. These products are redistributed by the action of water, wind, ice and gravity, in many cases providing parent material for another cycle of soil formation. Generally, transported parent material is more heterogeneous than that formed by the weathering of rock *in situ* (residual soils). Layering of a soil profile occurs when different parent materials are superimposed, and a fossil soil occurs when an existing soil is buried by the subsequent deposition of new parent material.

It is difficult to separate cause and effect in the case of organisms because the range of colonizing species influences the pace and direction of soil formation, but the soil that forms affects the competition for survival among the plants and other organisms. Human activity can exert a profound effect on the soil, for good or bad, through the clearing of natural vegetation, its use for agriculture, and by the addition of organic residues over a long period of time.

The importance of relief is obvious in old weathered landscapes of the tropics, where the concept of the catena evolved to describe the recurring suites of soil that occur on slopes. Slope has an important effect on soil drainage, which in turn influences the balance between oxidation and reduction of Fe and Mn compounds in the soil profile. The effect of relief on soil formation can also be seen in the gilgai country of Australia and the knob and kettle country of the prairies in western Canada.

Time is important because the value of a soil-forming factor may change in time (e.g. climatic change), and the extent of a pedogenic reaction depends on the length of time for which it has operated. A soil profile may attain a steady state if the rate of natural erosion matches the rate of weathering of the parent material. Individual pedogenic processes generally approach steady state at different rates, in some cases following a 'law of diminishing returns', in others a sigmoidal response over time. Soil–plant ecosystems developing on material exposed by the last retreat of Pleistocene ice sheets appear to have attained steady state within 1000–10,000 years. The zonal soils formed may be described as mature. On land surfaces that have been exposed for much longer (10^5–10^7 years), and where the rate of mineral weathering may be faster and leaching more severe, steady state soil–plant ecosystems are less likely to be achieved. Here parent material and relief determine the ultimate condition of the soil.

References

Avery B. W. (1990) *Soils of the British Isles*. CAB International, Wallingford.

Batey T. (1971) *Soil Profile Drainage*, in ADAS Advisory Papers, No. 9, Soil Field Handbook. Ministry of Agriculture, Fisheries and Food, London.

Birkeland P. W. (1999) *Soils and Geomorphology*, 3rd edn. Oxford University Press, New York.

Bouma J. & Hoosbeek M. R. (1996) The contribution and importance of soil scientists in interdisciplinary studies dealing with land, in *The Role of Soil Science in Interdisciplinary Research* (Eds R. J. Wagenet & J. Bouma) Soil Science Society of America Special Publication No. 45. Soil Science Society of America, Madison WI.

Duchaufour P. (1982) *Pedology, Pedogenesis and Classification* (Translated by T. R. Paton). Allen & Unwin, London.

Goldich S.S. (1938) A study of rock weathering. *Journal of Geology* 46, 17–58.

Holmes A. (1978) *Principles of Physical Geology*, 3rd edn. Nelson, London.

Hoosbeek M. R. & Bryant R. B. (1992) Towards the quantitative modeling of pedogenesis – a review. *Geoderma* 55, 183–210.

Jackson M. L., Tyler S. A., Willis A. L., Bourbeau G. A. & Pennington R. P. (1948) Weathering sequence of clay-size minerals in soils and sediments. 1. Fundamental generalizations. *Journal of Physical and Colloid Chemistry* 52, 1237–60.

Jenny H. (1941) *Factors of Soil Formation*. McGraw-Hill, New York.

Jenny H. (1980) *The Soil Resource, Origin and Behaviour*. Springer-Verlag, New York.

Milne G. (1935) Some suggested units of classification and mapping, particularly for East African soils. *Soil Research* 4, 183–98.

Minasny B. & McBratney A. B. (2001) A rudimentary mechanistic model for soil formation and landscape development. II. A two-dimensional model incorporating chemical weathering. *Geoderma* 103, 161–79.

Mitchell J. K. (1993) *Fundamentals of Soil Behavior*, 2nd edn. Wiley, New York.

Mohr E. C. J. & van Baren F. A. (1954) *Tropical Soils*. N. V. Uitgeveriz, The Hague.

Paton T. R., Humphreys G. S. & Mitchell P. B. (1995) *Soils, a New Global View*. UCL Press, London.

Phillips J. D. (1998) On the relations between complex systems and the factorial model of soil formation (with Discussion). *Geoderma* 86, 1–21.

Soil Survey Staff (1996) *Keys to Soil Taxonomy*, 7th edn. United States Department of Agriculture, Natural Resources Conservation Service, Washington DC.

Thornthwaite C. W. (1931) The climates of North America according to a new classification. *Geographical Review* 21, 633–55.

Thorp J. & Smith G. D. (1949) Higher categories of soil classification. *Soil Science* 67, 117–26.

Further reading

Amundson R., Harden J. & Singer M. (1994) *Factors of Soil Formation: A Fiftieth Anniversary Retrospective*. Soil Science Society of America Special Publication No. 33. Soil Science Society of America, Madison, WI.

Bland W. & Rolls D. (1998). *Weathering: an Introduction to Scientific Principles*. Edward Arnold, London.

Thomas M. J. (1994) *Geomorphology in the Tropics: a Study of Weathering and Denudation in Low Latitudes*. Wiley & Sons, Chichester.

Wysocki D. A., Schoeneberger P. J. & Hannan E. L. (2000) Geomorphology of landscapes, in *Handbook of Soil Science* (Ed. M. E. Sumner). CRC Press, Boca Raton, FL, E5–E39.

Young A. & Young R. (2001) *Soils in the Australian Landscape*. Oxford University Press, Melbourne.

Example questions and problems

1 If the suite of soils formed under a changing biotic influence (vegetation and organisms) is called a biosequence, what is the name given to a sequence in which
   (a) relief is the dominant soil-forming factor, and
   (b) parent material is the dominant factor?
   (c) What is another name for the suite of soils formed on the same parent material on a slope?

2 (a) What is the name of the concept of soil distribution that assumes climate has an over-riding influence on soil formation?
   (b) Which soil classification system derives from this broad concept? (Hint – see also Section 1.1).
   (c) Give an example where the assumption of climate as the dominant soil-forming factor is not justified.

3 Explain the difference in origin between intrusive and extrusive volcanic rocks.

4 Why is the mineral olivine weathered more readily than quartz.

5 Define what is meant by
   (a) an alluvial deposit,
   (b) colluvium, and
   (c) glacial till.

6 (a) What is loess?
   (b) Describe the process of solifluction.
   (c) Describe the process of saltation.

7 (a) Explain the concept of steady state in soil formation.
   (b) Suppose that the accumulation of organic matter (as organic C, C_t) in a soil forming on fresh parent material follows an exponential function of the type

$$C_t = C_m(1 - \exp(-kt)),$$

where C_m is the expected value of organic C in the soil at steady state and the incremental rate of increase k is 0.001 (1/yr). How many years will it take for the soil to reach 95% of its final steady state value?

Chapter 6

Hydrology, Soil Water and Temperature

6.1 The hydrologic cycle

A global picture

Water vapour enters the atmosphere by evaporation from soil, water and plants; it precipitates from the atmosphere as rain, hail, snow or dew. Precipitation P equals evaporation E over the entire Earth's surface; but for land surfaces only, the average annual P of 710 mm* exceeds the average annual E by 240 mm, the difference amounting to river discharge. If the part of the land surface that contributes little to evaporation – the deserts, ice caps and Tundra – is excluded from the calculation, the average rate of evaporation from the remainder is approximately 1000 mm per annum, which is not far short of the 1200 mm annual rate of evaporation from the oceans. Evaporation depends on an input of energy, the energy available being much greater at the Equator than at high latitudes. The capacity of the air to hold water also increases with temperature, so that the amount of precipitable water in the atmosphere is greatest at the Equator (40–50 mm) and least at the Poles (< 5 mm at the North Pole) (Penman, 1970). Although relatively small, the atmospheric source replenishes the much larger reserves of surface water, and water in the soil and regolith that sustains terres-

trial plant and animal life. Water in the soil and regolith may exist in an unsaturated or vadose zone (above the watertable or phreatic surface – see Fig. 5.14a), or in a saturated zone (below the watertable). Estimates of the size of these reserves are subject to much uncertainty, but are approximately 250 mm for soil water (range 50–500 mm), 400–1000 mm for surface water (lakes and rivers), and 15–45 m for groundwater.

Water balance on a local scale

A catchment (or watershed in the USA) is a part of the landscape that behaves as a discrete hydrologic unit. The processes involved in water cycling at a local scale can be illustrated by those in a segment of forested catchment (Fig. 6.1). Of the total precipitation on land, much is intercepted by the vegetation with the remainder falling directly on the soil. Water in excess of that required to wet the leaves and branches, the canopy storage capacity of usually 1–2 mm, drips from the canopy or runs down stems to the soil. Canopy drip (sometimes called throughflow), stem flow and direct rainfall constitute the net precipitation or net rainfall. Water retained by the canopy is lost by evaporation and is referred to as interception loss: it may be as much as one-quarter of the gross rainfall for broadleaf trees of temperate regions in summer.

Of the rain that reaches the soil surface, some may soak in while the remainder initially ponds on the surface and then runs off – this is called

* This the 'equivalent depth' of water, calculated as $1000 \times m^3$ water/m^2 of surface (refer to Box 4.6). Note that 1 mm = 1 L/m^2.

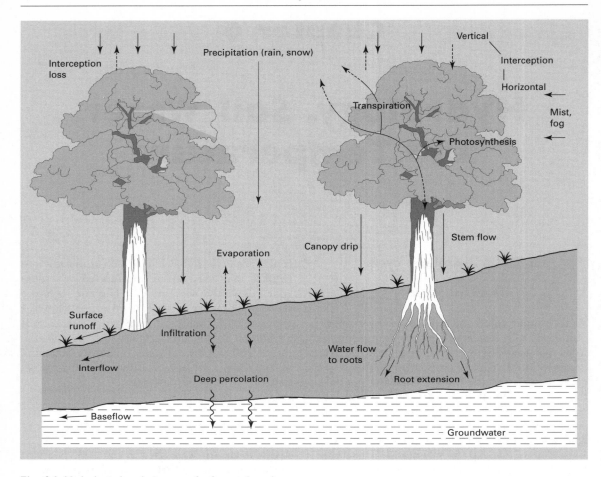

Fig. 6.1 Hydrological cycle in part of a forested catchment.

surface runoff or overland flow, which is likely to occur during periods of intense rainfall that do not last long. Movement of water into the soil from above is called infiltration. When the soil is thoroughly wet down to the watertable, water flows or percolates vertically to the groundwater. Subsurface lateral flow or interflow can also occur through soil on slopes, or when vertical flow into the subsoil is impeded, as occurs in soils with impermeable B horizons (Section 5.5), or with indurated layers such as an iron pan (Section 4.3).

Surface runoff generated by the process of 'in-filtration excess', described above, causes a rapid response in stream flow in a catchment. Normally, this runoff is augmented by 'saturation excess' flow from contributing areas – these are parts

of the catchment that quickly reach saturation during periods of significant rainfall, because they are normally wet and are fed by the convergence of both overland flow and interflow from higher areas. Water can also flow laterally in the groundwater and provides the bulk of a stream's baseflow, which is the relatively low volume flow that is, one hopes, maintained through dry periods. Surface runoff and interflow (collectively runoff R) make a much more variable contribution to stream flow than groundwater.

Radiant energy from the sun and atmosphere that is absorbed by the soil and vegetation is used to convert water into water vapour – the process of evaporation. Surfaces vary in their reflectance of the radiant energy (their albedo), as shown in Table 6.1, and the radiation energy balance

Table 6.1 Representative values of albedo for different surfaces. After Linacre, 1992.

Type of surface	Range of albedo (%)	Median value (%)
Tall grass (2 m) to short grass (0.3 m)	21–25	23
Field crops	15–24	20
Forests	10–18	15
Soil (wet) to soil (dry)	11–18	14
Water	4–13	7

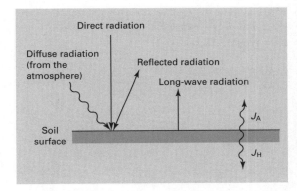

Fig. 6.2 Radiation energy balance and heat fluxes at the soil surface (redrawn from Linacre, 1992. From *Soils for Fine Wines* by Robert E. White, copyright © 2003 by Oxford University Press, Inc. Used by permission of Oxford University Press, Inc.).

Fig. 6.3 A standard raingauge.

at the surface is shown in Fig. 6.2. That part of the absorbed energy not dissipated through evaporation is partitioned between a transfer of heat to the air above (heat flux J_A), and deeper into the soil (soil heat flux J_H). The balance of these heat fluxes determines the soil temperature (Section 6.6). Slow upward movement of water occurs in response to evaporation from the soil surface. Plant canopies, however, offer a much larger surface for the evaporation of water, and deep-rooted plants have access to a potentially large volume of soil water. The decrease in water vapour pressure in the plant leaves provides the driving force for water to be sucked into the roots and through the plant, to be lost by evaporation. This is called transpiration. Soil water flows into the water-depleted zones around absorbing roots while the remainder is intercepted by the roots

as they grow through the soil. The combination of transpiration from vegetation and evaporation from bare soil surfaces is often referred to as evapotranspiration E_t.

The interaction of all these processes in a catchment, or another hydrologic unit such as an irrigated field, is expressed by the water balance equation (Box 6.1).

Precipitation P is measured with a rain gauge, the collecting surface of which should be a standard height (30 cm) above the soil surface because the 'rain catch' decreases with height above the land. An example of a standard rain gauge is shown in Fig. 6.3. Runoff R is measured in collecting ditches or small streams using weirs or flumes: an example of an H-flume is shown in Fig. 6.4. Measurement of evaporation is discussed in Section 6.5. A knowledge of runoff, deep drainage and stream flow in a catchment is important, first because stream flow provides water that can

Box 6.1 The water balance equation.

The water balance equation is an equation of mass conservation, that is

$$P = E_t + R + D + \Delta S, \tag{B6.1.1}$$

where P, E_t and R have been defined, and deep drainage D is defined as water percolating below the root zone, which may vary from 0.8–1 m deep under many annual species to 6 m deep under trees. The change in soil storage ΔS is measured relative to the soil's field capacity FC (Section 4.5). For catchments in which deep drainage losses are negligible, on an annual basis, the changes in soil storage ΔS (+ or −) tend to be small so that the difference between P and E is primarily runoff R. Over shorter periods, values of ΔS are nevertheless important for plant growth, and if ΔS is persistently negative, irrigation may be necessary to supplement rainfall.

Changes in ΔS (in mm) are calculated from changes in the volumetric water content θ (m^3/m^3) in the soil profile (Section 4.5). θ at different depths can be measured non-destructively with instruments that rely on measuring changes in the dielectric constant of the soil, using instruments such as a Time Domain Reflectometer (TDR) or a capacitance probe. Water has a very high dielectric constant compared with any other non-solid constituent in the soil, so changes in the soil dielectric constant can be related to changes in θ. Another instrument is the neutron probe, which is lowered into an aluminium tube, sealed at the base and permanently located in the soil. Fast neutrons are emitted from a radioactive source in the probe and are slowed down and reflected back to a detector, following collisions with H nuclei. By far the greatest source of variation in numbers of H nuclei in soil is due to changes in water content, so the neutron probe can be directly calibrated to give the value of θ. An example of a neutron probe is shown in Fig. B6.1.1. The reader is referred to specialist articles such as Or and Wraith (2000) for further details.

When P, E_t, R and ΔS are known, Equation B6.1.1 can be solved for appropriate time intervals to estimate the average drainage flux D, which is water that is potential recharge to groundwater. The size of the deep drainage flux is a contentious issue in many areas under agriculture, particularly where irrigation is used or the groundwater is saline and a rise in the watertable can cause problems for plant growth (Section 13.2).

Fig. B6.1.1 A neutron probe used for measuring soil water content *in situ*.

be conserved for human and animal use; second, because water in deep aquifers (permeable rocks that hold groundwater) is a valuable reserve from which to supplement surface supplies when E_t is consistently greater than P. The amount of rainfall that goes to runoff and drainage, and their proportioning, can be manipulated to some extent by land use and soil management in a catchment. It is therefore appropriate to discuss the properties of soil water and soil water movement in some detail.

Fig. 6.4 Surface runoff being measured in an H-flume.

6.2 Properties of soil water

Water quantity and water energy

The volumetric water content θ is an important variable indicating how much water a given volume of soil can hold. However, to understand what drives water movement in soil, we need to know about the forces acting on the water, which affect its potential energy. The energy status of soil water also influences its availability to plants. There is no absolute scale of potential energy, but we can measure *changes* in potential energy when useful work is done on a measured quantity of water, or when the water itself does useful work. These changes are observed as changes in the free energy of water, which gives rise to the concept of soil water potential. The definition of soil water potential ψ (psi) is given in Box 6.2.

Hydraulic head and pF

The hydrostatic pressure (in Pa) at the base of a column of water h metres high (equivalent to ψ in units of energy per unit volume) is given by

$$\psi = h \, \rho \, g, \tag{6.1}$$

where ρ is the density of water ($1000 \, \text{kg/m}^3$) and g is the acceleration due to gravity ($9.8 \, \text{m/s}^2$). Suppose that the pressure at the base of the water column is 1 bar ($10^5 \, \text{Pa}$), then the height h (m) of the column can be calculated by rearranging Equation 6.1 to give

$$h = \frac{\psi}{\rho g} = \frac{10^5}{1000 \times 9.8} = 10.2 \text{ m.} \tag{6.2}$$

When expressed in this form (energy per unit weight), the water potential is referred to as hydraulic head or simply as head (m). Hydrologists

Box 6.2 Free energy change and the definition of soil water potential.

Suppose a small quantity of water is added to a dry, non-saline soil at a constant temperature. This water is distributed so that the force of attraction between it and the soil is as large as possible, and the free energy of the water is reduced to a minimum. More water added to the soil is held by progressively weaker forces. Finally, with the last drop of water necessary to saturate the soil, the free energy of the soil water approaches that of pure water (no solutes) at the same temperature and pressure. The reference water must be 'free' in the sense that it is not affected by forces other than gravity. The energy status of soil water is then defined in terms of the difference in free energy between 1 mole of water in the soil and 1 mole of pure, free water at a standard temperature, pressure and a fixed reference height.

The free energy per mole of a substance defines its chemical potential. So for water in soil, we define the soil water potential ψ by the equation

$$\psi = \psi_w(\text{soil}) - \psi_w^\circ (\text{standard state}) = RT \ln (e/e^\circ), \quad \text{(B6.2.1)}$$

where ψ_w is the chemical potential (free energy/mol) of water, e/e° is the ratio of the vapour pressure of soil water to the vapour pressure of pure water at a standard temperature (298 K) and pressure (one atmosphere), R is the Universal gas constant (8.314 Joule/degree Kelvin/mol), and T is the absolute temperature (K). The reference height used for measuring soil water potentials is usually the soil surface. Note that for water at the standard temperature and pressure, and reference height, the maximum value of ψ is obtained at $e/e^\circ = 1$ for which $\psi = 0$ in saturated, non-saline soil. In unsaturated soil, ψ is negative because work must be done on any water extracted from the soil and the potential of the remaining water to do work is therefore lower.

The units of ψ in Equation B6.2.1 are energy per mole (Joule/mol), but to be compatible with earlier pressure measurements, ψ is often expressed as energy per unit volume (J/m^3) by dividing Equation B6.2.1 by V_w, the volume (m^3) per mole of water. The units of J/m^3 are equivalent to Newton/m^2 or Pascals (Pa), which are units of pressure. The analogy with pressure measurements goes further with the use of the term suction (implying negative pressure) to describe the action of forces on water as it is drawn into an unsaturated soil.

and soil scientists make use of this relationship and often express water potentials in terms of the height of an equivalent water column, because it is a very convenient unit for calculating head gradients in the field (Section 6.3).

In dry soils, the water potential, when expressed in head units in the old centimetre-gram-second system, was often a large and unwieldy number. It was therefore more convenient to use $\log_{10} h$ (cm), which defines the pF value of the soil water (compare the pH scale for H^+ ion activity). However, with the adoption of Système International (SI) units, soil pF values are now seldom used. Comparative values for the water potential of soil at various states of wetness, in values of ψ, head h and pF are given in Table 6.2.

Forces acting on soil water

The free energy of water in the soil is reduced in the following ways:

1 At low relative humidities (RH < 20%, or e/e° < 0.2 in Equation B6.2.1), water is adsorbed on clay minerals of small permanent charge as a monolayer in which the molecules are H-bonded to each other and to the surface. With an increase in RH, more water is adsorbed and the influence of the surface on the structure of the liquid water extends further into the solution; however, the bulk of the effect occurs in the first three molecular layers within approximately 1 nm of the surface.

2 At the charged surfaces of 2 : 1-type clay minerals, water is also strongly attracted to the adsorbed cations, the strength of attraction depending on whether the cations form inner- or outer-sphere complexes with the surface (Section 2.4). The electric field of the cation orientates the polar water molecules to form a hydration shell containing up to six water molecules for a monovalent cation, and up to 12 for a divalent cation such as Mg^{2+} (Fig. 6.5). The energy of hydration per mole of cation depends upon its ionic potential (Section 2.3) and polarizability, with energies for the common exchangeable cations falling in the order

$$Al^{3+} > Mg^{2+} > Ca^{2+} > Na^+ > K^+ \cong NH_4^+.$$

The greater the hydration energy of the cation, the more work is done in attracting water molecules

Table 6.2 Water potential, hydraulic head and pF for a soil of varying wetness.

Soil wetness condition	Water potential ψ (kPa)	Head h (m)	pF	Equivalent radius of largest pores that would just hold water (μm)
Soil very nearly saturated	−0.1*	0.01**	0	1500†
Water held after free drainage (field capacity)	−10	1.02	2	15
Approximately the water content at which plants wilt	−1500	153	4.18	0.1
Soil at equilibrium with a relative vapour pressure of 0.85 (approaching air-dryness)	−22,400	2285	5.36	0.007

* Calculated from Equation B6.2.1, assuming a temperature of 25°C.
** Calculated from Equation 6.2.
† Calculated from Equation 6.5.

Fig. 6.5 A hydrated monovalent cation.

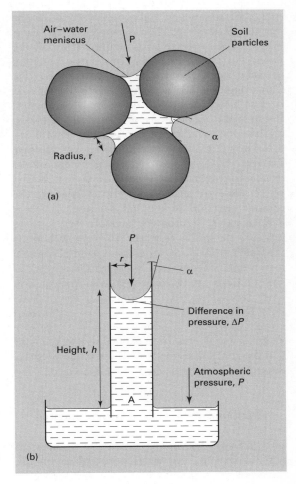

Fig. 6.6 (a) Water held under tension between soil particles. (b) The rise of water in a capillary tube.

and the larger the reduction in free energy of the water molecules attracted into the hydration shell. The relatively small amount of water present in dry soils (RH < 85%) is therefore associated mainly with clay surfaces and their exchangeable cations.
3 As water layers build up, water begins to fill the smallest pores and interstices between soil particles, forming curved air–water menisci (Fig. 6.6a). The behaviour of water between the particles is analogous to water in a capillary tube inserted into a free water surface (Fig. 6.6b). Water is sucked up into the capillary, of radius r, by surface tension forces acting all the way around the perimeter of the meniscus. At equilibrium, this upward force is balanced by the weight of water in the column, which is the product of the cross-sectional surface area and the pressure difference ($h\rho g$) between water at the top of the column (A) and the base (B); that is

$$\pi r^2 h \rho g = 2\pi r \gamma \cos \alpha. \qquad (6.3)$$

where γ (gamma) is the surface tension of water, α (alpha) is the angle of wetting, and π, r, ρ, g and h are already defined. Using Equation 6.1, we may write

$$\pi r^2 \psi = 2\pi r \gamma \cos \alpha, \qquad (6.4)$$

When α is very small, $\cos \alpha$ approaches 1, and Equation 6.4 can be simplified and rearranged to

$$\psi = \frac{2\gamma}{r}. \qquad (6.5)$$

Thus, the potential of water held in pores of radius r can be calculated, knowing the surface tension of water (0.075 N/m). The combined effects of adsorption and capillarity on the free energy of soil water are expressed through the matric potential ψ_m. For example, when water in an initially wet soil has drained from all the pores > 60 µm in diameter, the matric potential of water occupying the largest pores is −5 kPa. Equation 6.5 works well in mineral soils where α is close to zero and the soil does not repel water. But in sandy organic soils, α is often not near zero and the soil exhibits water-repellency (Box 6.3). 4 Solutes (ions and organic molecules) decrease the free energy of soil water, the extent of the decrease being measured by the osmotic potential ψ_s. Osmotic potential is numerically equal to the osmotic pressure of the soil solution, which is defined as the hydrostatic pressure necessary to just stop the inflow of water when the solution is separated from pure water by a semi-permeable membrane. The osmotic pressure OP (in Pa) of soil solutions can be calculated from the equation

$$OP = RTC, \qquad (6.6)$$

where R and T have been defined in Box 6.2, and C is the concentration (mol/m³) of solute particles (molecules and dissociated ions) in solution.

Components of the soil water potential

The variables ψ_m and ψ_s are called component potentials of the soil water potential ψ. In soil, these potentials are always ≤ 0 so that their contribution to ψ is always negative or zero. A positive contribution to ψ is made if the soil water is at a height above that chosen for the reference

Box 6.3 Non-repellent and water-repellent soils.

The angle of wetting or contact angle is measured from the liquid–solid interface to the liquid–air interface. At most soil mineral surfaces, the angle is close to zero and the water is said to 'wet' the surface (Fig. B6.3.1a). However, if these surfaces become coated with hydrophobic organic compounds, the cohesive forces between the water molecules become much stronger than the force of attraction between the water and the surface so that the contact angle increases markedly (Fig. 6.3.1b). The surface is then said to 'repel' water and the soil as a whole may become water-repellent. Such an effect can occur in sandy A horizons that undergo severe drying – water subsequently applied to the surface does not wet the surface, nor infiltrate, and tends to run off. Many soil surfaces exhibit some degree of water repellency when dried for a prolonged period, but the effect gradually disappears as the soil wets up again. Examples of water-repellent sandy soils are found in the southeast of South Australia and in southern Western Australia.

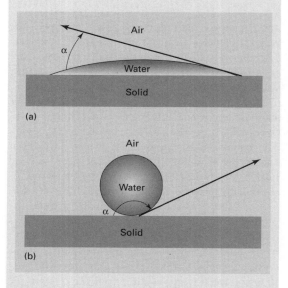

Fig. B6.3.1 Illustration of air–water–solid interfaces (a) small contact angle, water 'wets' the surface; (b) large contact angle, water is 'repelled' by the surface (after Jury et al., 1991).

level – this is expressed through a component called the gravitational potential ψ_g. In addition, soil water under a pressure greater than atmospheric pressure has a positive component – the pressure potential ψ_p.

Although it is technically feasible to increase the air pressure on a soil above normal atmospheric pressure, in practice variations in ψ_p due to changes in air pressure are negligible. The pressure potential ψ_p only becomes significant when the soil, or a zone in the soil profile, is saturated. For example, ψ_p at a depth A in Fig. 6.6b is directly proportional to the depth of water in the capillary tube. Obviously, ψ_m in saturated soil must be zero, so ψ_m and ψ_p can be considered as two subcomponents of a pressure-matric potential continuum for which:

- Above the watertable, $\psi_p = 0$ (at atmospheric pressure) and ψ_m is negative;
- at the watertable $\psi_m = \psi_p = 0$; and
- below the watertable, $\psi_m = 0$ and ψ_p is positive.

In summary, the soil water potential ψ is given by the equation

$$\psi = \psi_m + \psi_s + \psi_p + \psi_g. \tag{6.7}$$

In non-saline soils, Equation 6.7 simplifies to

$$\psi = \psi_m + \psi_p + \psi_g. \tag{6.8}$$

The measurement of soil water potential in the field is discussed in Box 6.4. The retention and movement of water in soil is discussed in terms of potentials and head gradients in the following sections.

Box 6.4 Field measurement of soil water potential.

A very useful instrument for measuring the combined expression of ψ_m and ψ_g in the field is the tensiometer, which gives rise to a tensiometer potential ψ_t. A tensiometer consists of a porous ceramic cup (with an air-entry value of at least 85 kPa) sealed to a length of polyvinyl chloride (PVC) tubing, at the top of which is a pressure-measuring device (Fig. B6.4.1a). The tube and cup are filled with water so that when the tensiometer is installed in the soil, the water in the tensiometer comes to equilibrium with water in the surrounding soil. The negative pressure of the soil water (less the osmotic effect of any solutes that can diffuse through the ceramic cup walls) is transmitted through the water column, to be read with a Bourdon gauge or pressure transducer at the top. Tensiometers of different length installed at depths in the soil measure ψ_g values of the soil water, relative to a reference level at the surface, as well as ψ_m values at those specific depths (Fig. B6.4.1b). This arrangement enables potential gradients for water movement to be measured, and also reveals any zones of saturation in the soil ($\psi_m = 0$). Any depth where the tensiometer potential gradient is zero defines the zero flux plane where no net upward or downward movement of water is expected.

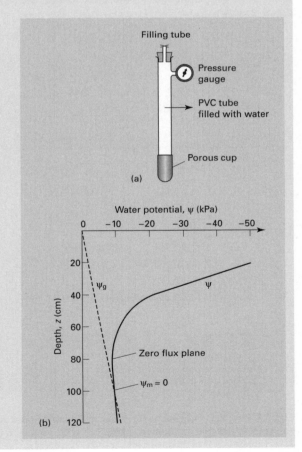

Fig. B6.4.1 (a) A vacuum gauge tensiometer. (b) A typical water potential profile in soil drying by evapotranspiration after being thoroughly wetted.

6.3 Infiltration, runoff and redistribution of soil water

Hydraulic head and Darcy's Law

For water movement through porous materials, provided the velocity is low enough that flow is laminar and not turbulent, the rate of flow is directly proportional to the driving force and inversely proportional to the resistance to flow. The driving force is determined by the water potential gradient, which as stated in Section 6.2, is most conveniently expressed in terms of hydraulic head h. Thus, the formal expression describing the rate of flow is given by Darcy's equation

$$J_w = -K \frac{dh}{dx}, \tag{6.9}$$

where J_w is the volume per unit time crossing an area A, perpendicular to the flow (called the water flux density or simply water flux), the pro-

portionality coefficient K is the hydraulic conductivity, which is also the reciprocal of the flow resistance, and dh/dx denotes the gradient in head per distance x in the direction of flow. (The negative sign accounts for the direction of flow being opposite to the direction of increasing head.) Water flow in soil can occur in the x, y, or z directions, but we shall confine ourselves here to considering movement in the z (vertical) direction.

For water movement in non-saline soils, the two important components of head h are the pressure head p (equivalent to ψ_m in head units) and the gravitational head z. For water below a watertable (i.e. in saturated soil) at a point A (Fig. 6.7a), the piezometric head h is given by the sum

$$h = p - z, \tag{6.10}$$

where p is positive and z is the depth of point A below the reference level (the soil surface). By convention, z is measured positively upwards.

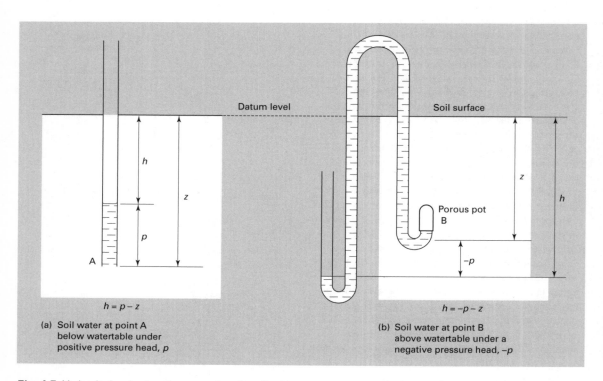

Fig. 6.7 Hydraulic head values for water in soil under (a) a positive pressure head p, or (b) a negative pressure head $-p$ (after Youngs and Thomasson, 1975).

Box 6.5 Factors affecting the hydraulic conductivity K.

The value of K depends on the amount of water in the pores and its viscosity, as well as the pore size distribution, tortuosity and surface roughness. Some insight into the effect of pore size (radius) on K is obtained from Poiseuille's Law, which describes the rate of water flow Q through a full cylindrical tube under a constant pressure gradient, i.e.

$$Q = \frac{\pi r^4 \Delta P}{8L\eta},$$ (B6.5.1)

where r is the tube radius, η the dynamic viscosity, and $\Delta P/L$ is the change in pressure over the length of tube L. Equation B6.5.1 shows that the volume of water flowing per unit time varies with the fourth power of the radius, and the water flux J_w ($= Q/\pi r^2$) varies with the radius squared. Hence in a soil, other factors being equal, water would move 100 times faster through water-filled pores of 1 mm radius than those of 0.1 mm radius.

As a soil desaturates ($\psi_m < 0$), water drains out of the largest pores first (Section 6.4) and the average size of pores still conducting water decreases. Water columns in the soil may become discontinuous so that the average pathway for water movement becomes more tortuous. The overall result is that hydraulic conductivity K

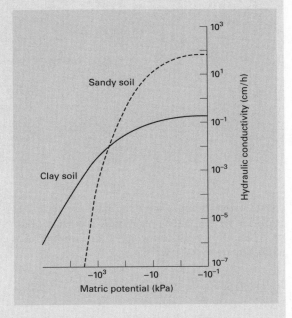

Fig. B6.5.1 Typical hydraulic conductivity–matric potential curves for a sandy soil and a clay soil.

decreases markedly from its maximum value at saturation (K_s). Examples of how K changes as the matric potential ψ_m decreases are shown in Fig. B6.5.1.

Alternatively, water may be held at a negative pressure head or suction ($-p$) in unsaturated soil (point B in Fig. 6.7b). Here the water in the manometer tube equilibrates with water in the soil surrounding the porous pot so that the head h is given by

$$h = -p - z.$$ (6.11)

The ideal condition for applying Darcy's Law is when both the head gradient and resistance to flow are constant over time, a condition attained during steady, saturated flow through a non-swelling soil. Under saturated conditions, the value of K defines the saturated hydraulic conductivity K_s (Box 6.5). Substituting for h in Equation 6.9 from Equation 6.10, we have

$$J_w = -K\left(\frac{d(p-z)}{dz}\right)$$ (6.12)

$$J_w = -K\left(\frac{dp}{dz} - 1\right).$$ (6.13)

Because the soil is saturated, the term dp/dz is zero and water moves solely under a gravitational head gradient so that

$$J_w = K_s.$$ (6.14)

Water flow in field soils also occurs under unsaturated conditions, and is usually not in steady state because of the dynamic cycle of wetting and drying, and water extraction by plants. In this case, for the reasons outlined in Box 6.5, K is not

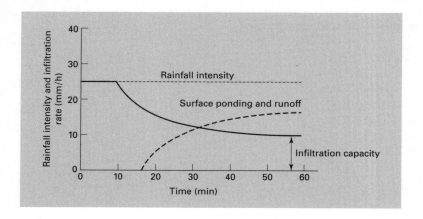

Fig. 6.8 Infiltration, surface ponding and runoff during a storm event.

constant but changes with the pressure head $-p$, and so is written $K(p)$ (meaning K is a function of p). Therefore, substituting for h in Equation 6.9 from Equation 6.11 we have

$$J_w = -K(p)\left(\frac{d(-p)}{dz} - 1\right). \quad (6.15)$$

If the pressure gradient $d(-p)/dz$ favours downward movement of water, it reinforces the gravitational head gradient, so that the overall gradient is then greater than 1 – this is the case for rainfall infiltrating into an initially dry soil (e.g. Fig. 6.9). However, if the pressure gradient favours upward movement of water, it acts against the gravitational head gradient and water may move up or down, depending on which component gradient is the greater. This situation occurs when soil is drying by evaporation after having been wet up by rain.

Stages of infiltration

Rain falling on the soil surface is drawn into the pores under the influence of both a pressure and gravitational head gradient, and if the rainfall intensity is less than the initial infiltration rate (IR) all the water is absorbed. In the early stages of wetting a dry soil, the pressure head gradient is predominant. However, as the depth of wet soil increases, the pressure head gradient decreases (see Equation 6.15) and the gravitational head gradient becomes the main driving force. The flux density J_w then approaches the saturated

hydraulic conductivity K_s, under which circumstances the soil may not be able to accept water rapidly enough to prevent ponding on the surface. For bare soil, structural changes in the surface following raindrop impact can also reduce the rate of acceptance so that the IR falls below the rainfall intensity within a few minutes to a few hours. Initially, ponding may be confined to small depressions, and this creates favourable conditions for preferential flow down macropores (see below). But once the surface detention of water reaches a threshold value, surface runoff occurs. This sequence of events for rainfall of moderately high intensity is illustrated in Fig. 6.8.

The wetting front

If rain continues to fall on the soil surface, and the soil is structurally homogeneous, water travels downwards at a constant rate J_w, which determines the infiltration capacity of the soil. As water penetrates more deeply, a zone of uniform water content, the transmission zone, develops behind a narrow wetting zone and well-defined wetting front. The graph of water content θ against depth for such a soil, illustrated in Fig. 6.9, shows a sharp change in water content at the wetting front. This effect occurs because water at the boundary takes up a preferred position of minimum potential energy in the narrowest pores for which the hydraulic conductivity is very low, and does not move at an appreciable rate until the large pores begin to fill.

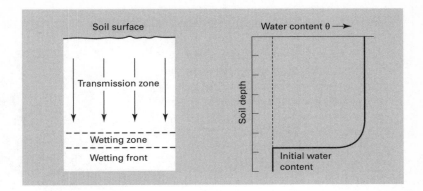

Fig. 6.9 Vertical infiltration into a homogeneous column of initially dry soil.

For a constant rainfall intensity J_w and soil water content θ, the mean pore water velocity $\bar{v}$ in the wet soil zone is given by

$$\bar{v} = \frac{J_w}{\theta}, \tag{6.16}$$

and the distance z travelled by the wetting front in time t is given by

$$z(t) = \frac{J_w t}{\theta}. \tag{6.17}$$

In practice, field soils are rarely structurally homogeneous due to the presence of old root channels, worm holes and gaps between aggregates, so the downward movement of water is more erratic than Fig. 6.9 would imply. A wetting front in a natural soil profile is shown in Fig. 6.10. Although the change in water content at the wetting front is sharp in this soil, there is also evidence of water

at some points moving ahead of the main transmission zone. At a local scale there will be a range of pore water velocities because some of the infiltrating water flows rapidly down channels and cracks (referred to as 'preferential' or 'bypass' flow), whereas the bulk of the water penetrates slowly into the micropores within aggregates ('matrix' flow). The implications of these variations in pore water velocity for the leaching of nitrate and other solutes are discussed in Section 10.2.

Within a well-developed transmission zone, the soil is nearly saturated (allowing for some trapped air pockets) and the pressure head gradient $d(-p)/dz$ is negligible. Therefore, Equation 6.15 simplifies to

$$J_w = K, \tag{6.18}$$

where K approaches the soil's saturated hydraulic conductivity (c. $0.95K_s$). Values of K_s range from < 1 to 500 mm/day, depending on the texture and structure of the soil. Methods for measuring K_s, or the field infiltration capacity, are described in Box 6.6.

Field capacity

When infiltration ceases, redistribution of water occurs at the expense of the initially saturated zone of soil. If the soil is completely wet to the watertable, or an artificial outlet such as a drain, drainage ceases when the pressure gradient acting upward balances the downward gravitational head gradient. Even when the soil is not completely wet depthwise, the drainage rate often becomes very small 1 to 2 days after rain or irrigation.

Fig. 6.10 Vertical infiltration into a soil in the field.

Box 6.6 Measuring soil hydraulic conductivity and infiltration capacity.

K_s can be measured in the laboratory by first saturating an intact soil core (usually from the bottom upwards to displace entrapped air), and then maintaining a small constant head of ponded water on the surface while measuring the rate at which water flows out at the base. The gradient in head through the core will be unity so that at steady state the drainage flux $J_w = K_s$ (from Equation 6.14).

A simple way to measure K_s in the field is to use a ring infiltrometer. Water is ponded to a known depth inside a ring infiltrometer – a metal ring 30–60 cm in diameter that is driven a short distance into the soil surface. When quasi-steady state infiltration is achieved, the rate of water loss from the ring is taken as an estimate of the infiltration capacity. In the previous discussion one-dimensional infiltration (vertical only) was assumed. Infiltration measured in this way on a dry heavy-textured soil will not be entirely one-dimensional (vertical) because considerable lateral flow occurs under a pressure gradient. The infiltrometer will therefore overestimate the field K_s, or infiltration capacity, unless an outer 'buffer ring' filled with water is used, as in the double-ring infiltrometer. The infiltration rate is measured from the inner ring only.

K can also be measured at small negative pressure heads (0 to −120 mm) in the field using a disc permeameter or tension infiltrometer (Fig. B6.6.1). Water infiltrates from a circular disc (radius r_o) covered by a porous membrane of air-entry pressure > 2 kPa. A thin layer of fine sand is spread on the soil surface under the disc to ensure good contact. The rate of water loss (Q) from the permeameter is measured when steady state infiltration is reached. The value of K is calculated from an approximate formula that allows for the lateral suction effect on infiltration (Wooding, 1968), that is

$$K = \frac{Q/\pi r_0^2}{1 + 4/\pi r_0 \alpha}. \qquad \text{(B6.6.1)}$$

Fig. B6.6.1 Disc permeameters in use for hydraulic conductivity measurements on a deep loamy soil on the North China Plain.

The value of α, the reciprocal of the capillary length, is 0.05–0.5 (1/cm) for soil textures ranging from clay loam to sand. The disc permeameter is used to obtain the relationship between K and matric potential for the initial part of the curves shown in Fig. B6.5.1. K values measured close to saturation can vary 10- to 100-fold within an area of 1 ha due to the effects of soil structure. A pressure head of −20 mm (−0.2 kPa) is recommended for measuring K_s, because at this small head, the effect of very large pores is eliminated and the measured K_s is less variable spatially.

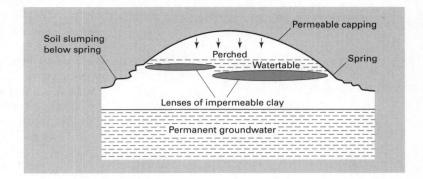

Fig. 6.11 A landscape profile showing perched watertables and spring lines.

This is possible because of the sharp fall in hydraulic conductivity when most of the macropores have drained of water. The water content attained after 2 days, in the absence of evaporation, defines the field capacity (FC), which corresponds to matric potentials between −5 and −10 kPa. For well-structured soils in Britain, the matric potential at FC is set at −5 kPa, whereas in Australia and the USA it is set at −10 kPa (Section 4.5). As is seen in Section 6.4, the concept of field capacity is useful for setting an upper limit to the amount of 'available' water in the soil; but it is imprecise for two reasons:

• The soil may remain wetter than FC due to frequent showers of rain for several days; and

• soils with a large proportion of micropores continue to drain slowly for several days after rain.

Steady state and transient flow

Flow under saturated or nearly saturated conditions is approximately in steady state, that is, the water flux into a given volume is equal to the water flux out. However, as the larger, more continuous pores drain and the soil becomes unsaturated, the resistance to flow increases and the value of K decreases accordingly. Flow then becomes non-steady or transient. Darcy's Law (Equation 6.15) can still be used to calculate water fluxes, provided that the pressure head gradient can be measured over short time intervals, and the $K(-p)$ function is known.

For drainage to a very deep watertable, the gradient $d(-p)/dz$ is much smaller than 1 and the water content throughout the profile adjusts so that $J_w \cong K(-p)$. Unsaturated flow of this kind

occurs in deep permeable soils below the root zone. Alternatively, flow is sometimes impeded by a layer of less permeable soil or parent material deep in the profile, resulting in localized saturation and the development of a perched water table (Fig. 6.11). The increase in pressure head immediately above the impeding layer induces lateral flow that may appear as a line of springs at another position in the landscape.

Non-steady state conditions

Soil water is rarely in steady state in the root zone. Intermittent rainfall and surface evaporation, combined with the variable plant uptake, cause the water content θ to fluctuate with time at any particular point in the soil. Flow can then be described only by combining Darcy's equation with an equation for the conservation of mass – the continuity equation. The latter relates the rate of change of θ in a small soil volume to the change in water flux into and out of that volume, allowing for any water removed by evapotranspiration E_t (Fig. 6.12). If we assume that flow is vertical only and that E_t is negligible, the continuity equation (in partial differential form) for the soil volume in Fig. 6.12 is written as

$$\frac{\partial \theta}{\partial t} = -\frac{J_w}{\partial z}. \tag{6.19}$$

Substituting for J_w from Equation 6.15 we obtain the Richards equation

$$\frac{\partial \theta}{\partial t} = \frac{\partial}{\partial z}\left\{K(p)\left(\frac{\partial(-p)}{dz} - 1\right)\right\}. \tag{6.20}$$

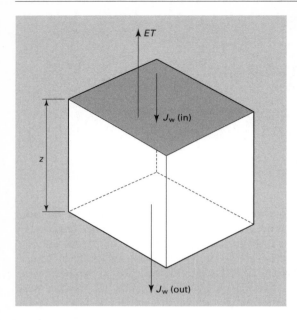

Fig. 6.12 Water flow through a soil volume under non-steady state conditions.

$$\frac{\partial(-p)}{\partial z} = \frac{\partial(-p)}{\partial \theta}\frac{\partial \theta}{\partial z}, \qquad (6.21)$$

where $d(-p)/d\theta$ is obtained from the soil water retention curve.

2 Since K is a function of $-p$ and $-p$ is a function of θ, K may be written as a function of θ, that is $K(\theta)$, and Equation 6.21 substituted into Equation 6.15 to give

$$J_w = -K(\theta)\left(\frac{\partial(-p)}{\partial \theta}\frac{\partial \theta}{\partial z} - 1\right). \qquad (6.22)$$

The expression $K(\theta)\dfrac{\partial(-p)}{\partial \theta}$ is called the soil water diffusivity $D_w(\theta)$.

3 Substitution of the expression for J_w from Equation 6.22 into Equation 6.19 gives the water content form of the Richards equation

$$\frac{\partial \theta}{\partial t} = \frac{\partial}{\partial z}\left(D_w(\theta)\frac{\partial \theta}{\partial z}\right) - \frac{\partial K(\theta)}{\partial z}. \qquad (6.23)$$

Equation 6.23 generally can only be solved numerically for specified initial and boundary conditions: examples are given in Dane and Topp (2002). Another example of the use of the continuity equation is in the modelling of solute leaching through soil (Box 6.7).

This equation cannot be solved in this form because it contains two unknowns, $-p$ and θ. The steps in overcoming this problem are as follows:
1 Use the soil water retention function (Section 6.4) to obtain an expression for $d(-p)/dz$ in terms of $-p$, θ and z, that is:

Box 6.7 The convection-dispersion model of solute transport.

Infiltration of rain and subsequent drainage (or evaporation) of water can lead to a redistribution of solutes in the soil. Net downward movement of solutes is called leaching (Section 5.3). On the assumption that mass flow of solution (convection) and molecular diffusion between regions of different concentration are the mechanisms involved in movement, the solute flux density J_s vertically downwards can be written as

$$J_s = J_v + J_d, \qquad (B6.7.1)$$

where the convective flux J_v is given by

$$J_v = J_w C \qquad (B6.7.2)$$

and the diffusive flux J_d is given by Fick's first law of diffusion

$$J_d = -D\theta\frac{\partial C}{\partial z}. \qquad (B6.7.3)$$

C is the concentration of the solute in the soil solution and D is its diffusion coefficient in the porous medium. Applying the principle of mass conservation to the solute flux through a defined volume of soil, we may write

Box 6.7 continued

$$\left\{\frac{\partial(\theta C)}{\partial t}\right\} = -\frac{\partial J_s}{\partial z}. \tag{B6.7.4}$$

Substituting for J_s from Equations B6.7.1, B6.7.2 and B6.7.3 gives

$$\theta\frac{\partial C}{\partial t} = -\frac{\partial}{\partial z}\left\{J_w C - D\theta\frac{\partial C}{\partial z}\right\},$$

which on rearrangement of terms gives

$$\frac{\partial C}{\partial t} = D\frac{\partial^2 C}{\partial z^2} - \frac{J_w}{\theta}\frac{\partial C}{\partial z}. \tag{B6.7.5}$$

In practice, if the substitution $J_w/\theta = \bar{v}$ (Equation 6.16) is made in Equation B6.7.5,

the value of D obtained by solving the equation is invariably found to be greater than the porous medium diffusion coefficient of the solute. This is due to the coupled effects of molecular diffusion and hydrodynamic dispersion on the solute as water flows through the heterogeneous pore space, a subject outside the scope of this book. For this reason, Equation B6.7.5 is written in the form

$$\frac{\partial C}{\partial t} = D_h\frac{\partial^2 C}{\partial z^2} - \bar{v}\frac{\partial C}{\partial z}, \tag{B6.7.6}$$

where D_h is called the hydrodynamic dispersion coefficient and Equation B6.7.6 is known as the convection-dispersion equation for solute transport.

6.4 Soil water retention relationship

Pore-size distribution

If a saturated soil core is placed on a porous plate attached to a hanging water column and allowed to attain equilibrium (Fig. 6.13a), the soil water content at the applied negative pressure can be measured. For pressures < -85 kPa, the soil is usually brought to equilibrium with an air pressure that is then equal but opposite to the negative pressure of water in the soil pores. A graph of θ against negative pressure (matric potential ψ_m) can be constructed, which is called the soil water retention curve, illustrated by curve A in Fig. 6.13b. The rate of change of θ with ψ_m is the differential water capacity.

Since the absolute value of ψ_m is inversely proportional to the radius of the largest pores holding water at that potential (Equation 6.5), the volume of pores (equivalent to $\Delta\theta$) having radii between r and $r - \delta r$, where δr is a very small decrement in r corresponding to a decrease in potential $\delta\psi_m$, can be determined from the slope of the water retention curve A, provided that shrinkage of the soil on drying is small. Furthermore, the larger the value of $\Delta\theta/\Delta\psi_m$, the greater

is the volume of pores holding water within the size class defined by δr. Therefore, the value of ψ_m when the specific water retention capacity is a maximum defines the most frequent pore-size class (through the relationship $\psi_m \propto 1/r$). This relationship is illustrated by curve B in Fig. 6.13b. The example chosen is typical of a well-structured soil showing an adequate volume of large pores between aggregates (macropores), which promote drainage and aeration, relative to small pores within aggregates (micropores) that hold water in the so-called available range (see below).

The analysis of soil water retention curves to give pore-size distributions is complicated by the phenomenon of hysteresis, as discussed in Box 6.8.

Available water capacity and plant available water

The available water capacity (AWC) defines the amount of water in a soil that is nominally available for plant growth. The upper limit is set by the field capacity (FC) and the lower limit as the value of θ at which plants lose turgor and wilt, that is, the permanent wilting point (PWP). When related to the soil water retention curve

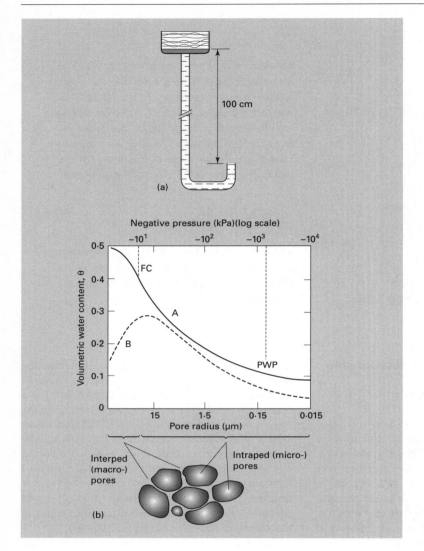

(a)

Negative pressure (kPa)(log scale)

FC

A

B

PWP

Interped (macro-) pores

Intraped (micro-) pores

(b)

Fig. 6.13 (a) Soil water under a negative pressure of −10 kPa (−100 cm head). (b) Graphs of the soil water retention curve (A) and differential water capacity (B) of a clay loam soil.

(Fig. 6.13b), these water contents are found to correspond to ψ_m values between −5 to −10 kPa for FC and −1500 kPa for PWP for a number of soil–plant combinations. In some intensive cropping systems, such as for vegetables, orchards and vineyards, the concept of readily available water (RAW) is also used. From experience, RAW is defined as the amount of water held between FC and ψ_m values of −40 to −60 kPa, depending on the age of the plants and the soil texture.

As pointed out in Section 6.3, FC is an inexact concept for many soils. Further, the onset of permanent wilting depends as much on the plant's ability to decrease the water potential in its tissues (to maintain a favourable gradient for water inflow from the soil) as it does on the soil water potential. For this reason, the choice of −1500 kPa for ψ_m at the PWP is arbitrary. Nevertheless, the AWC is a useful measure of the soil's reserve of available water, based primarily on its texture and structure. The total plant available water (PAW) is calculated by multiplying the AWC (expressed as an equivalent depth of water in mm per m depth of soil – see Box 4.6) by the plant's

Box 6.8 Hysteresis in soil wetting and drying curves.

The shape of the soil water retention curve depends on whether the soil is drying or wetting – a phenomenon called hysteresis, which occurs for two main reasons:

1 Large pores do not empty at the same pressure as that at which they fill, if access to the pore is restricted by narrow necks. Take, for example, a large pore of diameter D, which is connected to two smaller pores of diameter d_1 and d_2. On drying, the large pore will not empty until the negative pressure is great enough to break the meniscus across the larger of the two connecting pores, diameter d_2 (Fig. B6.8.1a). On wetting, water will advance into the large pore through pore d_1 and d_2 at roughly the same pressure because they are of similar size. However, the large pore will not fill completely until the negative pressure is small enough to sustain a meniscus across the distance D (Fig. B6.8.1b). Because $D > d_1$ or d_2, the absolute value of the negative pressure at which the large pore fills on wetting is less than that at which it empties on drying.

2 For a given pore, the contact angle between the water meniscus and the pore walls is greater when the meniscus is advancing (wetting) than when it is retreating (drying). This means that the radius of curvature is greater for the advancing (wetting) meniscus and that the absolute value of the pressure developed is correspondingly less for the same water content.

The combined effect of 1 and 2 is illustrated in Fig. B6.8.1c, which shows that the soil comes to equilibrium at a greater water content for a given matric potential on the drying than the wetting phase of the cycle. Note that the water content attained at zero potential at the end of the wetting phase may differ slightly from that at the start of the original drying phase because of air entrapment in the soil (Fig. B6.8.1c).

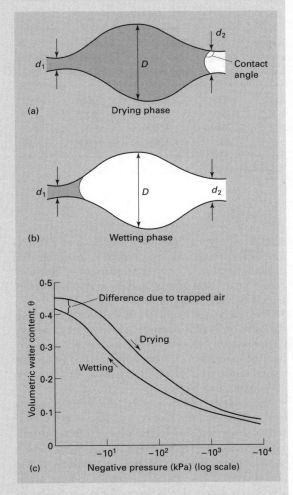

Fig. B6.8.1 (a) and (b) Differences in pore water content at similar negative pressures during wetting and drying. (c) Hysteresis in a soil water retention curve.

rooting depth (m). Some examples of PAW for different soil-rooting depth combinations are given in Table 6.3.

The concepts of AWC and PAW are useful for assessing the soil water available to crops under dryland farming, and for calculating the amount of water to be applied to crops under irrigation (Section 13.2). As a soil dries below FC, the difference between FC (mm) and the actual profile water content (mm) defines the soil water deficit (SWD). Theoretically, the maximum SWD attainable is determined by the PAW, on the assumption

Table 6.3 Plant available water PAW for different rooting depths in soils of varying texture.

Rooting depth (m)	Soil texture (mm)			
	Loamy sand	Sandy loam over clay (duplex profile)	Uniform or gradational clay loam*	Medium clay
1.0	100	160	200	140
1.5	150	240	300	210
2.0	200	320	400	280
3.0	300	480	600	420

* Soil with the best AWC and hence PAW.

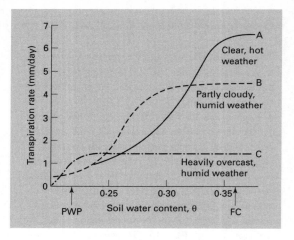

Fig. 6.14 The effect of atmospheric conditions and soil water content on plant transpiration rate (after Denmead and Shaw, 1962).

that all the water held between FC and PWP to the full rooting depth is equally available for growth (that is, a plant transpires at a constant rate over the whole range). This is approximately true if the following occur:

• The soil water retention curve is of a type that the bulk of available water is held at small negative pressures for which the hydraulic conductivity of the soil is relatively high – this holds for many sandy and sandy loam soils, but not for clay loams and clays.

• The evaporative power of the atmosphere is low enough that the rate of water movement from soil to root can satisfy the plant's transpiration. This proviso is demonstrated by Fig. 6.14, which shows that when the evaporative power of the atmosphere is low (curve C), almost all the available water is transpired before the plant begins to suffer water stress and the transpiration rate falls. By contrast, when the evaporative power of the atmosphere is high (curve A), the transpiration rate drops after only a small fraction of the AWC has been consumed (essentially the RAW has been consumed). The plant then shows signs of stress, such as loss of leaf turgor, at θ values well above the PWP. However, turgor is restored if cool, humid weather supervenes – this is a case of temporary wilting.

These observations illustrate the complexity of the interaction between soil properties ($\theta - \psi_m$ and $K - \psi_m$ relationships), plant properties (rooting depth, root morphology, salt accumulation, osmotic adjustment) and atmospheric factors affecting the plant's transpiration rate. The strong link between atmospheric factors and transpiration rate, which determines the rate at which the plant must withdraw water from the soil to remain turgid, focuses attention on the physically based measurements of evaporation discussed in Section 6.5.

6.5 Evaporation

The energy balance

The rate of evaporation from an open water surface (E_o) is measured using an evaporation pan, which is a shallow tank of water exposed to wind and sun. An example of the widely used Class A pan is shown in Fig. 6.15a. Note that the pan should be raised above the soil surface to allow air to circulate and minimize overheating of the water. Evapotranspiration E_t can be measured directly using weighing lysimeters, which are large tanks (approximately $1.5 \times 1.5 \times 1$ m deep) filled with soil and vegetation (Fig. 6.15b). Where such a lysimeter is installed in a field and planted to the same crop, pasture or even trees as in the surrounding area, the evaporative conditions should be the same. The soil in the lysimeter should be

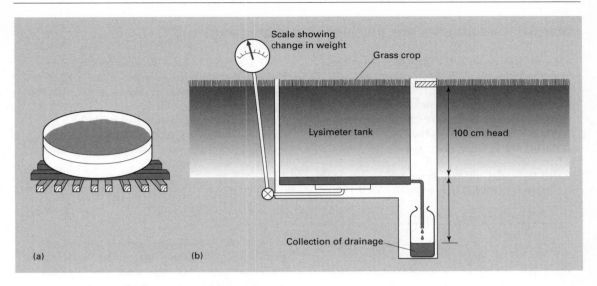

Fig. 6.15 (a) Class A evaporation pan. (b) A weighing lysimeter (after Pereira, 1973).

packed to reproduce the natural soil profile and have a collection vessel in which drainage D can be measured. If the depth of soil is 1 m, the minimum ψ_m attained at the surface due to drainage alone will be -10 kPa (corresponding to the FC). By measuring on a daily basis the input of rainfall P and irrigation I, and the change in weight (ΔW) of the lysimeter (allowing for any increase in crop mass), E_t for the crop and soil can be calculated from the water balance equation (Box 6.1)

$$E_t = P + I - D - \Delta W. \qquad (6.24)$$

In this equation all variables are expressed in mm of water.

When direct measurement is not possible, and especially for the estimation of E_t losses on a regional scale, the potential rate of evaporation from any surface – open water, plant or wet soil – can be calculated from meteorological data. The basic information required is:

• The net input of energy into the system; and
• the capacity of the air to take up and remove water vapour.

The main source of energy is solar radiation of which a fraction is directly reflected, depending on the albedo of the surface (see Table 6.1), and

some is lost as a return of long-wave radiation from the surface to the atmosphere. Of the net radiation absorbed R_N, part is dissipated by evaporation (component LE, where L is the latent heat* of vaporization and E the rate of evaporation), part by transfer of sensible heat* to the air (J_A) and part by transfer of sensible heat to the soil J_H (see Fig. 6.2); that is

$$R_N = LE + J_A + J_H. \qquad (6.25)$$

The transport capacity of the air is estimated from the vapour pressure gradient over the surface and the average wind speed, with an added empirical factor to account for surface roughness when evaporation from vegetation, rather than a free water surface, is considered. Equations for calculating the potential evapotranspiration rate (E_{to}) are outlined in Box 6.9.

For a fully vegetated surface, over a 24-h period, J_H in Equation 6.25 is small because the heat flux into soil during the day is approximately

---

* Latent heat of vaporization is 'hidden' heat absorbed when water changes from a liquid to vapour state; sensible heat is heat that can be felt.

Box 6.9 Calculation of the potential rate of evaporation.

A widely used equation for calculating the potential rate of evaporation is the Penman–Monteith combination formula (Smith, 1992), which is based on the energy balance (Equation 6.25) and an aerodynamic term for the turbulent transport of heat and water vapour from the evaporating surface to the air above. For a short, green grass crop (12 cm high), completely shading the soil and not limited for water, the evapotranspiration rate (called the reference rate E_{to}) is given by

$$E_{to} = \frac{0.408\Delta(R_N - J_H) + \gamma 900 U(e_a - e_d)/(T + 273)}{\Delta + \gamma(1 + 0.34U)},$$

(B6.9.1)

where Δ is the rate of change of the saturated vapour pressure with temperature, γ is the psychrometric constant, U is the horizontal wind speed at 2 m height, $e_a - e_d$ is the vapour pressure deficit above the canopy, and T is the average temperature (°C). The variables R_N, J_H, U, $(e_a - e_d)$ and T are measured at a 'weather station', such as the Bowen Ratio installation shown in Fig. B6.9.1. Measurements are made at short time intervals and aggregated for a day.

Another widely used equation, similar to the Penman–Monteith equation but which does not include a variable term for the vapour pressure deficit and wind speed, is that of Priestley and Taylor (1972)

$$E_{to} = [\alpha \frac{\Delta}{\Delta + \gamma}(R_N - J_H)]/L,$$

(B6.9.2)

where α is a dimensionless constant and the other terms have been defined. The constant α should be chosen for a particular region, but a value of 1.26 has been found to have wide applicability.

E_{to} values range from 6–7 mm/day in midsummer at low altitudes in southern Australia to < 1 mm/day in midwinter. On days of strong wind in summer, the value may rise to 8–10 mm/day.

Fig. B6.9.1 A field installation for measuring weather and other variables for E_t and the Bowen Ratio.

balanced by the heat flux out at night. Thus Equation 6.25 simplifies to

$$R_N = LE + J_A$$

i.e. $$\frac{R_N}{LE} = 1 + \frac{J_A}{LE}. \qquad (6.26)$$

The ratio of the sensible heat flux J_A to latent heat flux LE defines the *Bowen Ratio* ($J_A : LE$). Under humid temperate conditions, $J_A : LE$ is small and can be ignored so that to a reasonable approximation, Equation 6.26 becomes

$$\frac{R_N}{L} \cong E. \qquad (6.27)$$

Penman (1970) suggested that approximately 40% of the incoming solar radiation is dissipated in the evaporation of water from a green crop completely covering the soil, in which the supply of water is not limiting. Thus, given an integrated daily R_N value of 7.56 MJ/m² and an L value of 2.45×10^9 J/m³, substitution in Equation 6.27 gives an evaporation rate of 3.1 mm/day. This rate is an estimate of potential evapotranspiration E_{to}, which for the conditions specified, is comparable with the rate of transpiration under mild cloudy conditions shown in Fig. 6.14. However, as the soil water supply becomes limiting, the actual rate of evapotranspiration E_{ta} falls progressively below the potential rate as plants close their stomata and restrict the rate of transpiration from their leaves. Under drier conditions when plants suffer water stress, the leaf surfaces warm up and a much greater share of R_N is lost to the air as sensible heat (the Bowen Ratio increases). In this case, the actual rate of evapotranspiration can be obtained by solving Equation 6.25 for E, provided that J_A and J_H can be measured. Equipment for measuring E_{ta} using the Bowen Ratio approach is shown in Fig. B6.9.1.

Evaporation factors and coefficients

The reference evapotranspiration rate E_{to}, defined in Box 6.9, can be estimated from evaporation pan values (E_o) by multiplying by an appropriate pan coefficient (C_p). The value of the coefficient depends on the type of pan, its site and the weather

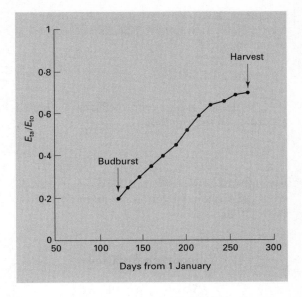

Fig. 6.16 Change in the ratio of E_{ta}/E_{to} during the development of a grapevine canopy from budburst to maturity (after L. E. Williams, personal communication).

conditions, but for a Class A pan situated on short grass in a relatively humid and low-wind environment, C_p is c. 0.85.

However, as pointed out above, the actual evapotranspiration rate E_{ta} of a crop falls below E_{to} as the soil dries out: it also falls when the plant canopy does not completely cover the soil surface. The extent of canopy cover is measured by the leaf area index (LAI), which is the leaf area per unit area (1 m²) of ground surface. Thus a relatively bare pasture may have an LAI < 1, whereas a healthy pasture completely covering the ground can have an LAI > 4. Similarly, the LAI of a cereal crop increases as it grows from seedling stage to maturity, as does the LAI of perennial crops such as grapevines developing from budburst to full leaf near harvest. If one assumes soil water is not limiting, the ratio $E_{ta} : E_{to}$ increases as the crop develops, as shown in Fig. 6.16, approaching 1 at maturity (if there is substantial advection of heat energy into an area of crop that is well supplied with water, E_{ta} may be 10–20% above E_{to} – this is an example of the 'oasis effect'). To calculate E_{ta} from either E_o or E_{to}, empirical coefficients have been derived that take account of changes in LAI as a crop grows and develops its canopy. If E_o values are used (adjusted by C_p), a crop factor C_f is used to calculate E_{ta} according to the equation

$$E_{ta} = \text{adjusted } E_o \times C_f. \tag{6.28}$$

If E_{to} values are used, a crop coefficient C_c is used according to the equation

$$E_{ta} = E_{to} \times C_c. \tag{6.29}$$

Examples of crop coefficients for a range of crops with and without irrigation are given by Howell *et al.* (1986).

Evaporation from soil

Stages of drying

The rate of evaporation from a soil surface is inversely related to the LAI, being a very small fraction of total ET when the LAI is large, to being the only component of ET when the soil is bare. Considering a bare soil, as in a recently cultivated field, while the surface remains wet, the rate of evaporation is determined by the energy balance between the soil and the atmosphere and the transport capacity of the air. Depending on the albedo, the rate can approach that of evaporation from an open water surface (commonly for wet soil $E_{soil} = E_{to} = 0.85\ E_o$). This stage is called constant-rate drying and may last for 1–3 days, depending on the soil type, latitude and season of the year.

With a fast initial evaporation rate from a soil, unaffected by a watertable, the rate of water loss soon exceeds the rate of supply from below and a surface layer, 1–2 cm deep, becomes air-dry. As the macropores empty of water first, the unsaturated conductivity for water flow across the dry zone from below decreases sharply and more than offsets the steep increase in matric potential gradient. The rate of evaporation is now controlled by soil properties and decreases with time, a stage known as falling-rate drying. The dry surface layer has a self-mulching effect because it helps to conserve moisture at depth, especially under conditions of high incident radiation in soils with a fine-granular or crumb structure.

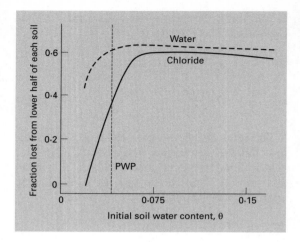

Fig. 6.17 Movement of water and chloride through soil blocks exposed to constant evaporative conditions (after Marshall, 1959).

Water movement in dry soil

Despite the self-mulching effect of an air-dry surface, if evaporation is prolonged and uninterrupted by rain, the zone of dry soil gradually extends downwards. This is because there is a slow flow of liquid water through soil wetter than the wilting point, to the drying surface, and there is also some diffusion of water vapour.

The flow of liquid water through relatively dry soil can be demonstrated by the simultaneous transfer of water and dissolved salts in a soil subjected to controlled, isothermal evaporative conditions. Fig. 6.17 shows the movement of water and Cl^- from the lower half of a series of soil blocks, initially at uniform water contents ranging between FC and PWP. Chloride moves with the water down to a water content only slightly greater than the PWP. Below that it is mainly water that moves, by vapour diffusion, in response to vapour pressure (VP) gradients created by differences in matric and osmotic potentials. The effect of matric and osmotic potentials on VP is normally small because, over the range 0 to $-1500\ \text{kPa}$, the relative VP (e/e^o in Equation B6.2.1) merely changes from 1.0 to 0.989. Only when the soil becomes air-dry and the relative VP of the soil water drops to $c.\ 0.85$ does the VP gradient under isothermal conditions become appreciable.

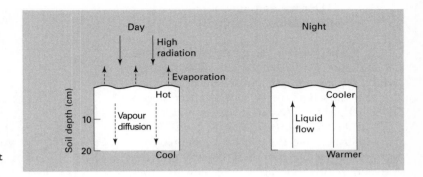

Fig. 6.18 Diurnal alternation of vapour and liquid water movement at the surface of a drying soil.

Temperature, however, has a large effect on VP, producing a threefold increase between 10 and 30°C.

A large diurnal variation in temperature can cause redistribution of salt within the top 15–20 cm of a dry soil through the alternation of vapour and liquid phase transfer of water (Fig. 6.18). As the surface heats up during the day, water vapour diffuses into the deeper, cooler layers and condenses. Transfer and condensation of water means heat transfer so that the temperature gradient, and hence VP gradient, gradually diminishes until the evening when, with concurrent surface cooling, the temperature differential is reversed. At this point, the increased water content and matric potential in the condensation zone induce a return flow of water to the surface, carrying dissolved salts that are deposited there when the water subsequently evaporates.

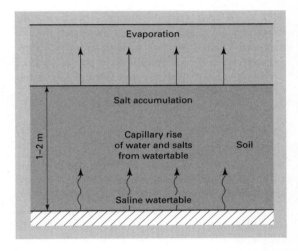

Fig. 6.19 Capillary rise of salts from saline groundwater.

Capillary rise

Soil water above the watertable is drawn upwards in continuous pores until the pressure gradient acting upwards is balanced by the gravitational head gradient acting downwards. This rise of water, known as capillary rise, depends very much on the pore-size distribution of the soil, being usually not greater than 1 m in sandy soils but as much as 2 m in some silt loams, as shown in Fig. 6.19. Thus, groundwater can be drawn from a watertable within 1–2 m of soil surface and lead to excessive evaporative losses. The actual rate of loss is governed by atmospheric conditions up to the point where the hydraulic conductivity of the soil becomes limiting. Accumulation of soluble

salts at the surface, as occurs in mismanaged irrigated soils, encourages capillary rise because the high osmotic pressure maintains an upward gradient even in moist soils, when the unsaturated hydraulic conductivity is large and appreciable water movement can therefore occur. The rise of groundwater often exacerbates the problem of soil salinity, which is discussed in Chapter 13.

6.6 Soil temperature

The effects of temperature on physicochemical and biological reactions in soil are discussed in Sections 5.2, 5.3 and 8.1. This section deals with the causes and nature of soil temperature variations.

Diurnal and annual temperature variations

Clearly the net radiant energy R_N and hence the soil heat flux J_H depend on the albedo of the surface, the time of day and season of the year. Not only does the type of surface affect albedo (Table 6.1), but it also varies with latitude (from c. 7% at the Equator to 56% at the North Pole), and with aspect and slope in the landscape. J_H is negative during the day (heat flowing into the soil) and positive at night (heat flowing from the soil to the air). This variation in flux causes the soil temperature to change in a sinusoidal way over a 24-hour period (diurnal variation); the longer term variation in heat flux between summer and winter causes a similar sinusoidal temperature variation of much longer period (annual variation). These diurnal and annual variations can be described by the equation

$$T(t) = T_a + A \sin\left(\frac{2\pi t}{\tau}\right), \quad (6.30)$$

where t is time, T_a is the average temperature, τ (tau) is the period of the wave (1 day or 1 year) and A is the wave's amplitude.

Variations in soil temperature are greatest at the surface and of decreasing amplitude with depth, as seen in Fig. 6.20. There is also a time lag between the attainment of either the maximum or the minimum temperature, which increases with depth. These effects are due to the nature of heat transport from the surface, given the time-dependent input as described by Equation 6.30 and the properties of the soil – in particular, its thermal conductivity λ_e and its heat capacity per unit volume C_h. The main processes involved in heat transport are conduction (involving molecular collisions) and convection of latent heat (involving the movement of water vapour – see Fig. 6.18), which can be described by the continuity equation for heat flow, that is

$$\frac{\partial T}{\partial t} = K_T \frac{\partial^2 T}{\partial z^2}, \quad (6.31)$$

where z is depth and K_T ($= \lambda_e/C_h$) is the thermal diffusivity.

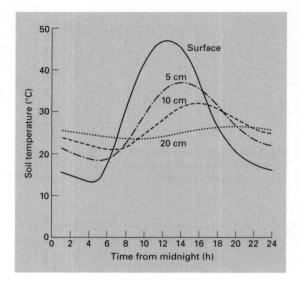

Fig. 6.20 Diurnal variations in soil temperature measured at different depths (after Jury et al., 1991).

The heat flux equation (6.31) can be solved with Equation 6.30 as an upper boundary condition, and the assumption that the amplitude of the temperature wave far below the surface is 0 (as $z \to -\infty$). The solution of Equation 6.31 for these conditions is

$$T(z,t) = T_a + A \exp\left(\frac{z}{z_d}\right) \sin\left(\frac{2\pi t}{\tau} + \frac{z}{z_d}\right). \quad (6.32)$$

Note that z is measured negatively downwards. The parameter z_d is called the damping depth and is given by

$$z_d = \left(\frac{K_T \tau}{\pi}\right)^{1/2}. \quad (6.33)$$

The damping depth is related to the lag time (Δt) for the temperature to change at depth z by the equation

$$\Delta t = \frac{z\tau}{2\pi z_d}. \quad (6.34)$$

These results show that:

• For a given period of the temperature wave, the time lag for temperature change is proportional to z and inversely proportional to $\sqrt{K_T}$;
• for given thermal properties, the damping depth $z_d \propto \sqrt{\tau}$, and
• all depths have the same average annual temperature.

The analysis confirms the observation that a rapidly changing surface temperature variation (diurnal) does not penetrate as deeply as a slowly changing temperature (over 1 year) of the same amplitude. Diurnal variations are normally < 5% below 30 cm, so this depth is commonly used to measure average soil temperatures.

The effect of mulches

The amplitude of diurnal variations in surface temperature is much reduced by vegetation and surface mulches through their effects on J_H, and the albedo and thermal conductivity of the surface. As well as having low thermal conductivity, light-coloured, granular plastic mulches transmit incoming short-wave radiation but prevent the loss of long-wave thermal radiation, thereby creating a 'greenhouse effect' at the soil surface (Section 3.1). However, grass or straw mulches in row crops can increase the chance of early morning frost in cool climates because of their insulating effect during the day. The greater albedo of such mulches compared with bare, dark soil also decreases the amount of radiant energy absorbed by the soil.

6.7 Summary

The interchange of water between the atmosphere and Earth's surface through precipitation P and evaporation E from soil, water and plants comprises the hydrological cycle. Within a defined catchment (or watershed)

P = E + Runoff R + Deep drainage D + Change in soil storage, ΔS.

On an annual basis, ΔS is usually small so that the sum of R and D determines the surplus water, which can be distributed to surface waters (streams, lakes and other storages) and groundwater. Nevertheless, changes in ΔS are important in the short term for plant growth and may decide crop success or failure.

The interaction between water and organo-mineral surfaces, exchangeable cations and dissolved salts reduces its potential energy. There is no absolute scale of potential energy, but we can measure *changes* in potential energy when useful work is done on a known quantity of water, or when the water does useful work. These changes are manifest as changes in the free energy per mole of water (i.e. its chemical potential μ_w), relative to pure water outside the soil at the same temperature and pressure and at a fixed reference height. The soil water potential ψ is therefore defined as

$$\psi = \mu_w(\text{soil}) - \mu_w^o(\text{standard state}) = RT \ln e/e^o,$$

where e/e^o is the relative vapour pressure of the soil water at temperature T, and R is the Universal gas constant. Potential ψ can be expressed as energy per unit volume, equivalent to a pressure in bars or Pascals; or as energy per unit weight, which is equivalent to hydraulic head in units of length. Head h and potential ψ are related through the equation

$$h = \psi/\rho g.$$

The value of ψ may be partitioned between several component potentials attributable to matric, osmotic, pressure and gravitational forces, respectively, as

$$\psi = \psi_m + \psi_s + \psi_p + \psi_g.$$

The matric potential ψ_m and pressure potential ψ_p comprise a continuum such that at normal atmospheric pressure in an unsaturated soil $\psi_p = 0$ and ψ_m is negative, whereas in saturated soil ψ_p is positive and $\psi_m = 0$.

Soil volumetric water content θ (m^3 water/m^3 soil) can be measured in the field by a neutron probe, time domain reflectometer (TDR), or a capacitance probe. Changes in the soil matric potential ψ_m (equivalent to a pressure head $-p$) down to −85 kPa (−850 cm head) are measured by a tensiometer. Under static equilibrium conditions, the relationship between θ and $-p$ defines the soil water retention curve, the slope of which (the differential water capacity, $d\theta/d(-p)$) depends on the pore-size distribution in non-swelling soils. Two points identified on this curve, the field capacity FC corresponding to matric potentials of -5 to -10 kPa, and the permanent wilting point PWP corresponding to a matric potential of -1500 kPa, set the approximate upper and lower limits for a soil's available water capacity (AWC). The AWC, expressed as the difference in θ ($\Delta\theta$) between FC and PWP can be converted to an equivalent depth of water d (mm per m depth of soil), according to the equation

$$d = 1000 \ \Delta\theta.$$

The plant available water (PAW) depends on the AWC and the rooting depth z of the plants, that is,

$$PAW = 1000 \ \Delta\theta z.$$

Rain or irrigation water at the soil surface infiltrates into dry soil at an initial rate determined by the rainfall intensity. However, the negative pressure gradient decreases as the soil wets, so that the infiltration rate eventually falls to a steady value equal to the soil's saturated hydraulic conductivity. Water is subsequently distributed through the soil profile in response to head gradients, and the flux of water J_w (volume per unit area per unit time) is given by Darcy's Law

$$J_w = -K\frac{dh}{dx},$$

where K is the hydraulic conductivity (the reciprocal of the resistance to flow) and dh/dx is the head gradient in the direction x. For vertical flow (in the z direction, measured positively upwards) in non-saline soils, the main driving forces are the gradients in pressure and gravitational heads, represented by the terms $-p$ and z, respectively. During saturated flow, the gradient $d(-p)/dz = 0$ and gravity is the driving force, so that $J_w = K_s$, the saturated hydraulic conductivity. As the soil desaturates K becomes a function of $-p$, and once the large pores have drained, K falls very sharply with an increase in the absolute value of $-p$. Even

in unsaturated soil, flow to a deep water table can be in steady state if $d(-p)/dz$ is constant; but normally flow under unsaturated conditions is transient and both $K(-p)$ and $d(-p)/dz$ must be known to calculate J_w. The change in water content in an unsaturated profile can be calculated by combining Darcy's Law with the continuity equation (describing the conservation of water mass) to give the Richards equation

$$\frac{\partial\theta}{\partial t} = \frac{\partial}{\partial z}\left\{K(p)\left(\frac{\partial(-p)}{\partial z} - 1\right)\right\}.$$

Solute transport through structurally homogeneous soils can be described by combining equations for mass flow and diffusion with a continuity equation, giving rise to the convection-dispersion equation.

Evapotranspiration from wet soil and vegetation well-supplied with water proceeds at a potential rate E_{to} that is determined by the net radiation input R_N and meteorological conditions. One measure of E_{to} for a short green grass crop completely covering the soil and not limited for water is called the reference E_{to}, and is calculated using the Penman–Monteith combination equation. When the supply of water is limiting, the actual evaporation rate E_{ta} is less than E_{to} because the rate of water uptake by the plants cannot match the evaporative demand of the atmosphere. Evaporation from bare wet soil proceeds at nearly the potential rate until the surface begins to dry (constant rate drying), but then begins to fall and becomes very slow once the top 1–2 cm of soil is air-dry. Vapour phase transport is insignificant in non-saline soils wetter than the PWP unless a temperature gradient exists between different layers in the soil.

Soil temperature is controlled by the net heat flux J_H, which is from the atmosphere into the soil during the day and in the opposite direction at night. The temperature at the surface fluctuates in a sinusoidal wave both diurnally (period 1 day) and annually (period 1 year). The phase lag in the wave increases with depth, but the amplitude of the temperature variation decreases. For diurnal temperature change, the variation below 30 cm is normally < 5%. Diurnal temperature variations in soil can be moderated by surface mulches.

References

Dane J. H. & Topp G. C. (Eds) (2002) *Methods of Soil Analysis. Part 4. Physical Methods.* Soil Science Society of America Book Series No. 5. Soil Science Society of America, Madison WI.

Denmead O. T. & Shaw R. H. (1962) Availability of soil water to plants as affected by soil moisture content and meteorological conditions. *Agronomy Journal* 54, 385–90.

Howell T. A., Bucks D. A., Goldhamer D. A. & Lima J. M. (1986) Management principles: irrigation scheduling, in *Trickle Irrigation for Crop Production.* (Eds F. S. Nakayama & D. A. Bucks). Elsevier, Amsterdam, pp. 241–79.

Jury W. A., Gardner W. R. & Gardner W. H. (1991) *Soil Physics*, 5th edn. Wiley & Sons, New York.

Linacre E. (1992) *Climate Data and Resources: A Reference and Guide.* Routledge, London.

Marshall T. J. (1959) *Relations between Water and Soil.* Technical Communication No. 50. Commonwealth Bureau of Soils, Harpenden, UK.

Or D. & Wraith J. M. (2000) Soil water content and water potential relationships, in *Handbook of Soil Science.* (Ed. M. E. Sumner). CRC Press, Boca Raton, FL, A53–A85.

Penman H. L. (1970) The water cycle. *Scientific American* 223, 98–108.

Pereira H. C. (1973) *Land Use and Water Resources in Temperate and Tropical Climates.* Cambridge University Press, Cambridge.

Priestley C. H. B. & Taylor R. J. (1972) On the assessment of surface heat flux and evaporation using large-scale parameters. *Monthly Weather Review* 100, 81–92.

Smith M. (1992) *Expert Consultation on Revision of FAO Methodologies for Crop Water Requirements.* Food and Agriculture Organization of the United Nations, Rome.

Wooding R. A. (1968) Steady infiltration from a shallow circular pond. *Water Resources Research* 4, 1259–73.

Youngs E. G. & Thomasson A. (1975) Water movement in soil, in *Soil Physical Conditions and Crop Production.* MAFF Bulletin No. 29. Her Majesty's Stationery Office, London, pp. 228–39.

Further reading

Evett S. R. (2000) Energy and water balances at soil–plant–atmosphere interfaces, in *Handbook of Soil Science.* (Ed. M. E. Sumner). CRC Press, Boca Raton, FL, A129–A182.

Marshall T. J., Holmes J. W. & Rose C. W. (1996) *Soil Physics*, 3rd edn. Cambridge University Press, Cambridge.

Rose C. W. (2004) *An Introduction to the Environmental Physics of Soil, Water and Watersheds.* Cambridge University Press, Australia.

Warrick A. W. (Ed.) (2001) *Soil Physics Companion.* CRC Press, Boca Raton, FL.

Example questions and problems

1 (a) Which of the component potentials of soil water are the most important in determining the movement of water through non-saline soils?

(b) Calculate the water potential ψ (in kPa) at the surface of a soil that is in equilibrium with air at a relative humidity of 80% and a temperature of 20°C. Use the equation

$$\psi = \frac{RT}{V_w} \ln\left(\frac{e}{e^\circ}\right)$$

where $R = 8.314$ J/degree Kelvin/mol and $V_w = 1.8 \times 10^{-5}$ m^3/mol.

2 Tensiometers are installed in a soil profile at intervals to a depth of 90 cm. The tensiometer potentials, read several days after rain has fallen, are given in the following table.

(a) Graph these values against the depths and fill in the blank cells in the table to show the gravitational and matric potentials. Take the soil surface as the reference height for gravitational potential.

Depth (cm)	10	30	50	70	90
Tensiometer potential (kPa)	−60	−34	−16	−5	−8
Gravitational potential (kPa)					
Matric potential (kPa)					

(Note – 1 kPa is equivalent to 10 cm head)

(b) Determine the depth of the zero flux plane in the soil profile.

  (c) In which directions is water expected to flow from this zero flux plane?

  (d) What is the cause of the low potential near the soil surface?

3 Runoff from a 20-ha grassland catchment in the Great Dividing Range of southern Australia was measured using an H-flume. In a storm in July (mid-winter), when 8 mm of rain fell in 20 min., the total volume of runoff measured over 3 h from the start of the storm was 1200 m^3.

  (a) Calculate the quantity of runoff in mm.

  (b) What was the runoff rate for the catchment in mm/h?

  (c) Would you expect any water to have been lost by evaporation during this event?

4 The manager of an irrigated cotton farm wants to know the water holding properties of his soil – a heavy black clay, so that he can apply water to avoid undue stress in the crop, while not wasting water. From neutron probe measurements, he determines that the average water content θ in the top metre of soil at field capacity (FC) is 0.42 m^3/m^3 and the permanent wilting point of the cotton plants is 0.28 m^3/m^3.

  (a) What is the AWC of the soil in mm water per m depth?

  (b) If the effective rooting depth of the crop is 60 cm, what is the PAW (in mm)?

  (c) If the manager's aim was to apply irrigation when the PAW had fallen to 50% of its maximum value, how much water would be needed to return the soil to FC?

5 In Q.4, the manager has potential evaporation (E_{to}) values from daily measurements at a weather station, and crop coefficients (C_c) for cotton at different stages of growth. The following table shows weekly E_{to} and corresponding C_c values for the crop from the seedling stage. Rainfall was also measured. Assuming the soil was at FC at the late seedling stage, complete the table by calculating the cumulative soil water deficit (SWD) and indicate when irrigation water should be applied to return the soil to FC, according to the criterion in Q.4(c).

Time from seedling stage (week)	1	2	3	4	5	6	7
Rainfall (mm)	0	5	0	8	0	2	0
E_{to} (mm)	28	24	35	38	42	40	48
C_c	0.25	0.40	0.52	0.65	0.78	0.90	1.02
Cumulative SWD (mm)							
Irrigate? (y/n)							

6 The saturated hydraulic conductivity K_s of the A horizon of a 'duplex' soil used for vegetable growing was 25 mm/h. Answer the following questions:

  (a) What is the head gradient for water flow in the saturated soil?

  (b) Given that the soil's porosity is 0.5, what is the average pore velocity for water moving through the saturated soil?

  (c) Suppose urea fertilizer (very soluble) was spread on the wet soil surface, and rain fell at 15 mm/h for 3 h soon afterwards. At what velocity would the dissolved fertilizer move through the soil?

  (d) To what depth would the peak concentration of dissolved fertilizer move after 3 h?

7 A 25-ha grassland catchment in western Victoria was instrumented to measure changes in the water balance. Changes in soil water content θ were measured with capacitance probes, and runoff was measured with an H-flume. In a storm in September, 15 mm of rain fell and produced a total runoff volume of 1250 m^3.

  (a) What was the runoff in mm?

  (b) The average SWD at the start of the storm was −20 mm. Assuming that evapotranspiration during the period of measurement was negligible, calculate how much rain (mm) was available to wet the soil.

  (c) What was the expected change in θ if the soil wetting was confined to the top 25 cm depth?

Chapter 7

Reactions at Surfaces

7.1 Charges on soil particles

Some definitions

Chapter 2 introduced the concepts of:
- *Permanent charge* on soil minerals arising out of the substitution of elements of similar size but different valency within the crystal structure; and
- *pH-dependent charge* arising through the reversible dissociation of H^+ ions from surface groups, according to their acid strength and the activity of H^+ ions in the soil solution.

The net permanent charge created by isomorphous substitution in clay minerals (mainly the 2 : 1 clays) is invariably negative and unaffected by the concentration and type of ions in the soil solution, unless the pH is so low that it induces decomposition of the crystal structure. With some of the hydrated oxides, there can be a small positive permanent charge due to the substitution of cations of higher valency within the structure. The result of isomorphous substitution is a mineral of constant surface charge density (σ_o) and variable electrical potential (ϕ).

On the other hand, reversible dissociation of H^+ ions from carboxyl and phenolic groups in organic polymers, or from O and OH groups in oxide surfaces and the edge faces of kaolinite crystals, results in a variable surface charge density (σ_H). H^+ and OH^- are called potential-determining ions for such surfaces, and σ_H measures the difference between the moles of H^+ and OH^- complexed by the surface. The electrical potential of the surface remains constant provided the pH of the solution does not change. Hence the expression 'constant potential' surface that is

sometimes used. When surfaces of constant and variable charge occur on the one mineral, such as kaolinite and some allophanes (Section 2.4), the resultant surface charge density is called the intrinsic surface charge density (σ_{in}), defined by

$$\sigma_{in} = \sigma_o + \sigma_H. \tag{7.1}$$

Methods of measuring or calculating σ_o, σ_H and σ_{in} are discussed by Sposito (1989). σ_{in} is an operationally defined parameter because its value depends on the experimental conditions under which it is measured.

The charged particle–solution interface

The existence of immobile charges on a mineral surface produces a change in the distribution of mobile ions in the soil solution in contact with the surface. A simple example is provided by a group of aluminosilicates called the zeolites, in which the permanent negative charge is distributed throughout the crystal structure to give a uniform volume charge. Ions dissolved in the solution permeating the mineral can diffuse through a network of interconnecting pores, undergoing frequent changes in electrical potential as they move from the influence of one fixed charge to another. However, when equilibrium is reached between the exchanger (the mineral) and the permeating salt solution, each ion experiences an average change in electrical potential on passing into or out of the exchanger, the value of which measures the Donnan potential. Within the exchanger cations are retained, or positively

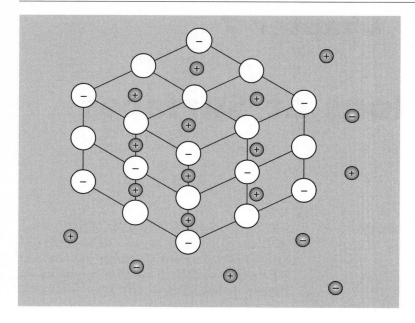

Fig. 7.1 Concentration of mobile cations (+) within the Donnan free space and exclusion of mobile anions (−).

adsorbed, and anions are excluded or negatively adsorbed. The volume from which anions are excluded is called the Donnan free space (Fig. 7.1). This model applies equally well to soil humic polymers and plant constituents, such as cell walls and cytoplasmic proteins, except that the charges on these are not permanent, but pH-dependent.

Non-uniform volume charge

Unlike the zeolites, the crystalline clay minerals have thin laminar structures (Section 2.3 and Box 2.2) in which the permanent negative charge acts as if it were spread over the planar surfaces of the crystals. The negative charge creates a surplus of cations (the counter-ions) and deficit of anions (the co-ions) in the solution immediately adjacent to the surface. The simplest case is one in which each surface charge is neutralized by the close proximity of a mobile charge of opposite sign – this parallel alignment of charges in two planes is called a Helmholtz double layer (Fig. 7.2).

However, the Helmholtz model is unrealistic in aqueous solutions. The high dielectric constant of water (meaning it is a poor conductor of electricity) diminishes the electrostatic force of attraction between the fixed and mobile charges. In addition, there is a diffusive force tending to draw the

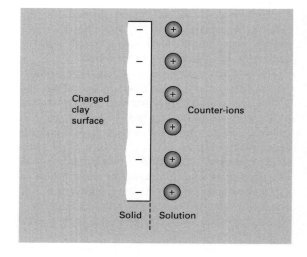

Fig. 7.2 The Helmholtz double layer.

cations away from the region of high concentration near the charged surface. At equilibrium, the electrostatic force across a plane of unit area near to and parallel to the surface is just balanced by the difference in osmotic pressure between the plane and the bulk solution far removed from the surface. The result is a diffuse distribution of cations and anions in solution (Fig. 7.3a), which together with the surface charge comprises the

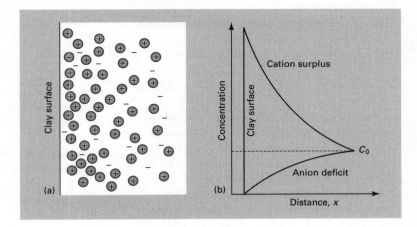

Fig. 7.3 (a) and (b) The ionic distribution at a negatively charged clay surface.

Gouy–Chapman double layer, so named after the two scientists who independently developed the mathematical description of the diffuse space charge. Such a diffuse double layer (DDL) can develop at surfaces of constant charge or variable charge.

Characteristics of the Gouy–Chapman double layer

The distribution of ions in solution in the direction at right angles to a charged planar surface (Fig. 7.3a) is given by the Boltzmann equation

$$C(x) = C_o \exp[-zF\phi(x)/RT], \qquad (7.2)$$

where $C(x)$ is the concentration of an ion of valency z (cation or anion) at a distance x from the surface where the potential is $\phi(x)$ (i.e. ϕ is a function of x); C_o is the ion concentration in the bulk solution, F is Faraday's constant and R and T are as defined in Box 6.2. It is implicit in Equation 7.2 that the potential in the bulk solution is zero. Thus, for a negatively charged surface, as x becomes smaller and the absolute value of $\phi(x)$ increases, the concentration $C(x)$ of cations increases exponentially relative to C_o. At the same time, the concentration of anions decreases exponentially, becoming vanishingly small as the surface is approached. The concentration profiles for cations and anions in solution near a negatively charged surface are illustrated in Fig. 7.3b.

At a surface of constant charge, the volume density of charge in the diffuse layer $\sigma(x)$ must remain constant irrespective of changes in the ionic composition of the solution. For each square cm of surface, the solid phase charge is neutralized by the surplus of cations over anions in solution, such that

$$\sigma(x) = \sigma(+) + \sigma(-), \qquad (7.3)$$

where $\sigma(+)$ and $\sigma(-)$ are the cationic and anionic charges in the diffuse layer (cmol charge (+) and (−) per cm², respectively). The effective thickness of the diffuse layer is measured by the parameter d_{ex}, which is the mean distance over which the co-ion concentration is depleted relative to the bulk solution. Because of the opposing forces of electrostatic attraction between unlike charges and the diffusion of ions from a region of high concentration to low concentration, d_{ex} is found to be inversely related to the charge on the counter-ion and the concentration C_o of the bulk solution. For solutions of mixed salts such as a soil solution, the property ionic strength (I) is a generic measure of ion concentrations, modified by the ionic charges, and we find I is related to d_{ex} according to

$$d_{ex} \propto \frac{1}{I^{1/2}}, \qquad (7.4)$$

that is,

$$d_{ex} \cong \frac{2}{(\beta C_o)^{1/2} z}, \qquad (7.5)$$

Box 7.1 Ionic strength and ion activity.

The ionic strength of a mixed salt solution is given by

$$I = 0.5\sum_{i}^{n} C_i z_i^2, \qquad (B7.1.1)$$

where C_i and z_i are the concentration and charge, respectively, of ion i in a mixture of ionic species $i = 1$ to n. The term 'concentration' refers to the mass of an ionic species per unit volume of solution (e.g. mol/L). An individual ion experiences weak forces due to its interaction with water molecules (the formation of a hydration shell), and stronger electrostatic forces due to its interaction with ions of opposite charge. Effectively, this means that the ability of the ion to engage in chemical reactions is decreased, relative to what is expected when it is present at a particular concentration with no interactions. This effect is accounted for by defining the activity a_i of ion i, which is related to its concentration by the equation

$$a_i = f_i\, C_i, \qquad (B7.1.2)$$

where f_i is the activity coefficient of the ion. Values of f_i range from 0 to 1. In very dilute solutions where the interaction effects are negligible, f_i approaches 1, and a_i is approximately equal to C_i. There is extensive theory on the calculation of activity coefficients, but all calculations make use of the ionic strength I, such as in the Debye–Huckel limiting law (Atkins, 1982)

$$\log_{10} f_\pm = -0.509\left|z^+ z^-\right| I^{1/2}, \qquad (B7.1.3)$$

which works well for values of I up to 0.01 M for a monovalent salt such as NaCl. Note that the value $f_\pm$ applies to both ionic species in solution, and that the absolute values of the charges z^+ and z^- are used in the formula.

where β, the DDL constant, has the value 1.084×10^{16} cm/cmol. The calculation of I from ionic concentration and charge, and the relationship between concentration and activity, are discussed in Box 7.1.

Calculated values of d_{ex} for different concentrations of NaCl and CaC1$_2$ solutions in contact with a clay surface are given in Table 7.1. The calculations show how the DDL is compressed as the charge on the cation increases or the solution concentration increases.

Modifications to the Gouy–Chapman double layer

The Gouy–Chapman theory works well for cation solutions of low to intermediate concentrations (0.0001–0.1 M) and surface charge densities of $1-4 \times 10^{-4}$ cmol charge/m^2. It breaks down once certain limits of cation valency and solution concentration are reached because:

• The ions are assumed to be point charges, which leads to absurdly high concentrations of cations in the inner region of the diffuse layer; and
• the theory ignores forces between the surface and counter-ions other than simple electrostatic

Table 7.1 Calculated diffuse layer thicknesses for different concentrations of 1 : 1 and 2 : 1 electrolytes.

Electrolyte concentration C_o (mmol/cm^3)	Effective diffuse layer thickness d_{ex} (nm)	
	NaCl	CaCl$_2$
0.1	1.94	1.0
0.01	6.2	3.2
0.001	19.4	10.1

forces. It predicts that cations of the same valency will be adsorbed with the same energy, whereas it is known from cation exchange measurements that Li$^+$ and K$^+$, or Ca^{2+} and Ba^{2+}, for example, are not held with the same tenacity by clays in contact with solutions of identical normality*.

These factors are accommodated in the Stern model – essentially a combination of the Helmholtz and Gouy–Chapman concepts – which splits the solution component of the double layer into two parts that are:

* A normal (N) solution is one containing 1 mmol of cationic or anionic charge per cm^3.

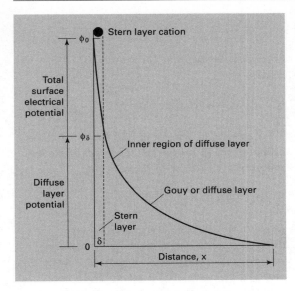

Fig. 7.4 Electrical potential gradients at a planar clay surface (after van Olphen, 1977).

1 A plane of counter-ions (cations) of finite size located less than 1 nm from the surface – the Stern layer, across which the electrical potential decays linearly; and
2 a diffuse layer of cations, across which the potential falls approximately exponentially, the inner surface of which abuts onto the Stern layer (Fig. 7.4). Allowance is made in the Stern model for a decrease in the dielectric constant of water due to its ordered structure very near the surface, and for specific adsorption forces between the surface and Stern layer cations (Box 7.2).

The Stern model predicts that an increase in ion concentration in solution not only compresses the double layer, but also increases the proportion of ions that reside in the Stern layer. The diffuse layer potential (ϕ_δ) decreases relative to the total surface potential (ϕ_o) and the repulsion of anions from the inner region of the diffuse layer is diminished. Cations in the diffuse part of the DDL are said to be non-specifically adsorbed.

Box 7.2 Specific adsorption of cations.

At a distance from the surface < 0.4 nm, the assumption of a plane of uniform surface charge does not hold and cations in the Stern layer experience a polarization force in the vicinity of each charged site. Depending on the balance of forces – the polarizing effect of the cation on the water molecules in its hydration shell compared with the polarization of the cation by the surface charge – the cation may shed its water of hydration to enter the Stern layer where it forms an inner sphere (IS) complex (Section 2.3) with surface groups of a siloxane trigonal cavity. (The plane of O atoms in the surface of a 2 : 1 phyllosilicate is called a siloxane surface). The cations most likely to do this belong to elements of Group I in the Periodic Table. These elements Li, Na, K, Rb and Cs form a monovalent cation series in which the ionic potential and hence hydration energy decreases with increasing atomic size. The small cation Li$^+$ tends to remain hydrated at the surface whereas the large Cs$^+$ ion tends to dehydrate and become strongly adsorbed.

The attractive force between the charged surface and solution cations, over and above the simple electrostatic force, is called a specific adsorption force and the cations are said to be specifically adsorbed. Cations held in the Stern layer are continually exchanging with cations in the diffuse layer, but at any instant the proportion of cations in the Stern layer is about 16% for Li$^+$, 36% for Na$^+$ and 49% for K$^+$ and the overall strength of cation adsorption increases in the order Li$^+$ < Na$^+$ < K$^+$ < Rb$^+$ < Cs$^+$.

Divalent cations formed by the Group II elements Mg, Ca, Sr, Ba and Ra can also form IS complexes at clay surfaces, but because of their much higher ionic potentials (and hydration energies) they tend to retain some of their hydration water. For example, when Ba^{2+} is adsorbed on vermiculite, a monolayer of water is retained between the crystal layers; but K$^+$, which has an almost identical ionic radius to Ba^{2+}, is adsorbed without any hydration water. The retention of hydration water results in the divalent

Box 7.2 *continued*

cations forming outer sphere (OS) complexes with siloxane surfaces (one layer of water molecules between the surface and the cation), which is the basis for clay quasi-crystal formation when Ca^{2+} is the predominant cation (Section 4.4). The strength of adsorption of the divalent cations of Group II increases in the order $Mg^{2+} < Ca^{2+} < Sr^{2+} < Ba^{2+} < Ra^{2+}$.

When these specific adsorption effects are accounted for, the net charge density on a clay particle (σ_p) may be written as

$$\sigma_p = \sigma_{in} + \sigma_{IS} + \sigma_{OS}, \tag{B7.2.1}$$

where σ_{IS} is the charge density due to ions forming IS complexes (other than H^+ and OH^-) and σ_{OS} is the charge density due to OS complexes.

Following from Equation 7.1, Equation B7.2.1 may be written

$$\sigma_p = \sigma_o + \sigma_H + \sigma_{IS} + \sigma_{OS}. \tag{B7.2.2}$$

For clays, σ_p is almost invariably negative and the balance of charge for the particle and its whole ionic atmosphere is made up by σ_d, such that

$$\sigma_p + \sigma_d = 0, \tag{B7.2.3}$$

where σ_d is the diffuse layer charge density comprising the surplus of cations *not* complexed with the surface, over the uncomplexed anions, according to Equation 7.3. Equation B7.2.3 is a statement of the charge balance for the surface and its adjacent solution.

Even the Stern model of the DDL is incomplete, because although inner sphere (IS) complexes (Stern layer cations) are accounted for, outer sphere (OS) complexes are not (Box 7.2). This shortcoming has led to the development of models based on the molecular properties of the interface (surface complexation models), as described by Sposito (1994).

Surfaces of variable charge

Oxide surfaces

Free oxides in soil (mainly Fe_2O_3, Al_2O_3 and MnO_2) exhibit pH-dependent charges due to the reversible adsorption of potential-determining H^+ ions. Of these oxides, the Fe and Al oxides can

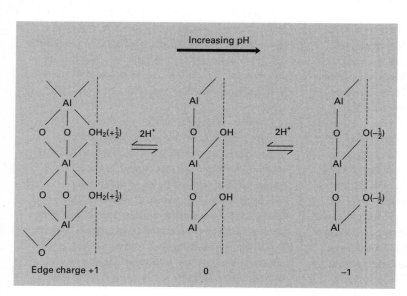

Fig. 7.5 Changes in surface charge on a hydrated Al_2O_3 edge face.

be abundant in many soils and quantitatively are the most important bearers of variable charge. The change in surface charge with pH change for a hydrated Al_2O_3 crystal is illustrated in Fig. 7.5. The reactive sites are OH groups that are co-ordinated to only one Al atom as shown and can change from O^- to OH to OH_2^+, depending on the acid strength of the site and the solution pH (see Box 3.6 for an explanation of 'acid strength'). Naturally, when the surface has a net positive charge, anions are attracted from the solution to form a double layer. The solution pH at which the surface charge is completely balanced by counter-ions or co-ions forming IS and OS complexes, so that the diffuse layer charge is zero, defines the point of zero charge (PZC) of the solid (Box 7.3).

Clay mineral edge faces

At the edge faces of clay crystals, the exposed O and OH groups are singly coordinated to Al and Si atoms in the alumina and silica sheets. Because the local structure differs from that of the free oxides, the acid strength of the O and OH groups differs also. The full negative charge of -2 per $0.4\,nm^2$ of edge area is developed only above pH 9 when the edge $\equiv$ Si-OH groups completely dissociate (see Fig. 2.16). The association of H^+ with the O and OH coordinated to Si and Al increases as the pH decreases from 9 to 5 in the manner illustrated in Fig. 7.6, the PZC of the edge face occurring $c.$ pH 6 for kaolinite. Thus, the edge faces of the clay minerals show a pH-dependent charge.

Organic matter

Although the charge of humified organic matter is completely dependent on pH due to the dissociation of carboxylic and phenolic groups, the charge is negative from pH 3 upwards and so augments the permanent negative charge of the

---

Box 7.3 Variably charged surfaces and the point of zero charge.

The PZC is the pH at which the net particle charge σ_p in Equation B7.2.3 is zero. If there is no net particle charge to be neutralized by ions in the diffuse layer, then σ_d must also be zero. It follows that when $\sigma_p = 0$, Equation B7.2.2 can be rearranged to give

$$\sigma_H = -\sigma_o - (\sigma_{IS} + \sigma_{OS}) \quad (pH = PZC). \quad (B7.3.1)$$

The value of σ_H changes with the pH of the solution in contact with the surface (pH_s), and the potential of such a surface (H^+ being the potential-determining ion) is related to the difference between PZC and pH_s by the Nernst equation

$$\phi_o = \frac{2.303RT}{zF}(PZC - pH_s). \quad (B7.3.2)$$

Thus, as pH_s decreases, the net proton surface charge σ_H increases and ϕ_o becomes more positive.

In soil, various inorganic cations and anions can form IS and OS complexes with oxide surfaces. The effect of these complexes on an oxide's PZC is resolved as follows. Noting that σ_o is a constant, we may rearrange Equation B7.3.1 to give

$$\sigma_H + \sigma_{IS} + \sigma_{OS} + \sigma_o(\text{constant}) = 0. \quad (B7.3.3)$$

Thus, we see that if the sum ($\sigma_{IS} + \sigma_{OS}$) in Equation B7.3.3 increases through net cation adsorption, σ_H must decrease and the pH at the PZC will increase (the acid strength of the surface weakens). Conversely, if ($\sigma_{IS} + \sigma_{OS}$) decreases through net anion adsorption, σ_H must increase and the pH at the PZC decreases. Thus, the formation of surface complexes changes the PZC of a pure mineral adsorbent in the same direction (+ or −) as the change in net surface charge through specific ion adsorption ($\sigma_{IS} + \sigma_{OS}$). This result has implications for the retention of nutrient ions and for particle to particle interactions (Section 7.4). The PZC for the common oxides found in soil ranges from 2–3 for silica, 7–8 for goethite and 8–9 for gibbsite, but the actual PZC of goethite and gibbsite can be $c.$ 1 pH unit lower because of the specific adsorption of oxyanions such as $H_2PO_4^-$.

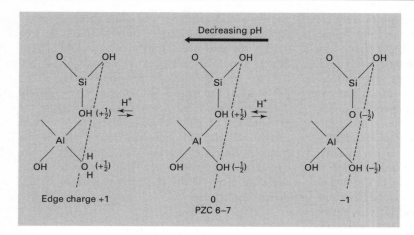

Fig. 7.6 Charge development at the edge face of kaolinite with change in pH.

crystalline clay minerals. In richly organic soils, the organic matter charge has the effect of considerably increasing the pH-dependence of the soil's cation exchange capacity, CEC. Charges on soil organic matter are discussed more fully in Section 3.4.

The significance of pH-dependent charges in soil

Edge charges are of greatest significance in the kaolinites, which have relatively few permanent charges (5–25 cmol charge/kg) and large edge : planar area ratios (1 : 10 to 1 : 5). For a kaolinite edge face area of 1–10 m²/g, the development of one positive charge amounts to 0.4–4 cmol charge/kg. Conversely, edge charges are less important in the 2 : 1 clay minerals that have high permanent negative charges (40–150 cmol charge/kg) and exist as smaller crystals, especially in the *c* direction, thereby offering a smaller edge : planar ratio than the kaolinites. The charge characteristics of soil clays may, however, be considerably modified by the presence of sesquioxides in variable amounts and spatial distribution.

The formation and occurrence of Fe and Al oxides are discussed in Section 2.4. With PZC values in the range pH 7–9, they are normally positively charged in soil and can occur as thin films adsorbed on the planar surfaces of clays, as illustrated by the Fe(OH)$_3$ film precipitated on kaolinite at pH 3 (Fig. 7.7). Al(OH)$_3$ films occur preferentially within the interlayer spaces of mica-

type clay minerals to form mixed layer minerals. Positive charge densities of 30–50 cmol charge/kg have been recorded for soil oxides, so that soils poor in organic matter that contain mainly

Fig. 7.7 Fe(OH)$_3$ deposit on a kaolinite cleavage face (courtesy of D. J. Greenland).

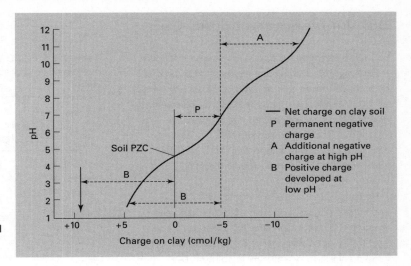

Fig. 7.8 A pH vs surface charge curve for an acidic red loam (Oxisol (ST) or Ferrosol (ASC)) (after Schofield, 1939).

kaolinite and sesquioxides can develop a net positive charge at low pH. Such behaviour is prevalent in highly weathered soils of the tropics, as exemplified by an acidic subsoil from South Africa, the pH *vs* charge curve of which is illustrated in Fig. 7.8. The presence of significant positive charge densities, wholly dependent on pH, has profound implications not only for the ability of the soil to retain cations, but also for the adsorption of anions, two topics that are discussed in the following sections.

7.2 Cation exchange

Exchangeable cations

A large proportion of the cations released by weathering and organic decomposition is adsorbed by clay and organic colloids, leaving a small concentration in the soil solution to be balanced by inorganic anions, and bicarbonate (HCO_3^-) generated from the respiration of soil organisms. These cations can be adsorbed as IS and OS complexes, and in the diffuse layer (DL) (Box 7.2). The IS and OS cations are immobilized on the surface for long times relative to the time taken for a DL cation to diffuse to the surrounding aqueous solution (Sposito 2000). However, as the external electrolyte concentration increases, the

distinction between DL and OS cations becomes less clear, so both categories of adsorbed cation are said to be exchangeable because they can readily exchange with cations in the bulk solution. In all but the most acidic or alkaline soils, the major exchangeable cations are Ca^{2+}, Mg^{2+}, K^+ and Na^+, in typical proportions of 80% Ca^{2+}, 15% Mg^{2+}, 5% ($Na^+ + K^+$) with small variable amounts of NH_4^+ substituting for any of these cations, depending on the extent of microbial nitrification (Section 8.3). Trace amounts of other cations such as Cu^{2+}, Mn^{2+} and Zn^{2+} are also present, although chelation with organic compounds and precipitation are the dominant mechanisms for the retention of these elements (Section 10.5).

Cation exchange capacity and 'base saturation'

Cation exchange capacity (CEC) is an expression of the negative charge per unit mass of soil (Section 2.5). However, CEC is measured by the amount of exchangeable cations (or a replacement 'index' cation) held per unit mass of soil, so it is an operationally defined quantity dependent on the conditions of measurement. Several methods have been developed to measure CEC that may give different results for the same soil, especially if some of the soil components have appreciable

Box 7.4 Measurement of soil CEC.

Soil CEC methods fall into two groups as follows:
• Exchangeable cations are displaced at a fixed pH by a concentrated salt solution, such as 1 M NH_4Cl or 1 M NH_4OOCH_3, both at pH.7. Displacement is effected by mass action (see cation exchange reactions). The 'index' cation (NH_4^+) is then displaced by another cation (e.g. Na^+) and the amount of NH_4^+ adsorbed (cmol charge/kg) is measured.
• Exchangeable cations are displaced by an unbuffered salt solution at no fixed pH and the amount of index cation that has been adsorbed (e.g. Ba^{2+} from a mixture of 0.1 M $BaCl_2$ and 0.1 M NH_4Cl solution), or remains in solution (Ag^+ from 0.01 M silver thiourea solution) is measured. Displacement is effected because of the high affinity of the index cation for the surface.

A pre-wash with aqueous ethanol may be used to remove soluble salts. The exchangeable cations are measured in the soil extracts (with or without a pre-wash); Al^{3+} and H^+ are usually measured separately by displacement in 1 M KCl.

CEC values determined by methods of the first group are widely reported in soil survey data and are used in some soil classifications. Representative values for several soil orders are given in Table B7.4.1. In soils of variable charge, CEC measured at

Table B7.4.1 CEC values of surface soils representative of several soil orders. After Sposito, 2000.

Soil order	CEC (cmol charge/ kg soil)	Soil order	CEC (cmol charge/ kg soil)
Alfisols (ST) or Chromosols/ Kurosols (ASC)	12	Mollisols (ST) or Dermosols (ASC)	22
Spodosols (ST) or Podosols (ASC)	11	Oxisols (ST) or Ferrosols (ASC)	5
Ultisols (ST) or Kurosols (ASC)	6	Vertisols (ST) or Vertosols (ASC)	37
Histosols (ST) or Organosols (ASC)	140	Aridosols (ST) or Tenosols/ Calcarosols (ASC)	16

pH 7 will over-estimate the effective CEC (ECEC) – that is, the CEC of the soil at its natural pH. For such soils, methods of the second group are preferred. Details of all these methods are given in specialist texts, such as Rayment and Higginson (1992) and Sumner and Miller (1996).

---

variable charge. Soluble salts can also inflate the results. For these reasons, it is most important to specify the method of measuring CEC (Box 7.4).

In calcareous soils, the sum of the cations ($\sum Ca^{2+}$, Mg^{2+}, K^+, Na^+) is invariably equal to the CEC because any deficit of cations on the exchange surfaces can be made up by Ca^{2+} ions from the dissolution of $CaCO_3$. In non-calcareous soils, however, $\sum Ca^{2+}$, Mg^{2+}, K^+, Na^+ is frequently less than the CEC, the difference being referred to as the exchange acidity (expressed as cmol H^+/kg soil). The ratio

$$\frac{\sum Ca^{2+}, Mg^{2+}, K^+, Na^+}{CEC} \times 100$$

has been called the per cent base saturation. The cations involved are called 'basic' simply to distinguish them from cations such as Al^{3+} that hydrolyse to release H^+ ions at comparatively low pHs (see below). Base saturation is crudely correlated with soil pH in many mildly acidic to neutral soils, ranging from c. 20 to >60% as the pH increases from 5 to >7.

'Acid clays' are Al-clays

Exchange acidity is not primarily due to H^+ ions. As the soil pH decreases below 5, increasing amounts of Al^{3+} ions can be displaced by leaching with strong salt solutions. Under acidic conditions

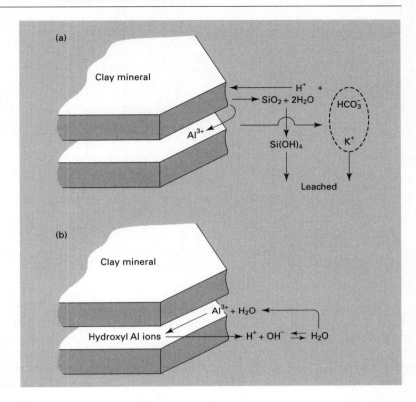

Fig. 7.9 (a) Primary processes in the weathering of a clay mineral in an acidic solution. (b) Secondary processes in acid clay weathering.

the weathering of clay minerals is accelerated, with the release of Al, SiO_2 and smaller amounts of Mg, K and Fe, and traces of Mn (Fig. 7.9). While the weak silicic acid ($Si(OH)_4$) formed is removed by leaching, Al, Mg, K and Mn are retained initially as the exchangeable cations Al^{3+}, Mg^{2+}, K^+ and Mn^{2+}. Hydrated ferric ions $(Fe(H_2O)_6)^{3+}$ are rapidly hydrolysed at pH < 3 and the Fe precipitated, in the pH range 4–5, as ferric hydroxide ($Fe(OH)_3$) or as an insoluble carbonate or phosphate compound, if the appropriate anions are present. Manganese precipitates as insoluble MnO_2 and, with the exception of illitic clay soils, K^+ is lost by leaching, so the result is an 'acid clay' dominated by exchangeable Al^{3+} with some Mg^{2+} (and H^+ ions produced by the hydrolysis of exchangeable Al^{3+}).

pH buffering capacity and titratable acidity

Large concentrations of exchangeable Al^{3+} greatly increase a soil's capacity to neutralize OH^- ions,

which is a measure of the soil's pH buffering capacity (buffering capacity is defined in Box 3.6). A hydrated Al^{3+} ion hydrolyses* in water according to the reaction

$$[Al(H_2O)_6]^{3+} + H_2O \leftrightarrow [Al(OH)(H_2O)_5]^{2+} + H_3O^+, \quad (7.6)$$

there being equal concentrations of the di- and trivalent Al ions at pH 5. The hydroxy-aluminium ions – abbreviated to $(AlOH)^{2+}$ – show a pronounced tendency to combine through bridging OH groups and build up into polymeric units of six and more Al atoms. The consequent release of H^+ contributes to the buffering capacity until eventually neutralization is complete and amorphous $Al(OH)_3$ precipitates on the external surfaces and in the interlayer spaces of the clay.

---

* Hydrolysis means 'the splitting of a water molecule'; the water molecule dissociates into H^+ and OH^- ions.

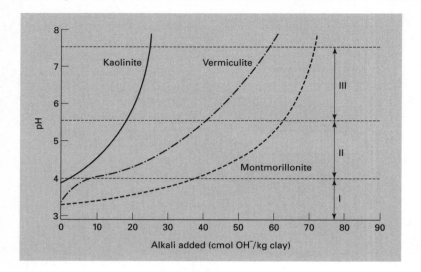

Fig. 7.10 pH–titration curves for kaolinite, vermiculite and montmorillonite in 0.1 M KCl.

Three stages are recognized in the neutralization of acid clays. These are illustrated in Fig. 7.10, where representative curves are shown for kaolinite, vermiculite and montmorillonite clays titrated in a neutral salt solution. Provided that sufficient time is allowed for near-equilibrium pH values to be attained, the three clays show markedly different buffering within the following ranges:

I $< pH\ 4$ exchangeable H^+ neutralized

II $pH\ 4$–5.5   exchangeable Al^{3+} hydrolysed

III $pH\ 5.5$–7.5   hydroxyl-Al ions hydrolysed.

Some 'weak' acidity due to H^+ ions released from hydroxy-Al polymers continues to be neutralized at $pH > 7.5$. The quantity of acid neutralized when a soil is equilibrated with a 0.1 M $BaCl_2$/triethanolamine solution at pH 8.2 is a measure of its titratable acidity (cmol H^+/kg). In practice, titration of a soil to a specific pH is used to estimate a soil's lime requirement (Section 11.3). In soils with much interlayer and surface hydroxy-Al coatings, the titratable acidity is always greater than the exchange acidity because part of the CEC is blocked by the hydroxy-Al polymers, which are not displaced in neutral salt solutions.

Aluminium is extensively hydrolysed when adsorbed by humified organic matter, the average charge per mole of adsorbed Al being +1. The Al^{3+} ions react preferentially with the stronger acidic groups on the organic matter, and H^+ ions

released by hydrolysis suppress the dissociation of weaker acidic groups. The apparent pH of Al-free organic matter is between 4 and 5, but rises to *c*. 6 when Al is added. Thus, the adsorption of Al decreases the acidity of soil organic matter and increases its capacity to adsorb anions such as phosphate (Section 7.3).

Cation exchange reactions

Exchange equations

The technique of using a strong solution of NH_4^+ ions to displace a soil's exchangeable cations and determine the CEC is an example of cation exchange. Cation exchange is a reversible process in which one mole of cation charge (or equivalent)* in solution replaces one mole of cation charge on the exchanger (clays and organic matter), as for example

$$(NH_4^+) + (Ca_{exchanger}) \leftrightarrow (NH_{4exchanger}) + \tfrac{1}{2}(Ca^{2+}),$$
$$(7.7)$$

where (NH_4^+) and (Ca^{2+}) are the activities of the ions in solution, and $(Ca_{exchanger})$ and $(NH_{4exchanger})$

---

* See Box 2.3 for an explanation of moles of charge and equivalents.

represent the activities of the ions on the exchanger. Following the law of mass action, the equilibrium constant for this exchange reaction can be written as

$$K_{ex} = \frac{(NH_{4exchanger})(Ca^{2+})^{1/2}}{(Ca_{exchanger})(NH_4^+)}. \qquad (7.8)$$

Much has been written about the validity of cation exchange equations, which all have to deal with the problem of measuring ion activities in the adsorbed state. One solution is to express an ion activity on the exchanger in fractional moles of charge, that is, moles of cation charge divided by the CEC (moles charge (+)/kg). If concentrations are substituted for ion activities in solution, Equation 7.8 can be rearranged to give the empirical Gapon equation

$$\frac{NH_{4exchanger}}{Ca_{exchanger}} = K_{NH_4,Ca} \frac{[NH_4^+]}{[Ca^{2+}]^{1/2}}. \qquad (7.9)$$

$K_{NH_4,Ca}$ is called the Gapon coefficient and square brackets indicate concentrations (mol/L). This equation assumes a limited range for the distribution of adsorption site affinities for the cations (Sposito, 2000), but has proved useful in interpreting some of the practical outcomes of cation exchange reactions in soils. For example, provided that the monovalent cation (Na^+ or K^+) amounts to < 25% of the exchangeable Ca^{2+} and Mg^{2+} combined, reasonably constant values of the Gapon coefficients $K_{Na–Ca,Mg}$ and $K_{K–Ca,Mg}$ are obtained for soils of similar clay mineral type from the equations

$$\frac{Na_{exchanger}}{Ca,Mg_{exchanger}} = K_{Na–Ca,Mg} \frac{[Na^+]}{[Ca^{2+} + Mg^{2+}]^{1/2}} \quad (7.10)$$

and

$$\frac{K_{exchanger}}{Ca,Mg_{exchanger}} = K_{K–Ca,Mg} \frac{[K^+]}{[Ca^{2+} + Mg^{2+}]^{1/2}}. \quad (7.11)$$

(Ca^{2+} and Mg^{2+} have sufficiently similar adsorption affinities, relative to the monovalent cations, to be regarded as interchangeable). Cation exchange equations can describe how the proportions of adsorbed cations respond to changes in concen-

tration in the soil solution, or alternatively, how well the adsorbed cations buffer changes in soil solution concentrations caused by leaching or plant uptake.

An equation similar to Equation 7.10 has been of great value in predicting the likely 'sodium hazard' of irrigation waters (Section 13.2), and Equation 7.11 has been applied to the study of potassium availability in soil.

Soil K availability

Equation 7.11 may be rewritten as

$$\frac{exchangeable\ K}{CEC - exchangeable\ K} = K' \frac{K}{\sqrt{Ca + Mg}} \quad (7.12)$$

where the quantities on the left side are in cmol charge/kg soil, and K' is the Gapon constant. When exchangeable K is small (< 10% of the CEC), Equation 7.12 reduces to the simple form

$$exchangeable\ K = K'' \frac{K}{\sqrt{Ca + Mg}}, \qquad (7.13)$$

where K'' includes the Gapon constant and the CEC of soil. Usually the change in exchangeable K (ΔK) is plotted against the activity ratio (AR) of $K/(Ca + Mg)^{1/2}$ to give a quantity/intensity (Q/I) relation for the soil's exchangeable potassium (Fig. 7.11). However, the reactions of K in soil are complex, especially in soils containing much partially-expanded clay mineral. As described in Section 2.4, K^+ ions form IS complexes in the unexpanded interlayer spaces of illite and micaceous clays and are therefore non-exchangeable. (NH_4^+, having a similar ionic potential to K^+, may also be held as a non-exchangeable cation in this way.) However, with weathering and the depletion of K in solution, the clay crystals slowly exfoliate from the edges exposing the interlayer K to exchange by other cations in solution (Fig. 7.12). The exposed interlayer sites at the crystal edges (which have been called 'wedge-shaped' sites) have a higher affinity for K^+ than the planar surfaces, so that the whole crystal exhibits a changing affinity for K^+ relative to Ca^{2+} and Mg^{2+}, as first the edge sites and then the planar surface sites are occupied. This change is reflected in a gradual

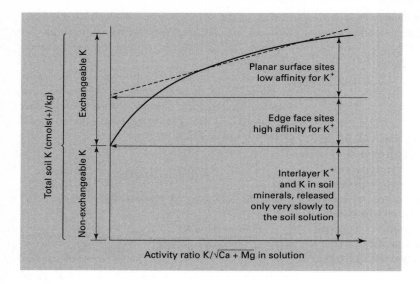

Fig. 7.11 Schematic Q/I relationship for K in a micaceous clay soil (after Beckett, 1971).

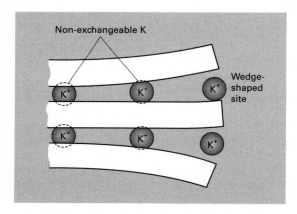

Fig. 7.12 The weathering edge of a micaceous clay crystal.

decrease in the slope of the Q/I plot, and hence in the soil's buffering capacity for K, going from small to large AR values.

The ratio law

Gapon Equation 7.11 indicates if the proportion of K^+ to $(Ca^{2+} + Mg^{2+})$ on the exchanger is constant, the ratio $K^+/\sqrt{Ca^{2+} + Mg^{2+}}$ in solution is also constant. Schofield (1947, p. 257) was the

first to formulate such observations into a general hypothesis called the ratio law which states that

> when cations in solution are in equilibrium with a larger number of exchangeable cations, a change in the concentration of the solution will not disturb the equilibrium if the concentrations of all the monovalent ions are changed in the one ratio, those of all the divalent ions in the square of that ratio, and those of all the trivalent ions in the cube of that ratio.

For 'concentration' one may also read 'activity'. The constancy of cation activity ratios depends on the effective exclusion of the accompanying anions such as Cl^- from the adsorption sites, since only then will the activities of the exchangeable cations be unaffected by changes in the Cl^- concentration of the bulk solution. Chloride exclusion is determined by the thickness of the double layer (Equation 7.5), so that the ratio law is valid only if:
• There is a preponderance (> 80%) of negative over positive charge on the surface;
• the bulk solution concentration is not too high.

As theory predicts, the upper limit of concentration for which the ratio law applies varies with the valency of the cation, being approximately 0.1 M for Na^+ and K^+, 0.01 M for Ca^{2+} and 0.002 M for Al^{3+}. Despite these limitations, the

Box 7.5 Measurement of soil pH.

Soil pH is commonly measured after 10 g air-dry soil are shaken in 50 mL of distilled water and the tips of a glass electrode and calomel reference electrode are inserted into the supernatant solution. For soils containing predominantly negatively charged clays, dilution of the soil solution by distilled water increases the absolute value of the surface potential and changes the distribution of H^+ ions between the DDL and the bulk solution. The proportion of H^+ ions in the DDL relative to the bulk solution increases so that the measured pH, which is the bulk solution pH, is higher than that of the natural soil. This effect is especially noticeable in saline soils. An alternative method is to shake the soil with 0.01 M $CaCl_2$ solution until equilibrium is attained. This solution, containing the most abundant exchangeable cation in many soils, also has an ionic strength I approximating that of the soil solution. In this soil suspension, both H^+ and Ca^{2+} exchange with other cations in the DDL, but the ratio of $H^+/\sqrt{Ca^{2+}}$ in solution remains relatively constant, even when the ratio of soil to equilibrating liquid changes. Thus, pH measured in 0.01 M $CaCl_2$ is a more accurate reflection of the natural soil pH, as is demonstrated by the comparison between soil pH measured in water, in 0.01 M $CaCl_2$, and in a solution that has been shaken with successive samples of the soil to achieve equilibrium (Table B7.5.1).

Many soil analysis laboratories, particularly in Australia, have now adopted pH measured in 0.01 M $CaCl_2$ at a soil : liquid ratio of 1 : 5 as the preferred method of pH measurement. Because $pH(CaCl_2)$ is generally 0.6–0.8 units lower than $pH(H_2O)$ at the same soil : liquid ratio, the solution in which soil pH is measured should always be specified.

Table B7.5.1 pH measurements on a basaltic red loam. After White, 1969.

Soil : liquid ratio (g/mL)	pH in H_2O	pH in 0.01 M $CaCl_2$	pH in an equilibrium solution*
1 : 2	5.08	4.45	4.45
1 : 5	5.29	4.45	4.45
1 : 10	5.43	4.46	4.45
1 : 50	5.72	4.52	4.45

* 0.01 M $CaCl_2$ solution after successive equilibrations with four separate samples of the soil.

ratio law is useful in interpreting ion exchange reactions in soils, such as those occurring during leaching by rainwater or when saline water is used for irrigation (Section 13.2). Another example concerns the measurement of soil pH (Box 7.5).

7.3 Anion adsorption and exchange

Adsorption sites and exchange mechanisms

Protonated hydroxyl groups on the edge faces of crystalline clay minerals, or on allophane and sesquioxide surfaces, can adsorb anions from solution according to the reaction

$$M - OH_2^+ + A^- \leftrightarrow M - OH_2^+ A^-, \qquad (7.14)$$

where M is a metal cation (e.g. Al^{3+} or Fe^{3+}). The anion A^- may exchange with other anions in the soil solution, in accordance with the ratio law. For example, the ratio of adsorbed $H_2PO_4^-$ to SO_4^{2-} will vary with the activity ratio $H_2PO_4^-/\sqrt{SO_4^{2-}}$ in solution, provided that the concentration of cations at the surface is vanishingly small. However, when the positive sites are only isolated spots on planar surfaces of clay minerals, or the soil solution concentration is low enough for the diffuse layer at planar surfaces to envelop the positive sites at the edges, anion adsorption is determined by the characteristics of the double layer at the negatively charged surfaces (Fig. 7.13a). Because of the magnitude of the electrostatic force of repulsion between the surface negative charge and the anions, the SO_4^{2-} concentration at the inner surface of the diffuse layer is less than

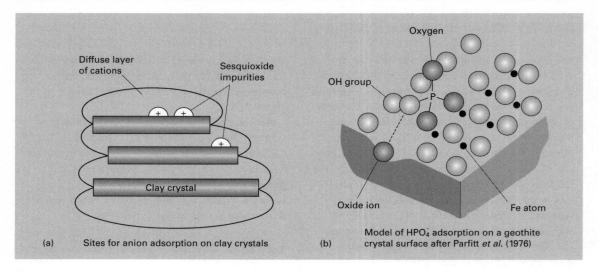

Fig. 7.13 (a) and (b) Physical models of the interactions between clays, sesquioxides and anions in solution.

the $H_2PO_4^-$ concentration, and as the soil solution becomes more dilute, the SO_4^{2-} concentration in the bulk solution increases relatively more than the $H_2PO_4^-$ concentration. Under these conditions the cation-anion activity products $(Ca^{2+}) \times (SO_4^{2-})$ and $(Ca^{2+}) \times (H_2PO_4^-)^2$ measured in the bulk solution should be constant.

Equation 7.14 is an example of simple anion exchange, or non-specific adsorption, and is typical of anions such as Cl^- and NO_3^-. The anion exchange capacity (AEC) of sesquioxide surfaces at pH 3.5 is in the range 30–50 cmol charge (–)/kg sesquioxide (Section 2.5), but the soil's AEC will generally be less than one-tenth of this value (Zhang and Zhao, 1997). Oxyanions such as $H_2PO_4^-$, HCO_3^-, SO_4^{2-}, MoO_4^{2-}, $H_3SiO_4^-$ and $B(OH)_4^-$ are primarily adsorbed by a ligand exchange mechanism in which an OH or OH_2^+ group in the mineral surface is displaced by an O of the oxyanion to form an IS complex – this is an example of specific adsorption. The reactive groups are charge-unsatisfied OH groups (i.e. OH coordinated to only one Al atom in the mineral – Section 7.1), which in gibbsite, for example, occur on the edge faces of the plate-like crystals. However, with poorly crystalline and amorphous oxides and allophanes the valence-unsatisfied OH groups are more widely distributed, so the extent of specific adsorption of anions such as $H_2PO_4^-$ is much greater.

Ligand exchange

Examples of two possible ligand exchange reactions between $H_2PO_4^-$ ions and a metal oxide surface are

$$M-OH + H_2PO_4^- \leftrightarrow M-O-\overset{\overset{\displaystyle OH}{|}}{\underset{\underset{\displaystyle OH}{|}}{P}} = O + OH^- \quad (7.15)$$

and

$$M-OH_2^+ + H_2PO_4^- \leftrightarrow M-O-\overset{\overset{\displaystyle OH}{|}}{\underset{\underset{\displaystyle OH}{|}}{P}} = O + H_2O. \quad (7.16)$$

In these reactions, the anion forms an ionic-covalent bond with the metal cation (M) by displacing either an OH or OH^{2+} group and entering into the metal's coordination shell. The anion is therefore very strongly adsorbed, a condition sometimes described as chemisorbed. Depending on the oxide and the ambient pH, both types of site M – OH and M – OH^{2+} occur on the surface, in which case when $H_2PO_4^-$ is adsorbed, the solution

Plate 1.4 Profile of an Alfisol (ST) or Chromosol (ASC) showing well-developed A, B and C horizons.

Plate 2.4 Boulders and stones covering the soil surface in a vineyard in the central Rhone Valley, France.

Plate 3.5 Basidomycete fungal colony on rotting wood.

Plate 3.10 Moist dark earthworm casts on the surface of a soil under pasture.

Plate 5.5 Deep, uniform, coarse-textured soil formed on granite in south-central Chile. The scale is 15 cm long.

Plate 5.9 'The 'mottled zone' of an ancient laterite (Oxisol (ST) or Ferrosol (ASC)) buried under more recent alluvium in southern New South Wales.

Plate 5.13 Litter layer colonized by fungi under Acacia and Eucalypt trees.

Plate 9.4 An Fe-humus podzol (Spodosol (ST) or Podosol (ASC)) developed on a well-drained parent material.

Plate B9.4.1 Red Brown Earth showing a sharp texture contrast between the A and B horizons.

Plate 9.11 A highly dissected laterite profile in the Kimberley region of northwest Australia.

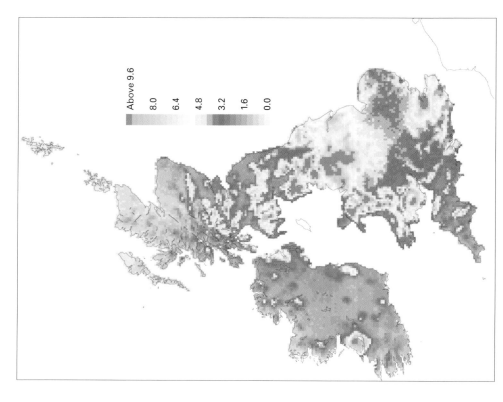

Plate B10.3.1(b) Total S deposition (kg/ha/yr) for the UK in 2010 predicted by the Hull acid rain model (HARM). Figure published with permission of the National Expert Group on Transboundary Air Pollution (2001).

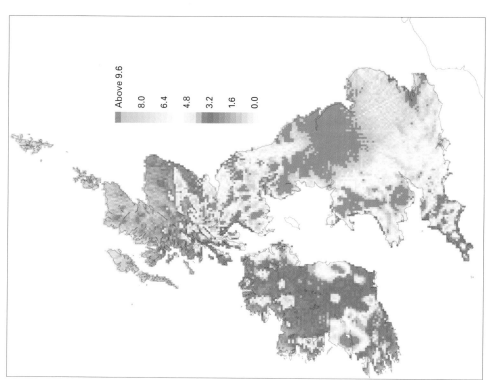

Plate B10.3.1(a) Total S deposition (kg/ha/yr) for the UK in 1999 predicted by the Hull acid rain model (HARM). Figure published with permission of the National Expert Group on Transboundary Air Pollution (2001).

Plate 11.3(a) Grapevine leaves showing Fe deficiency (courtesy of Scholefield Robinson Horticultural Services, Netherby, South Australia).

Plate 11.3(b) Leaves of *Glycine javanica* showing K deficiency (normal leaf in the lower-centre).

Plate 13.2 An area of dryland salinity in southwest Western Australia. Note the bare ground and salt-tolerant vegetation in the foreground.

Plate B15.2.1 A map of soil classes for a catchment in the Hunter Valley, Australia, based on a DEM and a biophysical model to predict soil distribution. Courtesy of the Australian Centre for Precision Agriculture, The University of Sydney.

pH can increase *and* the net positive charge on the surface decrease. The latter effect decreases the PZC of the surface.

Experimentally, decreases in surface charge from 1 to 0 have been observed per mole of anionic charge adsorbed on pure Fe and Al oxides. Studies of P adsorption on goethite using IR spectroscopy suggest that an HPO_4^{2-} ion displaces two singly coordinated OH groups of adjacent Fe atoms to form a binuclear HPO_4 bridging complex (Fig. 7.13b). Phosphate can also be adsorbed by ligand exchange on hydroxy-Al polymers that form on clay surfaces between pH 4.5 and 7. Such polymers may constitute the positively charged sites on planar surfaces as shown in Fig. 7.13a. These compounds are most stable between pH 5 and 6.5, so that P adsorption in acid soils that have a predominance of phyllosilicates in the clay fraction often shows a maximum in this pH range (Section 10.3).

Ligand exchange is the principal mechanism by which $H_2PO_4^-$, MoO_4^{2-}, $H_3SiO_4^-$ and $B(OH)_4^-$ are adsorbed, up to large amounts well in excess of a soil's simple AEC. HCO_3^- and SO_4^{2-} are intermediate in that they can be adsorbed by this mechanism, or by simple electrostatic attraction as in Equation 7.14. Phosphate and other anions non-specifically adsorbed at positive sites on oxide surfaces can be desorbed by decreasing the ion's concentration in the ambient solution, but anions coordinated to the surface can be desorbed only by raising the pH or introducing another anion with a greater affinity for the metal of the oxide. Consequently, the adsorption of oxyanions such as $H_2PO_4^-$ on oxide surfaces is greater, the lower the solution pH.

7.4 Particle interaction and swelling

Attractive and repulsive forces

The interaction between clay-size particles in soil, especially their response to changes in soil water content, depends on the balance between attractive and repulsive forces at their surfaces. The attractive forces between clay particles are of two main kinds:
• *Electrostatic*, as between negatively charged planar faces and positively charged edges; or between

K^+ ions and the adjacent siloxane surfaces in the interlayers of illite crystals; or between partially hydrated Ca^{2+} and Mg^{2+} ions and the siloxane surfaces of adjacent vermiculite crystals;
• van der Waals' forces between individual atoms in two particles. For individual pairs of atoms the force decays with the sixth power of the inter-atomic distance, but the total force between two particles is the sum of all the inter-atomic forces which results not only in an appreciable force between macromolecules, but also in less rapid decay in force strength with distance.

The origin of the attractive forces is such that, apart from the effect of the exchangeable cation species, they vary little with changes in the solution phase. On the other hand, the repulsive forces that develop due to the interaction of diffuse layers are markedly dependent on the composition and concentration of the soil solution, as well as on the surface charge density of the clay. As shown in Section 7.1, the thickness of the diffuse layer is controlled by the charge on the counter-ion and the concentration of the bulk solution. The model of the double layer presented so far assumes no impediment to the diffuse distribution of ions at the charged surface – the concentration in the mid-plane between two clay crystals is equal to the concentration in the bulk solution. This is reasonably true of dilute clay suspensions in salt solutions, but in concentrated suspensions and in natural soils the particles are close enough for the diffuse layers to interact, when the balance between forces of attraction and repulsion determines the net attractive force between the particles. If the net force is attractive (Fig. 7.14a), the particles remain close together and are described as flocculated. It also follows that soil microaggregates, of which the clay particles form a part, will remain stable. Conversely, if the net force is repulsive (Fig. 7.14b) the particles move farther apart until a new equilibrium position is attained under the constraint of an externally applied force, or the particles move far enough apart to exist as separate entities in the solution – a state described as deflocculated.

The excess concentration of ions at the mid-plane between two particles where the diffuse layers overlap gives rise to an osmotic or swelling pressure between the two particles. Swelling pressures of $1–2 \times 10^3$ kPa may develop for particle

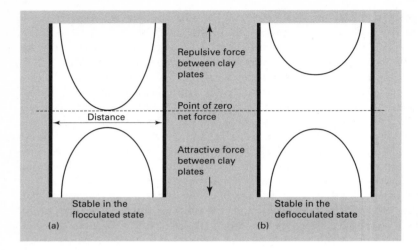

Fig. 7.14 (a) and (b) The interaction of attractive and repulsive forces between clay crystals in salt solutions.

separations from < 1 nm upwards, depending on the type of clay mineral, the interlayer cations and the concentration of the bulk solution.

This interpretation of clay interactions relies heavily on an understanding of colloid behaviour in dilute suspensions. Nevertheless, clays in soil are normally in very close contact and the water : solid ratio, even in wet soil, is much less than in a suspension. As a result, additional short-range forces dependent on the surface charge density of the clay, ion–ion correlations and the structure of water at surfaces come into play in natural soils. These have been reviewed by Quirk (1994) and are responsible for some unusually stable microstructures, as described below.

Flocculation–deflocculation and swelling behaviour

Face-to-face flocculation

The stacking of single layers of Ca-montmorillonite in roughly parallel alignment with much overlapping to form quasi-crystals, and of thicker crystals of Ca-illite or Ca-vermiculite to form domains, is briefly described in Section 4.4. Effective surface areas for Ca-montmorillonite of 100–150 m^2/g compared with 750 m^2/g for the completely dispersed clay suggest that each quasi-crystal consists of 5–8 layers (Fig. 7.15). Anions are totally excluded from the interlayer regions

Fig. 7.15 Face-to-face flocculation of Ca–saturated montmorillonite.

and a diffuse layer develops only at the external surfaces of the quasi-crystal. The interlayer cations appear to occupy a mid-plane position and provide a strong attractive force that stabilizes the interlayer spacing, irrespective of the external solution concentration.

Swelling behaviour

Except in atmospheres drier than *c.* 50% RH (Section 6.2), Ca-montmorillonite holds at least two layers of water molecules in the interlayer spaces due to the hydration of the cations. The *d* spacing is then 1.4–1.5 nm (Section 2.4). Additional water is taken up as the soil solution becomes more dilute, but only to the extent of expanding the *d* spacing to 1.9 nm, which is maintained even in distilled water. This behaviour is

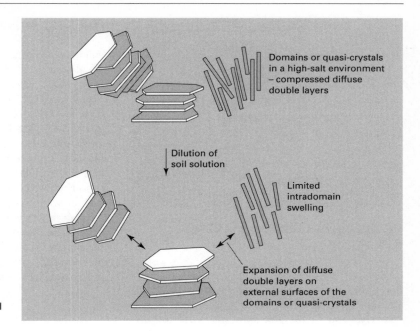

Fig. 7.16 Swelling responses of a Ca-saturated 2 : 1 clay domains or quasi-crystals to dilution of the soil solution.

typical of intracrystalline swelling in Ca-montmorillonite or intradomain swelling in Ca, Mg-illite or vermiculite.

As observed in Section 4.4, however, quasi-crystals and domains exhibit long-range order over distances up to 1 µm. Diffuse double layers at the outer surfaces of the quasi-crystals (and domains) lead to swelling and the creation of wedge-shaped spaces where water can be held by surface tension. Swelling of this kind, described as intercrystalline swelling, is illustrated in Fig. 7.16: it induces particle realignments that release the strains imposed by the crystal packing and bending during the previous drying cycle. Conversely, the large surface tension forces that develop during a drying phase contribute to the net attractive force holding the particles in close contact.

The swelling behaviour of montmorillonite and vermiculite is entirely changed when divalent cations are replaced by monovalent cations such as Na$^+$ and Li$^+$ (Table 7.2). For Na-montmorillonite, once the external salt concentration decreases below 0.3 M NaCl, repulsive forces prevail and the swelling pressure forces the layers apart to distances (up to 4 nm) that can be predicted from DDL theory. Swelling is a precursor to eventual

Table 7.2 Intracrystalline swelling of 2 : 1 clay minerals saturated with different cations. After Quirk, 1968.

Clay mineral	Moles of charge per unit cell	Exchangeable cation	Swelling in dilute solution (nm)
Mica	−2.0	K, Na	none
Vermiculite	−1.3	Ca, Mg, Na	1.4–1.5
		K	1.2
		Cs	1.2
		Li	>> 4
Montmorillonite	−0.67	Ca, Mg	1.9
		K	1.5 and >> 4
		Cs	1.2
		Na, Li	>> 4

deflocculation that occurs at solution concentrations ≤ 0.01 M in NaCl and even lower concentrations for Li$^+$.

Edge-to-face flocculation

Kaolinite crystals are larger than those of montmorillonite and the hydrous micas, and they

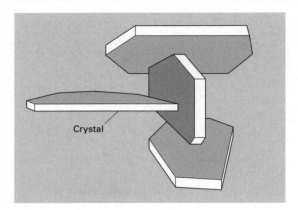

Fig. 7.17 Edge-to-face flocculation of kaolinite at low pH.

7.5 Clay–organic matter interactions

Much of the organic matter in soil is complexed in some way with crystalline clay minerals and free oxides. Knowledge of these interactions is therefore of great importance in understanding virtually every process – physicochemical and biological – that occurs in soil. Broadly, organic compounds may be divided into cationic, anionic and non-ionic compounds of small (< 1000 g) to large (> 10,0000 g) molar mass. Because of the bi-polarity of charge on clays at pH < 6, organic cations are adsorbed mainly on the planar faces and anions at the edges, except that small organic anions can bond with exchangeable cations on the planar faces. Uncharged organic molecules interact with the surfaces through polar groups and non-specific van der Waals' forces.

Electrostatic interactions

Cations

Many organic cations are formed by protonation of amine groups, as in the alkyl* amines ($R - NH_2$) and amino acids ($RCHNH_2COOH$). Protonation is enhanced by the greater acidity of water molecules hydrating metal cations adsorbed on clay surfaces. The greater the ionic potential of the cation M and its polarizing effect on the $M - OH_2$ bond formed with the hydrating water molecules, the greater is the tendency for a H^+ ion to be expelled from one of these molecules. The resultant protonation of a neutral organic molecule can be represented by

$$M(H_2O)_n^{m+} + R - NH_2 \leftrightarrow$$
$$MOH(H_2O)_{n-1}^{(m-1)+} + R - NH_3^+. \qquad (7.17)$$

On dry clay surfaces the polarizing effect of the cations is concentrated on fewer water molecules, so that the potential to dissociate is increased and the protonation of adsorbed organic molecules correspondingly enhanced. Any displacement of the metal cations by organic cations, such as the alkyl amines, decreases the affinity of the clay

flocculate face-to-face in concentrated solutions at high pH. Leaching and a decrease in the concentration of the soil solution cause the diffuse layers to expand, and the increased swelling pressure disrupts the face-to-face arrangement. If there is a concurrent decrease in soil pH, the edge faces become positively charged and the mutual attraction of double layers at planar and edge faces encourages edge-to-face flocculation, as shown in Fig. 7.17.

Of the two mechanisms, face-to-face flocculation is the norm for montmorillonitic and micaceous clays (2 : 1 clays), provided that di- and trivalent exchangeable cations predominate on the surfaces. In acidic soils, this kind of flocculation can be enhanced by the presence of hydroxy-Al polymers on the planar surfaces that act as an 'electrostatic bridge' between the crystals. However, edge-to-face flocculation is more common in kaolinites because of the positive edge charge at a pH < PZC for the edge faces (c. pH 6). At higher pHs (6–8), flocculation may be aided by positively charged sesquioxide particles (PZC of 7–9) forming 'bridges' between negatively charged edge faces. Edge-to-face flocculation is insensitive to changes in the solution concentration, but is weakened by the adsorption of oxyanions such as $H_2PO_4^-$ at kaolinite edge faces. In the same way, specific adsorption of phosphate and organic anions by sesquioxides may actually reverse the surface charge, rendering the oxide negatively charged and ineffective as a clay bridging agent.

---

* A hydrocarbon group such as CH_3, C_2H_5, . . . R.

surface for water, making the soil more water-repellent and diminishing the swelling propensity of the clay.

Organic cation adsorption takes place on the internal and external surfaces of 2 : 1 expanding clays. The adsorption affinity increases with the chain length of the compound due to van der Waals' forces between the adsorbed molecules (a co-adsorption phenomenon), so that adsorption in excess of the CEC may be achieved. Proteins adsorbed in interlayer spaces are less susceptible to microbial attack; conversely, extracellular enzymes, which are functional proteins, often show reduced biochemical activity in the adsorbed state.

Anions

Organic compounds are also adsorbed due to the electrostatic attraction of $-COO^-$ groups to positively charged sites on clay and oxide edge faces, and allophane surfaces. In addition, these carboxyl groups may undergo ligand exchange with OH and OH_2^+ groups in the surfaces, which provides a much stronger form of adsorption than the former.

Organic anions and uncharged polar groups can form IS and OS complexes with cations adsorbed on negatively charged organic or mineral surfaces. The former complex involves direct coordination of the anion with the metal cation, following the displacement of a hydrating water molecule (the cation-bridge shown in Fig. 7.18). The OS complex is a weaker bond formed between the anion (or polar molecule) and the metal cation without the displacement of a water molecule (the water-bridge

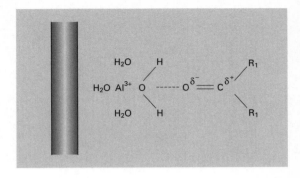

Fig. 7.19 Organic molecule forming a 'water-bridge' with an exchangeable cation.

of Fig. 7.19). Adsorption of organic molecules by cation-bridging is more likely when the exchangeable cations are monovalent (because these cations lose hydration water more easily), whereas water-bridging is favoured when the cations are di- or trivalent.

Phenolic compounds such as tannins, which are active constituents of the litter of many coniferous trees, form coordination complexes with adsorbed Fe and Al. The compound must contain at least three phenolic groups, of which two coordinate the metal cation and the third dissociates to confer a negative charge on the whole complex. Thus, these compounds in small concentrations can be effective deflocculants of clays. However, there are also long-chain polyanions that are attracted to clay particle edge faces, under acidic conditions, attaching to the particles in the manner of a 'string of beads' (Fig. 7.20). Under

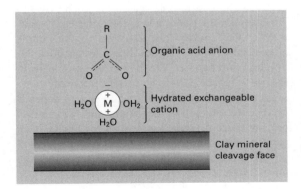

Fig. 7.18 Organic acid molecule forming a 'cation-bridge' (after Greenland, 1965).

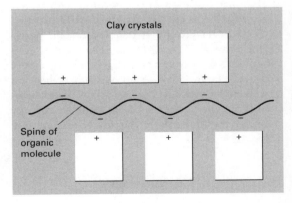

Fig. 7.20 'String of beads' arrangement of an organic polyanion and clay crystals.

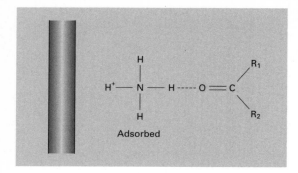

Fig. 7.21 Hydrogen-bonding between two organic compounds at a clay surface.

these conditions, these compounds are particularly effective flocculants of clays, and have been applied successfully, if expensively, as soil conditioners to improve soil structure (Section 11.4).

Van der Waals' forces and entropy stabilization

Alcohols, sugars and oligosaccharides of small molar mass are not adsorbed in competition with water unless present in high concentrations relative to the water. Very polar compounds can interact with clay surfaces by forming water-bridges, as discussed above, or by direct H-bonding to another organic compound that is already adsorbed (Fig. 7.21).

Alcohols (e.g. polyvinyl alcohol) and polysaccharides of large molar mass are strongly adsorbed at clay surfaces, even in competition with water. The attraction occurs partly through H-bonding to O and OH groups in the surface and partly through van der Waals' forces. The greatest contribution to the adsorption energy probably comes from the large increase in entropy when an organic macromolecule displaces many water molecules from the surface – the increase in entropy stabilizes the molecule in the adsorbed state (see Table 4.1).

7.6 Summary

Retention and exchange of cations and anions, soil acidity and pH buffering capacity, water adsorp-tion, swelling, flocculation and deflocculation of clay particles are some of the many soil properties determined by reactions at clay mineral, oxide and organic surfaces.

Apart from their large specific surface areas, clay mineral, oxide and organic surfaces are important because of the charge they bear, which is:
• Either a *permanent charge* (usually negative) due to isomorphous substitutions in the crystal lattice; or
• a *pH-dependent (variable) charge* due to the reversible dissociation of H^+ ions from OH and OH_2^+ groups at clay crystal edge faces and oxide surfaces, or from COOH and ⬡–OH groups of organic matter.

Planar clay surfaces give rise to an electrical double layer, comprising Stern layer cations more tightly adsorbed at the surface than the ions forming a diffuse layer extending out into the bulk solution. The diffuse layer results from a balance between a diffusive force between the region of high ion concentration near the charged surface and the lower concentration in the bulk solution, and the opposing electrostatic force of attraction. The thickness of this diffuse layer therefore varies with the cation charge and the bulk solution concentration.

Cations in the Stern layer lose their hydration shell and form inner-sphere (IS) complexes with the surface. The tendency to do this is greatest for monovalent cations whose overall strength of adsorption falls in the order $Cs^+ > Rb^+ > K^+ > Na^+ > Li^+$. Of higher hydration energy, divalent cations form mainly outer-sphere (OS) complexes in which one layer of water molecules is retained between the cation and the surface. The overall strength of adsorption falls in the order $Ra^{2+} > Ba^{2+} > Sr^{2+} > Ca^{2+} > Mg^{2+}$. In addition to the simple electrostatic force, dipole–dipole attraction, H-bonding and van der Waals' forces act on ions in the Stern layer, leading to specific adsorption.

Proton dissociation from surface groups depends on the acid strength of the groups and the solution pH. For example, sesquioxide surfaces change from net positively charged at low to neutral pH, to negatively charged at high pH. The pH at which the diffuse layer charge at any surface is zero defines the point of zero charge (PZC) of the surface.

Cation exchange capacity (CEC) is an operationally defined quantity that measures the net negative charge on a surface by replacement of the exchangeable cations with an 'index' cation. Some methods use a strong salt solution, such as M NH_4Cl at pH 7; others use an index cation of high affinity for the surface at a lower concentration (e.g. 0.1 M $BaCl_2$) at the soil pH. The former methods overestimate CEC in acidic soils with pH-dependent charge; the latter methods are preferred for such soils because they measure the effective CEC (ECEC). The difference between the CEC and the sum of exchangeable cations ΣCa^{2+}, Mg^{2+}, K^+, Na^+ defines the exchange acidity (comprising Al^{3+} and H^+). Ca^{2+}, Mg^{2+}, K^+ and Na^+ are referred to as 'basic' cations to distinguish them from the 'acidic' cations Al^{3+} and H^+. 'Acid clays' are primarily Al^{3+}-clays. Hydrolysis of exchangeable Al^{3+} and polymerization of the hydroxy-Al products release H^+ ions and account for much of the pH buffering capacity of a soil over the pH range 4–7.5. Titration of a soil to pH 8.2 gives a measure of its titratable acidity.

Exchange between cations in the double layer and the soil solution may be described by the Gapon equation, e.g.

$$\frac{K_{exchanger}}{Ca,Mg_{exchanger}} = K_{K-Ca,Mg} \frac{K^+}{\sqrt{Ca^{2+} + Mg^{2+}}}.$$

Changes in ion concentrations in solution that occur on leaching a soil, or in a soil influenced by saline water, may be explained by similar equations, the right-hand-side of which exemplify the ratio law. This law predicts that when cations in solution are in equilibrium with a larger number of exchangeable cations, the activity ratio $a^+/(a^{n+})^{1/n}$ remains constant over a limited range of solution concentration.

Inorganic and organic anions can be adsorbed at pH-dependent surface sites when these are positively charged (non-specific adsorption). However, oxyanions such as $H_2PO_4^-$, $H_3SiO_4^-$, MoO_4^{2-} and $B(OH)_4^-$ are primarily adsorbed by ligand exchange in which the O of the anion displaces an OH or OH_2^+ group from the coordination shell of a metal cation in the surface. This results in a very strong bond and the anion is said to be chemisorbed. Organic anions also interact with cations adsorbed at negatively charged surfaces by forming cation-bridges (an IS complex formed by displacing a hydrating water molecule) or water-bridges (an OS complex formed with the cation). Large organic molecules are held by van der Waals' forces and stabilized by an entropy increase from the large number of water molecules displaced.

The repulsive force developed as the diffuse layers of adjacent clay crystals overlap is manifest as a swelling pressure, which varies with water content and can disrupt the microstructure of clay domains. Partially hydrated Ca^{2+} ions forming IS complexes, located at the mid-plane between two opposing surfaces, predispose to especially stable quasi-crystals of montmorillonite or domains of vermiculite. Hydroxy-Al polymers can also act as electrostatic 'bridges' between planar clay surfaces in 2:1 clay minerals. These arrangements are examples of face-to-face flocculation of clay particles, which depends on the type of cation and the bulk solution concentration. On the other hand, crystals may also flocculate edge-to-face through positive and negative charge interaction, as in the case of kaolinite at pH < 6. Depending on the sign of the surface charge, organic cations and anions can be important in linking together clay minerals and sesquioxides and enhancing clay flocculation, a necessary precursor of stable microaggregate formation.

References

Atkins P. W. (1982) *Physical Chemistry*, 2nd edn. Oxford University Press, New York.

Beckett P. H. T. (1971) Potassium potentials – a review. *Potash Review*, Subject 5, pp. 1–41.

Greenland D. J. (1965) Interaction between clays and organic compounds in soils. 1. Mechanisms of interaction between clays and defined organic compounds. *Soils and Fertilizers* 28, 415–25.

Parfitt R. L., Russell J. D. & Farmer V. C. (1976) Confirmation of the surface structures of goethite (α-FeOOH) and phosphated geothite by infrared spectroscopy. *Journal of the Chemical Society, Faraday Transactions* 72, 1082–87.

Quirk J. P. (1968) Particle interaction and soil swelling. *Israel Journal of Chemistry* 69, 213–34.

Quirk J. P. (1994) Interparticle forces: a basis for the interpretation of soil physical behaviour. *Advances in Agronomy* 53, 121–83.

Rayment G. E. & Higginson F. R. (1992) *Australian Laboratory Handbook of Soil and Water Chemical*

Methods. Australian Soil and Land Survey Handbook. Inkata Press, Melbourne.

Schofield R. K. (1939) The electric charges on clay particles. International Society of Soil Science. British Empire Section, Cambridge, pp. 1–5.

Schofield R. K. (1947) A ratio law governing the equilibrium of cations in the soil solution. *Proceedings of the 11th International Congress of Pure and Applied Chemistry (London)* 3, 257–61.

Sposito G. (1989) *The Chemistry of Soils*. Oxford University Press, New York.

Sposito G. (1994) *Chemical Equilibria and Kinetics in Soils*. Oxford University Press, New York.

Sposito G. (2000) Ion exchange phenomena, in *Handbook of Soil Science* (Ed. M. E. Sumner). CRC Press, Boca Raton FL, B241–B263.

Sumner M. E. & Miller W. P. (1996) Cation exchange capacity and exchange coefficients, in *Methods of Soil Analysis. Part 3 Chemical Methods*. (Ed. D. L.

Sparks). Soil Science Society of America, Madison WI, pp. 1201–29.

Van Olphen H. (1977) *An Introduction to Clay Colloid Chemistry*, 2nd edn. Wiley Interscience. New York.

White R. E. (1969) On the measurement of soil pH. *Journal of the Australian Institute of Agricultural Science* 35, 3–14.

Zhang X. N. & Zhao A. Z. (1997) Surface charge, in *Chemistry of Variable Charge Soils* (Ed. T. R. Yu). Oxford University Press, New York, pp. 17–63.

Further reading

McBride M. B. (1994) *Environmental Chemistry of Soils*. Oxford University Press, New York.

Sparks D. L. (1995) *Environmental Soil Chemistry*. Academic Press, San Diego.

Zumdahl S. S. & Zumdahl S. A. (2003) *Chemistry*, 6th edn. Houghton Mifflin Company, Boston.

Example problems and questions

1 (a) What is the ratio of silica to alumina sheets in a crystal of (i) kaolinite, (ii) montmorillonite.

   (b) Does kaolinite have a large cation exchange capacity (CEC) compared with montmorillonite and illite (give a reason for your answer)?

   (c) (i) Identify the position of K^+ ions in the crystal structure of illite, and (ii) what happens to this K as the crystal weathers?

   (d) What other cation can substitute for K^+ in the crystal structure of illite?

2 (a) Distinguish between the origin of the permanent and variable charge on clay crystals.

   (b) Describe the change in charge on the edge faces of clay minerals as the soil solution pH changes from pH < 6 through the neutral range (6–7) to pH > 7.

   (c) How does an increase in pH of the soil solution affect the tendency of kaolinite clay to flocculate or deflocculate?

3 A 10 g sample of montmorillonite clay from a black earth was leached with 100 mL of M NH_4Cl to replace the existing exchangeable cations with NH_4^+ ions. The leachate was analysed for the displaced cations Ca^{2+}, Mg^{2+}, K^+ and Na^+. The results were as follows.

Cation	Leachate concn (mg/L)	mmol of cation in 100 mL of leachate	mmol of cation charge per g of clay	Exchangeable cation (cmol charge (+)/kg clay)
Ca^{2+}	1600			
Mg^{2+}	192			
K^+	156			
Na^+	69			

   (a) Complete the unfilled columns in this table (note – the molar masses of Ca, Mg, K and Na are 40, 24, 39 and 23 g, respectively).

   (b) Calculate the CEC of this clay on the assumption that the sum of the four cations is equal to the CEC.

4 (a) Write an equation for the first step in the hydrolysis of hydrated Al^{3+} ions.

   (b) A sample of vermiculite was leached with 0.1 M HCl to create a sample of H^+-saturated clay. After being washed to remove excess acid, the clay was titrated with 0.05 M KOH to a pH > 7.5. Identify the buffering ranges and the neutralization reactions that occur as the H^+-saturated clay is titrated.

(c) Suppose that 5 g H^+-saturated vermiculite, which has a CEC at pH 8.2 of 150 cmol charge/kg, was titrated with 0.05 M KOH. What volume of KOH solution would be required for complete neutralization of the clay?

5 (a) The pH of a Sodosol was measured in distilled water (pH_w) at a soil : solution ratio of 1 : 2 and 1 : 5. Indicate which ratio should give the higher pH reading.

(b) The pH of this soil was also measured in 0.01 M $CaCl_2$ solution at a 1 : 5 ratio (pH_{Ca}). Give an estimate of the likely difference between pH_w and pH_{Ca} measured at the 1 : 5 ratio.

(c) Calculate the ionic strength I of the 0.01 M $CaCl_2$ solution using the formula

$$I = 0.5 \sum_{i}^{n} C_i z_i^2.$$

6 A Ferrosol has a content of 10% sesquioxides, which have an AEC at the soil pH of 20 cmol charge/kg. If soluble P fertilizer is cultivated into the soil at the rate of 20 kg P/ha, and all the P is chemisorbed by sesquioxides in the top 15 cm, calculate

(a) How many mmoles of P are adsorbed per kg of sesquioxide (assume 2×10^6 kg soil per ha-15 cm)?

(b) If the P is adsorbed as $H_2PO_4^-$, and for each mole adsorbed 0.5 moles of OH^- ion is released into the solution, calculate the new AEC of the sesquioxides in the soil. (Molar mass of P = 31 g).

7 (a) Consider a sample of Ca^{2+}-saturated montmorillonite. Indicate the expected d spacing for the montmorillonite quasi-crystals when placed (i) in 0.1 M $CaCl_2$ solution, and (ii) distilled water.

(b) Suppose this sample of montmorillonite clay is filtered, washed briefly with distilled water, and re-suspended in a solution of 1 M NaCl. Write an equation for the reversible reaction between the exchangeable Ca^{2+} and Na^+ ions in solution.

(c) What would you expect to happen to the d spacing of the clay after this treatment?

8 (a) If a sample of Na-montmorillonite clay is placed in distilled water, what would you expect the clay to do (give a reason for your answer)?

(b) Suppose that $CaCl_2$ solution is added to the clay in water (10 mL of 1 M $CaCl_2$ per 100 mL) and the suspension is shaken. What would you expect the clay suspension to do (give a reason for your answer)?

Chapter 8

Soil Aeration

8.1 Soil respiration

Types of respiration

The importance of soil organisms in promoting C turnover is outlined in Chapter 3. Aerobic respiration involves the breakdown or dissimilation of complex C molecules, O_2 being consumed and CO_2, water and energy for cellular growth being released. Under such conditions the respiratory quotient R_q, defined as

$$R_q = \frac{\text{volume of } CO_2 \text{ released}}{\text{volume of } O_2 \text{ consumed}}, \qquad (8.1)$$

is equal to 1. When respiration is anaerobic, however, R_q approaches infinity because O_2 is no longer consumed but CO_2 continues to be evolved. The respiration rate R_r of the soil organisms is measured as the volume of O_2 consumed (or CO_2 released) per unit soil volume per unit time.

Strictly speaking, soil respiration refers to the respiration of soil organisms, but in most situations it is impractical to separate the respiration of soil organisms from the respiration of living plant roots. Further, the respiration of soil microorganisms is greatly stimulated by the abundance of carbonaceous material (mucilage, sloughed-off cells and exudates) in the soil immediately around a root. This zone under the influence of a root – extending up to a few mm from the root surface – is called the rhizosphere. The stimulus to soil respiration provided by the presence of a crop is demonstrated in Table 8.1, where respiration rates for cropped and fallow soil in the south of England are compared for the months of January (mean soil temperature of 3°C at 30 cm depth)

Table 8.1 Soil respiration as influenced by crop growth and temperature. After Currie, 1970.

| | Respiration rate, R_r (L/m²/day) | | | |
| | January (mid-winter) | | July (mid-summer) | |
	Fallow	Cropped	Fallow	Cropped
O_2 consumed	0.5	1.4	8.1	16.6
CO_2 released	0.6	1.5	8.0	17.4
R_q	1.20	1.07	0.99	1.05

and July (mean soil temperature of 17°C at 30 cm depth). This soil had an unusually deep A horizon so that the respiration rates were about twice those normally recorded in the field.

Measurement of respiration rates

An instrument for measuring respiration rates is called a respirometer, and one designed specifically for field measurements is illustrated in Fig. 8.1. The soil water content is regulated by controlled watering and drainage; temperature is recorded and the circulating air monitored for O_2, the deficit being made good by an electrolytic generator. Carbon dioxide is absorbed in vessels containing sodium bicarbonate and measured by titration. Such measurements show that the respiration rate depends on:
• Soil conditions, such as organic matter content, O_2 supply and water content;
• cultivation and cropping practices; and
• environmental factors, principally temperature.

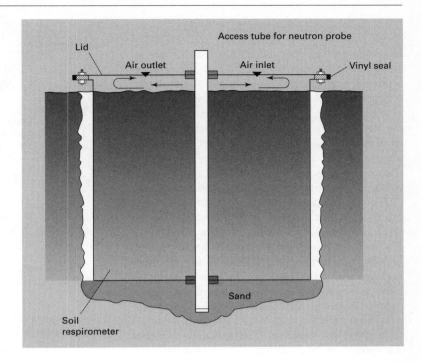

Fig. 8.1 Diagram of a soil respirometer (after Currie, 1975).

The effect of temperature on respiration rate is expressed by the equation

$$R_r = R_o Q^{T/10}, \tag{8.2}$$

where R_r and R_o are the respiration rates at temperature T and 0°C respectively, and Q is the magnitude of the increase in R_r for a 10°C rise in temperature, called the Q-10 factor, which is approximately 2. Temperature change is the cause of large seasonal fluctuations in soil respiration rate in temperate climates, as illustrated in Fig. 8.2.

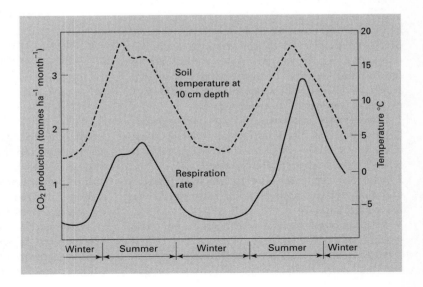

Fig. 8.2 Seasonal trends in soil temperature and respiration rate in southern England.

Cycles of respiratory activity

In temperate regions, the seasonal maximum in R_r lags 1–2 months behind that of soil temperature. Superimposed on this seasonal pattern is the effect of weather, because within any month of the year the daily maximum in R_r may be up to twice the daily minimum, a fluctuation strongly correlated with surface soil temperatures. Finally, there are minor variations in R_r caused by the diurnal rise and fall of soil temperature. The causes of soil temperature variations are discussed in Section 6.6.

These trends are modified by other factors: for example, at Rothamsted in southern England the respiratory peak is consistently lower during a dry summer than a wet one, and the rate tends to be higher in spring than in autumn for the same average soil temperatures. The latter effect is attributable to two factors, primarily a greater supply of organic residues at the end of the quiescent winter period than after a summer of active decomposition, together with spring cultivations that break up large aggregates and expose fresh organic matter to attack by micro-organisms.

8.2 Mechanisms of gas exchange

In section 4.5, the average composition of the air in a range of soils was given as 78% N_2, c. 20% O_2, 1% Ar and 0.1–1.5% CO_2. The soil air composition is buffered against change by the much larger volume of the Earth's atmosphere. There is also a large reservoir of HCO_3^- in the oceans. Buffering is achieved by the dynamic exchange of gases as O_2 travels from the atmosphere to the soil and CO_2 moves in the reverse direction, a process called soil aeration. Three transport processes are involved in soil aeration:
1 Dissolved O_2 is carried into the soil by percolating rainwater. The contribution is small owing to the low solubility of O_2 in water (0.028 mL/mL at 25°C and 1 atmosphere pressure);
2 mass flow of gases due to pressure changes of 0.1–0.2 kPa created by wind turbulence over the surface; and
3 diffusion of gas molecules through the soil pore space.

Gaseous diffusion is normally far more important than mass flow in either the liquid or gas

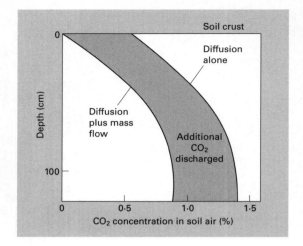

Fig. 8.3 Concentration *vs* soil depth profiles for CO_2 in the soil air under a surface crust (after Currie, 1970).

phase (1 and 2). This is illustrated by the gas concentration profiles shown in Fig. 8.3, for a soil with a thin surface crust, where diffusion alone decreases the concentration of CO_2 in the soil air at a depth of 1 m to < 1.5%. The effect of mass flow, on the other hand, is merely to reduce the CO_2 concentration by a further 0.5%, virtually all of this decrease occurring across the surface crust where mass flow due to atmospheric pressure changes is effective.

Gas exchange by diffusion

The rate of gas diffusion, or specifically the diffusive flux J_g, is given by the equation

$$J_g = -D\frac{\partial C}{\partial x}, \tag{8.3}$$

where J_g is the volume of gas crossing a unit area perpendicular to the direction of diffusion in unit time, $\partial C/\partial x$ is the change in gas concentration C (mL/mL of air-space) per unit distance x, and D is the gas diffusion coefficient (*cf.* Equation B6.7.3 for solute diffusion). Values of D for CO_2 and O_2 in water and air are given in Table 8.2, from which it can be seen that for the same concentration gradient, the diffusive flux is potentially 10,000 times faster through air-filled pores than

Table 8.2 Diffusion coefficients (cm^2/s) of O_2 and CO_2 in air and water.

	Oxygen	Carbon dioxide
Air	2.3×10^{-1}	1.8×10^{-1}
Water	2.6×10^{-5}	2.0×10^{-5}

water-filled pores. Further, since D_{O_2} is slightly larger than D_{CO_2}, when the soil R_q is 1, O_2 diffuses into the soil faster than CO_2 can diffuse out. A pressure difference builds up and mass flow of N_2 (the most abundant gas in the pore space) occurs to eliminate it. Nevertheless, under most steady-state aerobic conditions it is sufficiently accurate to equate the diffusive flux of O_2 into the soil with that of CO_2 out.

Air-filled porosity and tortuosity

Values of D for O_2 and CO_2 through the air-filled pores are always less than D in the outside air because of the restricted volume of the pores and the tortuosity of the diffusion pathway. These effects are expressed in the equation

$$D_e = \alpha \varepsilon D_o, \qquad (8.4)$$

where D_e is the effective diffusion coefficient in the soil, and D_o is the diffusion coefficient of the gas in the bulk air. The term ε is the air-filled porosity (Section 4.5) and α is a dimensionless impedance factor, which ranges from 0 to 1 and decreases as the pathway becomes more tortuous.

The value of α varies with ε in a complex way that depends on a soil's structure, texture and water content. Some of these effects are illustrated in Fig. 8.4. For example, the change in α in dry sand (Fig. 8.4a) is limited by the extent to which the pore-size distribution of the sand can be altered by changing its packing density, or by the addition of smaller particles. However, when the sand is wet up (Fig. 8.4b), water films fill the narrowest necks between pores so that the shape as well as the size distribution of the pores is altered. When water is added to dry soil aggregates so that as θ increases, ε decreases as shown in Fig. 8.4c, and α initially increases as the intra-aggregate pores fill and the tortuosity and roughness of the gas diffusion path decreases. However, once the larger inter-aggregate pores begin to fill, α decreases markedly as it does in wet sand, approaching a minimum value at $\varepsilon \approx 0.1$, a value corresponding to the average proportion of discontinuous or 'dead-end pores' in the soil.

We conclude that in aggregated soils, the intra-aggregate pores make little contribution to gas diffusion and the major pathway is through the pores between aggregates, most of which are drained at water contents less than field capacity. This being so, the value of α will reflect management practices that change the proportion of macropores in the total porosity. Soil structural deterioration through continuous cultivation, for example, may not necessarily result in a marked decrease in total porosity, but it is usually associated with a disproportionate decrease in the macroporosity. Consequently, the value

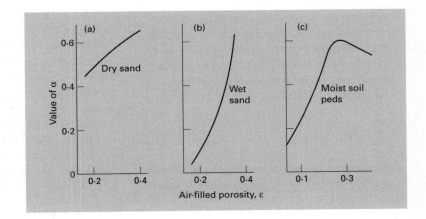

Fig. 8.4 The relation between impedance factor α and air-filled porosity ε for wet and dry sand and soil aggregates (after Currie, 1970).

of α, which may be as high as 0.6 in a well-structured soil, may fall as low as 0.1 in an over-cultivated arable soil.

Profiles of O_2 and CO_2 in field soils

In the field, the concentrations of O_2 and CO_2 in the soil pores change in space and time due to variations in the air-filled porosity, tortuosity and soil respiratory activity. Examples of changes in the O_2 partial pressure in the soil air during summer and winter for soils of contrasting texture are given in Fig. 8.5. A quantitative description of O_2 and CO_2 movement under transient conditions in the field requires a combination of the diffusive flux equation with a continuity equation that includes a term for soil respiration (Box 8.1).

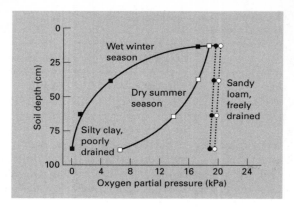

Fig. 8.5 Changes in O_2 partial pressure in contrasting soils with depth and season of the year. Solid symbols, winter; open symbols, summer.

Box 8.1 Calculation of changes in O_2 and CO_2 concentrations with soil depth.

For transfer of O_2 or CO_2 through the soil pore space in the vertical direction z, we may write the flux equation

$$J_g = -D_e \frac{\partial C}{\partial z}, \tag{B8.1.1}$$

which, when combined with a requirement for mass conservation, gives the continuity equation

$$\varepsilon \frac{\partial C}{\partial t} = -\frac{\partial J_g}{\partial z} \pm R_r, \tag{B8.1.2}$$

where R_r is the rate of CO_2 production ($+R_r$) or O_2 consumption ($-R_r$). The gas concentration C is usually expressed in mL gas/mL air and R_r in mL gas/cm^3 soil/s, which accounts for the appearance of ε (cm^3 air/cm^3 soil) in Equation B8.1.2. Combining Equations B8.1.1 and B8.1.2 gives

$$\varepsilon \frac{\partial C}{\partial t} = D_e \frac{\partial^2 C}{\partial z^2} \pm R_r. \tag{B8.1.3}$$

Assuming that steady-state conditions are attained (i.e. $\partial C/\partial t = 0$) and that R_r and D_e do not change

substantially with depth, Equation B8.1.3 can be integrated to give an expression relating the change in O_2 concentration (ΔC) in the gas phase to the depth z over which that change occurs, that is

$$\Delta C = \frac{R_r z^2}{2 D_e}. \tag{B8.1.4}$$

Substitution of appropriate values for R, z, and D_e in this equation gives a change in O_2 concentration consistent with field measurements. For example, for a moist well-aggregated soil in which ε = 0.2 and α = 0.4 (Fig. 8.4 (c)), we may substitute $R_r = 2 \times 10^{-7}$ mL gas/cm^3 soil/s, z = 100 cm, and $D_e = 0.0184$ cm^2/s, and obtain the result that the change in O_2 between the surface and 1 m depth is 0.054 mL/mL air space. By Dalton's law of partial pressures, this change corresponds to a change in O_2 partial pressure of 0.054 bars (5 kPa). The O_2 partial pressure would fall from c. 21 kPa at the surface to 16 kPa at 1 m depth, a result that is intermediate between a freely drained sandy loam and poorly drained silty clay in summer (Fig. 8.5).

Gas diffusion through water-filled pores

Gas diffusion is more complicated in soils that are poorly drained and possibly waterlogged for considerable periods of the year. In this case diffusion occurs through air and water to varying degrees. The cytochrome oxidase enzyme system, which catalyses the reduction of O_2 to H_2O in plants and micro-organisms, has a very high affinity for O_2, as indicated by a K_M value of 2.5×10^{-8} M (this defines the concentration of dissolved O_2 at which the rate of the reaction is 50% of the maximum rate). However, in stirred bacterial suspensions it is found that aerobic respiration can only be sustained at dissolved O_2 concentrations $> 4 \times 10^{-6}$ M, which indicates that diffusion of O_2 to the enzyme site within the bacterial cell is the rate-limiting step in respiration. The concentration of O_2 in water in equilibrium with the normal partial pressure of O_2 in the atmosphere at 25°C is

$$0.21 \times 0.028 = 0.0059 \text{ mL } O_2/\text{mL water,}$$

which is equivalent to 2.6×10^{-4} M*. Thus, soil micro-organisms should continue to respire aerobically provided that the O_2 partial pressure in their *immediate* surroundings is approximately one-sixtieth or more of that in the atmosphere, that is ≥ 0.35 kPa. Although O_2 partial pressures above this value may be maintained in the few air spaces remaining in a waterlogged soil, the respiring organisms mostly live in the water-filled pores *within* aggregates. Thus, O_2 must diffuse to the organisms through water, a slow process due to the low solubility and low diffusion coefficient of O_2 in water (Table 8.2). In many soils, therefore, anaerobic pockets develop at the centre of aggregates larger than a certain size when these aggregates remain wet for any length of time.

Figure 8.6 shows the effect of respiration rate on the minimum size of a wet aggregate in which anaerobic conditions can develop at its centre. This relationship is obtained by integrating Equation B8.1.3 for appropriate initial and boundary conditions (Box 8.2).

* The volume of a gas at 0°C and 1 atmosphere pressure is 22.4 L/mol.

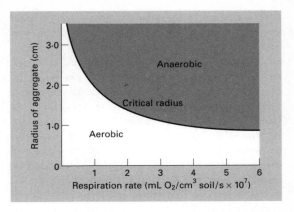

Fig. 8.6 Critical radius for the onset of anaerobic conditions at aggregate centres (after Greenwood, 1975).

Box 8.2 Estimating the critical aggregate size for the onset of anaerobic conditions.

Assuming that the aggregates are spherical (so the radius r is substituted for z) and steady-state conditions apply, Equation B8.1.3 can be integrated to give

$$r = \sqrt{\frac{6 \Delta C D_e}{R_r}}, \qquad \text{(B8.2.1)}$$

where ΔC is the difference in O_2 concentration between the aggregate surface and its centre, D_e is the effective diffusion coefficient of O_2 through the water phase and R_r is the respiration rate inside the aggregate. If the partial pressure of O_2 in the air space at the aggregate surface is assumed to be 21 kPa, while at the centre of the aggregate it is zero, Equation B8.2.1 can be solved for a range of R values to give the minimum radius of aggregate, the centre of which will just be anaerobic. When D_e has a value of 1.5×10^{-5} cm²/s, the results shown in Fig. 8.6 are obtained for a range of R_r values.

Provided there is some continuity of air-filled pores ($\varepsilon > 0.1$) and respiration rates are low ($< 1 \times 10^{-7}$ mL/cm^3 soil/s), anaerobic conditions are unlikely to occur unless the sites of active respiration are separated from the gas phase by more than 2 cm of water-saturated soil. However, as respiration rates rise, anaerobic conditions are likely in aggregates of radius 1 cm or more. This is probably a common occurrence in clay soils in Britain and other humid temperate regions in spring, and predisposes to denitrification, as discussed in Section 8.4. Because the solubility of CO_2 is much greater than that of O_2 (0.759 mL/mL compared to 0.028 mL/mL respectively), its diffusive flux through water is larger with the result that CO_2 partial pressures do not rise to more than *c.* 1 kPa, even at the centre of wet aggregates. Nevertheless, once the pores between aggregates fill with water, as in a waterlogged soil, the CO_2 partial pressure can rise to 10 kPa and more, and may have detrimental effects on the growth of plants and micro-organisms.

8.3 Effects of poor soil aeration on root and microbial activity

Plant root activity

The picture emerging from Section 8.2 is one of a mosaic of aerobic and anaerobic zones in heavy-textured soils during wet periods of the year, as illustrated in Fig. 8.7. The anaerobic zones exert

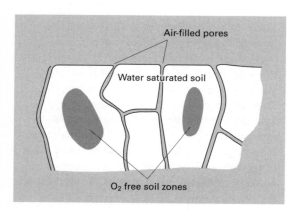

Fig. 8.7 Distribution of water and oxygenated zones in a structured clay soil (after Greenwood, 1969).

no serious ill effect (other than through the reduction of NO_3^-, as discussed later) because most of the roots grow through the larger pores (> 100 μm diameter) where O_2 diffusion is rapid. Oxygen can also diffuse for limited distances through plant tissues from regions of high to low O_2 concentration, an attribute best developed in aquatic species and paddy rice (*Oryza sativa*), because of large intercellular air spaces in which the value of D_{O_2} approaches 1×10^{-1} cm^2/s. Fig. 8.8 shows a healthy young rice crop growing in a completely puddled and waterlogged soil.

However, when wet weather coincides with temperatures high enough to maintain rapid respiration (typically in spring), the whole root zone may become anaerobic. Even if sustained for only a day, this condition can have serious effects on the metabolism and growth of sensitive plants. Species intolerant of waterlogging experience accelerated glycolysis and the accumulation of endogenously produced ethanol and acetaldehyde. Cell membranes become leaky, and ion and water uptake is impaired. The characteristic symptoms are wilting, yellowing of leaves and the development of adventitious roots at the base of the stem. The production of endogenous ethylene (C_2H_4) may be stimulated and cause growth abnormalities, such as leaf epinasty – the downward curvature of a leaf axis.

Soil microbial activity

Soil aeration determines the balance between aerobic and anaerobic metabolism, which has profound effects on the energy available for microbial growth and the type of end-products. Some general points about aerobic and anaerobic organisms are summarized in Box 8.3.

Aerobic processes: nitrification

Nitrification entails the biological oxidation of inorganic N forms to NO_3^-. However, the principal nitrifying organisms are chemoautotrophic bacteria that derive energy for growth solely from the oxidation of NH_4^+ produced by saprophytic decay organisms, or supplied from fertilizers and in rain. The oxidation occurs in two steps:

Fig. 8.8 Healthy young rice plants in a completely puddled and waterlogged soil.

Box 8.3 Aerobic and anaerobic organisms compared.

Aerobic organisms:

• Are generally beneficial to soil and plants;
• the heterotrophs break down complex C compounds to produce much microbial biomass, humus, water and CO_2 (Section 3.2);
• mineralize essential elements such as N, P and S during decomposition;
• specialized groups of autotrophs, such as the nitrifying bacteria and sulphur-oxidizing bacteria, produce NO_3^- and SO_4^{2-};
• some rhizosphere-dwelling species of bacteria may produce growth regulators, including gibberellic acid and indolacetic acid, which stimulate plant growth.

Anaerobic organisms:

• Are much less numerous than aerobes, comprising about 10% of the total soil population, but some aerobes are facultative anaerobes (i.e. can grow anaerobically when the O_2 supply is limiting);
• generate less metabolic energy per mole of substrate during anaerobic respiration, so that cell growth is slower than for aerobic respiration;
• form more varied end-products, including CO_2, H_2O and reduced C compounds such as organic acids, C_2H_5OH, CH_4, C_2H_4 and other hydrocarbon gases;
• use alternative electron acceptors (NO_3^-, MnO_2, $Fe(OH)_3$ and SO_4^{2-}) and produce reduced inorganic forms, some of which are undesirable (Section 8.4).

$$NH_4^+ + \frac{3}{2}O_2 \rightarrow NO_2^- + 2H^+ + H_2O + \text{energy}$$

$$(8.5)$$

carried out by species of the genera *Nitrosomonas*, *Nitrosospira*, *Nitrosococcus*, *Nitrosovibrio* and *Nitrosolobus*, and

$$NO_2^- + \frac{1}{2}O_2 \leftrightarrow NO_3^- + \text{energy} \qquad (8.6)$$

carried out by species of the genera *Nitrobacter*, *Nitrococcus* and *Nitrospira* (Paul and Clark, 1996). Note that although the first reaction in nitrification is the conversion of NH_3 to NH_2OH on the surface of the ammonia mono-oxygenase (AMO) enzyme (Box 8.4), the NH_4^+ ion is by far the predominant form of reduced N in soil, as governed by the reversible equilibrium

$$NH_4^+ \leftrightarrow NH_3 + H^+, \qquad (8.7)$$

Box 8.4 Pathways of nitrification and nitrification inhibition.

Chemoautotrophic bacteria such as *Nitrosomonas europea*, *Nitrosospira briensis*, *Nitrosolobus multiformis* and *Nitorosovibrio tenuis* (found in soil) and *Nitrosococcus oceanus* (found in seawater) oxidize NH_3 to NO_2^- via hydroxylamine (NH_2OH), producing small amounts of NO and N_2O as by-products. NO_2^- is oxidized to NO_3^- by *Nitrobacter* species (e.g. *Nitrobacter winogradsky*). These autotrophic organisms are all obligate aerobes.

The enzyme catalysing the NH_3 oxidation step, ammonia mono-oxygenase (AMO), is inhibited by a range of organic compounds, which forms the basis of attempts in ecological studies and practical agriculture to limit the rate of conversion of NH_3/NH_4^+ to NO_3^-, and hence conserve mineral N. The gas acetylene (C_2H_2) is effective at low partial pressures (up to 0.1 kPa), competing for the enzyme's functional site and inactivating it. Nitrapyrin (2-chloro-5-trichloromethyl pyridine), commercial name N-serve, is widely used in agriculture, usually dissolved in an organic solvent because of its low solubility in water. Effective concentrations are 10–50 mg/kg, but the compound is decomposed in soil with half-life of about 2 weeks at 20°C, and is less effective in organic-rich soils. Obtaining a uniform distribution through the soil can also be a problem.

There is also a heterotrophic pathway for nitrification carried out by some fungi, Actinomycetes and bacteria (all organisms which do not depend on the energy of NH_3 oxidation for growth, nor which fix CO_2 as the first step in carbohydrate synthesis). The substrate is generally amino-N, although some organisms may oxidize NH_4^+. The contribution of heterotrophic organisms to overall nitrification is not well known: it is more important in soils of pH < 5.5 (in water), and those high in organic matter and undisturbed, such as forest soils and soils of permanent grasslands, where it may comprise up to 20% of the total nitrification. Heterotrophic nitrification is not inhibited by C_2H_2 or nitrapyrin.

---

which has a pK value of 9.5. Hence, we write the first step of nitrification as shown in Equation 8.5.

These energy-releasing reactions are coupled to the reduction of CO_2 and cell growth. Because CO_2, which is the sole source of C for these organisms, is plentiful, the growth rate and hence bacterial numbers are limited primarily by the supply of oxidizable substrates (NH_4^+ and NO_2^-). For example, the demand for NH_4^+ created by the more numerous soil heterotrophs restricts the rate of NH_4^+ oxidation by the slower growing nitrite-oxidizers, and the absence of detectable NO_2^- in well-aerated soils indicates that NO_2^- supply is potentially rate-limiting for the growth of *Nitrobacter*.

Growth of the micro-organisms and rate of nitrification are also influenced by temperature, moisture, pH and O_2 supply as follows:
• The temperature optimum lies between 30 and 35°C and nitrification is very slow at temperatures < 5°C;

• the optimum moisture content is *c.* 60% of field capacity. *Nitrosomonas* species are more susceptible to dry conditions than *Nitrobacter*, but sufficient bacteria survive short periods of desiccation for increased nitrification to occur during the flush of decomposition that follows the rewetting of an air-dry soil (Fig. 10.6);
• the growth of *Nitrosomonas* is more sensitive to low pH than *Nitrobacter*, which might be expected from reactions 8.5 and 8.6. Growth rates measured in pure culture or in laboratory soil perfusion experiments are at a maximum at pH $\geq$ 7.6 for *Nitrosomonas*, and pH 6.6–7.6 for *Nitrobacter*. Nitrifying organisms in field soils also produce less NO_3^- per unit time at low pH, and hence grow more slowly. But measurements of the short-term nitrification rate at non-limiting NH_4^+ concentrations suggest that soil nitrifiers adapt to the prevailing pH such that the optimum pH for nitrification is close to the soil pH (Fig. 8.9). This relationship also suggests that the optimum pH for nitrifiers in any soil is unlikely to exceed 6.6.

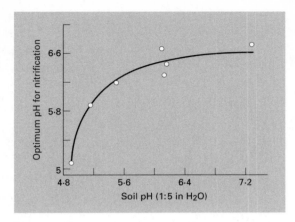

Fig. 8.9 pH optimum for nitrifying organisms in a range of acid to neutral soils. The pH optimum was based on short-term nitrification rates measured under non-limiting NH_4^+ supply (after Bramley and White, 1991).

Anaerobic processes: fermentation and reduction products

The metabolic pathway for carbohydrate oxidation in plants and micro-organisms is the same, in the presence and absence of O_2, as far as the key intermediate, pyruvic acid ($CH_3COCOOH$). This is called glycolysis. Normally, reduced coenzymes in the cell are reoxidized as electrons are passed along a chain of respiratory enzymes to the terminal acceptor O_2. However, in the absence of O_2, as in a waterlogged soil, there are two possible alternatives:

• Other organic compounds serve as final electron acceptors and are reduced in turn – the process of fermentation; or

• inorganic compounds such as NO_3^-, MnO_2, $Fe(OH)_3$, and SO_4^{2-} are reduced (Section 8.4).

As indicated in Fig. 8.10, several organic acids of low molecular weight – the volatile fatty acids (VFAs) – are formed as intermediates in carbohydrate fermentation. The more abundant VFAs are acetic (CH_3COOH) and butyric (C_3H_7COOH) acids, the concentrations of which in the soil solution can sometimes be > 1–10 mM for several weeks in flooded soils. Even in normally aerobic soils, when unfavourable conditions develop (through surface compaction and excess water), crop residues may ferment to produce localized concentrations of soluble phytotoxins, usually acetic acid, which are high enough to inhibit seed germination and seedling establishment.

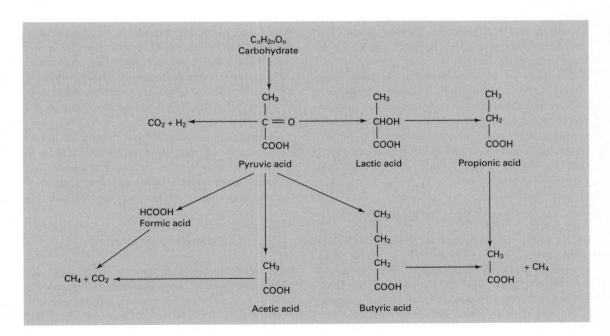

Fig. 8.10 Products of the fermentation of carbohydrates (after Yoshida, 1975).

In addition to the VFAs, a range of low molecular weight hydrocarbon gases – methane (CH_4), ethylene (C_2H_4) and ethane (C_2H_6) – is produced during the first few days of waterlogging. However, as anoxic conditions (anaerobiosis) are prolonged, the acids and hydrocarbon gases are dissimilated to the final end products, mainly CH_4, CO_2 and some H_2, by the obligate anaerobes that multiply slowly in a completely waterlogged soil. Thus, although some CH_4 is produced during the initial stages of anaerobiosis, the bulk is produced by the reduction of CO_2 or dissimilation of CH_3COOH as reducing conditions intensify, according to the reactions

$$CO_2 + 4H_2 \rightarrow CH_4 + 2H_2O \qquad (8.8)$$

and

$$CH_3COOH \rightarrow CH_4 + CO_2. \qquad (8.9)$$

As a result of these reactions, the ratio of CO_2 to CH_4 emitted from waterlogged soils is variable. Methane is a radiatively active gas accounting for *c.* 15% of the Enhanced Greeenhouse Effect (Box 3.1). About 70% of these emissions are anthropogenic, with most of the remainder coming from natural wetlands. Paddy rice soils contribute *c.* 15–20% of the anthropogenic CH_4 emissions, which is approximately the same as from cattle and other ruminants (Aulakh *et al.*, 2001). Less N is immobilized in a given time under anaerobic conditions because decomposition of the substrate is slower and the growth yield of the micro-organisms is lower (Section 3.3). Consequently, more N appears as NH_4^+ during anaerobic conditions, and putrefaction products, such as volatile amines, thiols and mercaptans, may also be produced.

Physiological effects of anaerobiosis

Of the range of fermentation products – VFAs, hydrocarbon gases, CO_2 and H_2 – produced in poorly aerated and waterlogged soils, only acetic and butyric acids and C_2H_4 are likely to have adverse effects on root activity and plant growth. It appears that an anaerobic to aerobic interface may be necessary for C_2H_4 production in soil. Very low concentrations of C_2H_4 (~0.1 mg/L

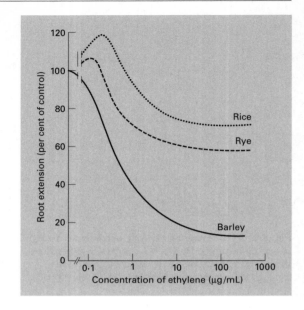

Fig. 8.11 The effect of ethylene on root growth (after Smith and Robertson, 1971).

in the gas phase) stimulate root elongation, but higher concentrations, which may persist for several weeks in wet, poorly drained soil, or beneath the smeared plough layer in a clay, can inhibit the root growth of sensitive crops such as barley (Fig. 8.11). Some aerobic micro-organisms metabolize C_2H_4, so the C_2H_4 concentration in soil reflects the balance between microbial production, diffusive losses and decomposition. This explains why very little C_2H_4 accumulates in well-aerated soil where it is destroyed or lost as rapidly as it is formed.

Methane is also metabolized by aerobic micro-organisms in soil, so even though waterlogged soils (such as paddy rice soils) are important sources of CH_4 that is released to the atmosphere, other well-drained soils can be significant sinks for the gas.

Indices of soil aeration

From the foregoing it can be seen there are potentially several ways in which the aeration status of a soil can be assessed. These are outlined in Box 8.5. However, once the O_2 partial pressure in the bulk soil air has fallen below *c.* 0.4 kPa (Section 8.2), and anaerobic respiration

Box 8.5 Ways of assessing a soil's aeration status.

• Measurement of the rate of O_2 diffusion through the soil. A popular method is based on the reduction of O_2 at the surface of a platinum (Pt) electrode inserted into the soil, to which a fixed voltage is applied relative to a reference electrode. The resultant electron transfer causes a current flow that is proportional to the rate of O_2 diffusion to the Pt electrode. Problems can occur through poor contact between the electrode and surrounding soil.
• Alternatively, the concentration of O_2 in a sample of air extracted from the soil can be measured by gas-liquid chromatography (GLC). Obviously, where aerobic and anaerobic zones exist in close proximity (as in Fig. 8.7), such a measurement is only a spatial average for a volume of soil in which the exact distribution of the O_2 concentration remains undefined.
• The detection of fermentation and putrefaction products by their odour (especially the volatile amines and mercaptans), and where possible, by quantitative analysis using GLC (measuring the VFAs, CH_4 and C_2H_4).
• By noting any visual evidence of chemical changes in the soil profile, such as the solution and reprecipitation of Mn as small black specks of MnO_2, or the appearance reddish brown mottles in a background of bluish-grey reduced Fe compounds.

(reduced) state, or *vice versa*, through the transfer of electrons. One such reversible reaction is

$$H_2 \text{ (gas)} \leftrightarrow 2H^+ + 2e^-, \quad (8.10)$$

where e^- stands for a free electron. The molecules of H_2, from which electrons are removed, are said to be oxidized. The half-reaction (8.10) must be coupled with another reversible half-reaction in which, for example, electrons are added to molecular O_2, i.e.

$$\tfrac{1}{2}O_2 \text{ (gas)} + 2e^- \leftrightarrow O^{2-}. \quad (8.11)$$

Overall, electrons flow from the species of lower electron affinity (H_2) to higher electron affinity (O_2), and the resultant coupled reaction is written as

$$\tfrac{1}{2}O_2 \text{ (gas)} + 2H^+ + 2e^- \leftrightarrow H_2O \text{ (liquid)}. \quad (8.12)$$

The general form of a redox reaction is therefore

$$\text{Ox} + m\text{H}^+ + n\text{e}^- \leftrightarrow \text{Red}. \quad (8.13)$$

The equilibrium constant K_e for reaction (8.13) is written as

$$K_e = \frac{\text{(Red)}}{\text{(Ox)(H}^+)^m(\text{e}^-)^n}. \quad (8.14)$$

In this expression, (Ox) and (Red) are the activities of the oxidized and reduced species, respectively, m and n are stoichiometric coefficients, and the free electron is assigned an activity (e^-).

We may take $-\log_{10}$ of the terms in Equation 8.14 and rearrange to give

$$\text{pe} = \frac{1}{n}[\log K_e + \log \text{(Ox)} - \log \text{(Red)} - m\text{pH}], \quad (8.15)$$

where pe is the negative logarithm of the free electron activity, sometimes called the electron potential (Bartlett, 1999). If K_e, (Ox), (Red), pH, m and n are known, pe can be calculated. Large positive values of pe (low electron activity) favour the existence of oxidized forms, or those that are electron acceptors. However, in heterogeneous soil systems, it is virtually impossible to assign values to all these parameters and variables.

becomes dominant, the redox potential is a useful measure of the intensity of the ensuing reducing conditions. Measurement of redox potential is discussed in the next section.

8.4 Oxidation–reduction reactions in soil

Redox potential

Oxidation–reduction or redox reactions are those in which a chemical species (a molecule or ion) passes from a more oxidized to a less oxidized

Box 8.6 An expression for the redox potential.

The equation for the free energy change (ΔG) in a reversible reaction, such as reaction (8.13), is

$$\Delta G = \Delta G^0 + RT \ln \frac{(\text{Products})}{(\text{Reactants})}. \qquad (B8.6.1)$$

By combining Equation B8.6.1 with the equation for electrical work done for a given free energy change,

$$\Delta G = -nFE, \qquad (B8.6.2)$$

where n is the number of electrons transferred through an electrical potential difference E, and F is Faraday's constant, we may write

$$E = E^0 - \frac{RT}{nF} \ln \left\{ \frac{(\text{Red})}{(\text{Ox})(\text{H}^+)^m} \right\}. \qquad (B8.6.3)$$

That is

$$E_h = E^0 + 2.3 \frac{RT}{nF} \left\{ \log_{10} \frac{(\text{Ox})}{(\text{Red})} - m\text{pH} \right\}. \qquad (B8.6.4)$$

In Equation B8.6.4, E_h is the equilibrium redox potential and E^0 is the standard redox potential of the half-reaction $\text{H}^+ + \text{e}^- \leftrightarrow \frac{1}{2}\text{H}_2$ (gas). E^0 is zero when the pressure of H_2 gas at 25°C is 1 atmosphere and the activity of H^+ ions in solution is one (i.e. pH = 0). It is clear that oxidizing systems will have relatively positive E_h values and reducing

systems have low positive to negative E_h values. It also follows that for a given ratio of (Ox)/(Red), the lower the pH the higher the E_h value at which reduction occurs.

In practice, it is more realistic to refer the measured E_h to pH 7, rather than pH 0, in which case the standard redox potential is defined by

$$E_7^0 = E^0 - \left(2.3 \frac{RT}{F} \right) \frac{m}{n} \text{pH}. \qquad (B8.6.5)$$

Therefore, substituting for E^0 in Equation B8.6.4 gives

$$E_h = E_7^0 + \frac{0.059}{n} \log_{10} \frac{(\text{Ox})}{(\text{Red})}. \qquad (B8.6.6)$$

Equation B8.6.6 shows that the change in E_h as the ratio (Ox)/(Red) changes is inversely related to n, the number of electrons transferred in the redox reaction; also that the point of maximum poise occurs when (Ox) = (Red) and $E_h = E_7^0$. To measure the redox potential, a Pt electrode and suitable reference electrode (usually a calomel electrode) are placed in the wet soil and the electrical potential difference recorded. Aerobic soils have E_h values of 0.3–0.8 volts (V), but reproducible values are only obtained in anaerobic soils, in which the potential ranges from c. 0.4 to −0.4 V. E_h values measured on soils that have been dried, sieved and rewetted are meaningless.

Alternatively, the oxidation–reduction status of the system can be measured by the redox potential, which is the difference in electrical potential between the two half-reactions of the coupled reaction (8.13), as explained in Box 8.6. Note that when (Ox) = (Red), Equation 8.15 simplifies to

$$\log K_e = n\text{pe} + m\text{pH}, \qquad (8.16)$$

so that in a soil at redox equilibrium, the sum of $n$pe and mpH tends to remain constant. If (Ox) increases (more electron acceptors present), pe increases and pH decreases; alternatively, if

(Red) increases (more electron donors present), pe decreases and pH increases. Thus, Equation 8.16 describes the condition of maximum poise (or electron buffering) of the redox system. As can be seen from Equation B8.6.6, maximum poise also occurs when the redox potential $E_h = E_7^0$.

Sequential reductions in soil deprived of oxygen

Once part or all of the soil has become anaerobic, substances other than O_2 are used as terminal

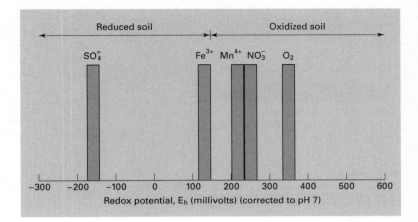

Fig. 8.12 Threshold redox potentials at which the oxidized species shown become unstable (after Patrick and Mahapatra, 1968).

acceptors for the free electrons produced during respiration. A redox reaction will proceed spontaneously in the direction of decreasing free energy $(-\Delta G)$, but the rate at which it proceeds may be very slow if the activation energy is high or if the redox half-reactions are physically not well coupled. In soil, redox reactions are coupled through various micro-organisms growing as facultative or obligate anaerobes (Section 8.3). The following sequence of reactions, involving different inorganic compounds as terminal acceptors for the electrons produced during respiration, may occur

$$2NO_3^-(soln) + 12H^+(soln) + 10e^- \leftrightarrow$$
$$N_2(gas) + 6H_2O \text{ (liquid)} \qquad (8.17)$$

$$MnO_2(solid) + 4H^+(soln) + 2e^- \leftrightarrow$$
$$Mn^{2+}(soln) + 2H_2O \text{ (liquid)} \qquad (8.18)$$

$$Fe(OH)_3(solid) + 3H^+(soln) + e^- \leftrightarrow$$
$$Fe^{2+}(soln) + 3H_2O \text{ (liquid)} \qquad (8.19)$$

$$SO_4^{2-}(soln) + 10H^+(soln) + 8e^- \leftrightarrow$$
$$H_2S(gas) + 4H_2O \text{ (liquid)}. \qquad (8.20)$$

The approximate E_h value at which each oxidized form becomes unstable is shown in Fig. 8.12. Each reaction poises the soil in a particular E_h range until most of the oxidized form has been consumed and then E_h drops to a lower value as the reducing power intensifies. Note that reactions (8.17) and (8.20), in which a gas of low water-solubility is produced, are essentially irreversible.

Generally, the boundary between oxidized and reduced soil is set in the ferric (Fe^{3+})/ferrous (Fe^{2+}) redox potential range.

Denitrification

The first reaction (8.17) summarizes the process of denitrification that is carried out by facultative anaerobic bacteria, predominantly *Pseudomonas* and *Bacillus* species, once the partial pressure of O_2 has fallen to a low level (< 0.4 kPa). The complete pathway for NO_3^- reduction is

$$NO_3^- \leftrightarrow NO_2^- \rightarrow NO \text{ (gas)} \rightarrow N_2O \text{ (gas)} \rightarrow N_2 \text{ (gas)}.$$
aerobic anaerobic

Only trace amounts of nitric oxide (NO) are produced, and the main products are nitrous oxide (N_2O) and dinitrogen (N_2), with different groups of micro-organisms being involved in NO_2^- reduction and N_2O reduction. Laboratory incubations under anaerobic conditions indicate a temperature optimum of c. 40°C for denitrification, being very slow at temperatures < 10°C. Temperature and pH affect the ratio of $N_2O : N_2$ evolved during denitrification in closed systems. At low temperatures and pH < 5, the ratio of $N_2O : N_2$ is 1 or greater, but at temperatures of 25°C or greater and pH > 6, most of the N_2O is reduced to N_2. In the field, however, both gases can escape from the site of formation by diffusion and some N_2O, which is about as soluble as CO_2, is lost in the drainage water from very wet soils. The very variable ratio of $N_2O : N_2$ produced

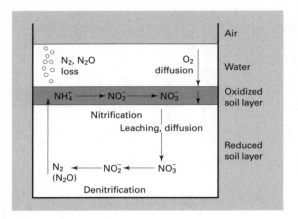

Fig. 8.13 Nitrification and denitrification in the oxidized and reduced zones of a flooded soil (after Patrick and Mahapatra, 1968).

during denitrification makes difficult the measurement of field losses of N by this pathway (Section 10.2).

Denitrification losses can be serious in agricultural soils, especially those exhibiting a mosaic of aerobic and anaerobic zones (Fig. 8.7), and where the supply of C substrates and NO_3^- as a terminal electron acceptor is not limiting. An extreme situation can arise in paddy rice soils that are completely waterlogged during the early period of crop growth (Fig. 8.8). The sequence of processes occurring in the oxidized and reduced soil layers is illustrated in Fig. 8.13. The NO_3^- produced aerobically diffuses or is leached into the anaerobic zone where it is denitrified. Accordingly, N fertilizer in such systems is best applied in an NH_4^+ or urea form and placed directly in the anoxic zone below the soil surface.

More intensive reducing conditions

Reactions 8.18 and 8.19 are characteristic of gleying, a pedogenic process discussed in Section 9.2. The former reaction is less effective in poising the soil than either NO_3^- or Fe^{3+} reduction because MnO_2 is very insoluble and relatively few microorganisms use it as a terminal electron acceptor. Nevertheless, in very acid soils, waterlogging may lead to high concentrations of exchangeable Mn^{2+}, which cause manganese toxicity in susceptible

plants. Anaerobiosis and Fe^{3+} reduction generally increase the phosphate concentration in the soil solution, especially in acid soils where phosphate is specifically adsorbed on Fe_2O_3 particles and coatings on clay surfaces (Section 7.3), or precipitated as very insoluble Fe phosphates. The effect is complicated, however, by the inevitable rise in pH on prolonged waterlogging ($3H^+$ ions are consumed for each electron transferred in reaction 8.19). At pH > 6, Fe^{2+} compounds become less soluble and a metastable iron oxyhydroxide containing both Fe^{3+} and Fe^{2+} iron precipitates ($Fe_3(OH)_8$) according to the reaction

$$2Fe(OH)_3 + Fe^{2+} + 2OH^- \leftrightarrow Fe_3(OH)_8. \qquad (8.21)$$

With an increase in CO_2 partial pressure, $FeCO_3$ may also precipitate and the pH eventually stabilizes between pH 6.5 and 7. Freshly precipitated $Fe_3(OH)_8$ presents a highly active surface for the re-adsorption of phosphate and certainly all the dissolved P is readsorbed when aerobic conditions return and ferrihydrite precipitates (Section 2.4).

At very low E_h values (Fig. 8.12), especially in the presence of plentiful C substrates, SO_4^{2-} is reduced due to the activity of obligate anaerobes of the genus *Desulphovibrio*. As the concentration of weakly dissociated H_2S builds up in the soil solution, ferrous sulphur (FeS) precipitates and slowly reverts to iron pyrites (FeS_2), according to the reaction

$$FeS(solid) + S(solid) \leftrightarrow FeS_2(solid). \qquad (8.22)$$

Iron pyrites is a mineral characteristic of marine sedimentary rocks and of sediments currently being deposited in river estuaries and mud flats. Under the intense reducing conditions that lead to sulphur accumulation, the roots of paddy rice plants become blackened and impaired in their function.

8.5 Summary

During respiration by soil organisms and plant roots, complex C molecules are dissimilated to provide energy for cellular growth. When conditions are aerobic, one mole of CO_2 is released for each mole of O_2 consumed and the respiratory quotient R_q is 1. Respiration continues aerobically, at a rate determined by soil moisture, temperature and substrate availability, provided that O_2 from the atmosphere moves into the soil fast enough to satisfy the organisms' requirements, and CO_2 can move out. This dynamic exchange of gases is called soil aeration.

Diffusion of gases through the air-filled pore space is the most important mechanism of soil aeration. The rate of diffusion or diffusive flux J_g is given by

$$J_g = -D_e \frac{\partial C}{\partial z},$$

where C is the gas concentration, z is depth and D_e is the effective diffusion coefficient, which takes account of the effect of differences in air-filled porosity ε and the tortuosity α of the gas diffusion pathway. Given the possibility that CO_2 and O_2 fluxes vary with time and depth in the soil, the flux equation must be combined with a continuity equation to calculate O_2 and CO_2 concentration profiles in soil; that is

$$\varepsilon \frac{\partial C}{\partial t} = D_e \frac{\partial^2 C}{\partial z^2} \pm R_r,$$

where R_r, the soil respiration rate, is positive when measured as CO_2 release and negative when measured as O_2 uptake. The respiration rate approximately doubles for each 10°C rise in temperature T from the value R_o at 0°C to 30°C, according to the Q-10 equation

$$R_r = R_o Q^{T/10}.$$

Temperature change causes most of the large variation in R_r on the daily and annual time scales.

Sandy soils with ε values > 0.1 m³ air/m³ soil maintain O_2 partial pressures close to the atmospheric value of 21 kPa. Similarly, well-structured clay soils at field capacity can maintain O_2 partial pressures close to 20 kPa in their macropores, but sites of active microbial respiration within aggregates are likely to become anaerobic if separated from air-filled pores by more than 1 cm of water-saturated soil in spring, or > 2 cm in winter. The reason for this is that the diffusion coefficient D for both O_2 and CO_2 in water are 10,000 less than in air, and gas exchange through water-filled pores is very slow. Slow diffusion is especially true of O_2 because of its low water-solubility; whereas diffusion of CO_2 via water-filled pores is faster because of its higher water-solubility. Thus, even well-drained clay soils may exhibit a mosaic of aerobic and anaerobic zones that is particularly conducive to denitrification, if the temperature is favourable and C and NO_3^- supplies adequate. Denitrification is carried out by facultative anaerobes: the switch from aerobic to anaerobic respiration occurs at O_2 partial pressures < 0.35 kPa.

Once the O_2 supply is exhausted, which may happen within 1–2 days of a soil's becoming completely waterlogged, bacteria that can respire either aerobically or anaerobically (facultative anaerobes) continue to grow, and obligate anaerobes begin to multiply. Most plants are intolerant of waterlogging, but some such as marsh plants and paddy rice survive because of O_2 diffusion through their large intercellular air spaces. During anaerobic respiration, carbohydrates are incompletely oxidized or fermented to volatile fatty acids (VFAs) – which can attain localized concentrations high enough to be toxic to seedlings – and low molecular weight hydrocarbon gases, including methane (CH_4) and ethylene (C_2H_4). As anaerobiosis continues, the VFAs are further dissimilated, the end products being CO_2, CH_4 and H_2.

In the absence of O_2, other substances serve as electron acceptors in microbial respiration, and sequential reduction of NO_3^-, MnO_2, Fe^{3+} compounds, and SO_4^{2-} occurs, with some overlapping, as reducing conditions intensify. The reducing power of the soil system can be defined by the electron potential pe (the negative logarithm of the free electron activity), but is usually measured by the redox potential E_h, which for the coupled reversible redox reaction

$$Ox + mH^+ + ne^- \leftrightarrow Red$$

is given by the equation

$$E_h = E_7^0 + \frac{0.059}{n} \log_{10} \frac{(Ox)}{(Red)},$$

where (Ox) and (Red) are the activities of the oxidized and reduced forms, respectively, and E_7^0 is the redox potential of the standard state at pH 7.

References

Aulakh M. S., Wassman R. & Rennenberg H. (2001) Methane emissions from rice fields – quantification, mechanisms, role of management and mitigation options. *Advances in Agronomy* **70**, 193–260.

Bartlett R. J. (1999) Characterizing soil redox behavior, in *Soil Physical Chemistry*, 2nd edn. (Ed. D. L. Sparks). CRC Press, Boca Raton, FL, pp. 371–96.

Bramley R. G. V. & White R. E. (1991) The variability of nitrifying activity in field soils. *Plant and Soil* **126**, 203–8.

Currie J. A. (1970) Movement of gases in soil respiration, in *Sorption and Transport Processes in Soils*. Society of Chemical Industry Monograph 37, pp. 152–69.

Currie J. A. (1975) Soil respiration, in *Soil Physical Conditions and Crop Production*. MAFF Bulletin 29. Her Majesty's Stationery Office, London, pp. 461–8.

Greenwood D. J. (1969) Effect of oxygen distribution in soil on plant growth, in *Root Growth*. (Ed. W. J. Whittington). Butterworths, London, pp. 202–21.

Greenwood D. J. (1975) Measurement of soil aeration, in *Soil Physical Conditions and Crop Production*. MAFF Bulletin 29. Her Majesty's Stationery Office, London, pp. 261–72.

Patrick W. H. & Mahapatra I. C. (1968) Transformation and availability to rice of nitrogen and phosphorus in waterlogged soils. *Advances in Agronomy* **20**, 323–59.

Paul E. A. & Clark F. E. (1996) *Soil Microbiology and Biochemistry*. Academic Press, San Diego.

Smith K. A. & Robertson P. D. (1971) Effect of ethylene on root extension of cereals. *Nature New Biology* **234**, 148–9.

Yoshida T. (1975) Microbial metabolism of flooded soils, in *Soil Biochemistry*, Volume 3. (Eds. E. A. Paul & A. D. McLaren). Marcel Dekker, New York, pp. 83–122.

Further reading

Bartlett R. J. & Bruce B. R. (1993) Redox chemistry of soils. *Advances in Agronomy* **50**, 152–208.

Rowell D. L. (1988) Flooded and poorly drained soils, in *Russell's Soil Conditions and Plant Growth*, 11th edn. (Ed. A. Wild). Longman Scientific and Technical, Harlow, pp. 899–926.

Example questions and problems

1 (a) What is the main mechanism for gas exchange between the atmosphere and soil air?

(b) (i) What is the normal range of CO_2 concentration (percent by volume) in the air spaces of a well-aerated soil? (ii) Express this concentration range as a partial pressure.

(c) Which gas, O_2 or CO_2, is the more soluble in water?

2 (a) The mean daily respiration rate (R_r) of a fallow arable soil at 0°C was measured with a field respirometer as 0.25 L O_2 consumed/m² of soil surface. If the Q-10 factor for this soil is 2.2, estimate the mean daily respiration rate for the same soil in summer when the mean daily soil temperature is 20°C. (Hint – use $R_r = R_o Q^{T/10}$.)

(b) The production of CO_2 was also measured in this soil at 0°C, amounting to 2.1 L/m² over 1 week. Calculate the respiratory quotient for the soil.

3 (a) Which two soil properties interact to change the effective diffusion coefficient for gases in the soil air relative to their diffusion coefficients in bulk air?

(b) How does good soil aggregation improve the rate of gas diffusion and soil aeration?

(c) What is the minimum value of air-filled porosity for good aeration (give the units)?

4 The average flux of CO_2 from a grassland soil in summer was measured as 3 L/m² soil surface/day.

(a) Given an air-filled porosity of 0.2, and impedance factor of 0.5, calculate the effective diffusion coefficient for CO_2 gas in this soil (D for CO_2 in bulk air is 0.18 cm²/s).

(b) Given that the CO_2 concentration in the air above the soil was 0.0365%, and assuming diffusion through air-filled pores is the dominant mechanism for gas transport, calculate the average CO_2 concentration (in mL/mL) in the soil air at 10 cm depth. (Hint – assume steady state conditions and equate CO_2 release at the surface to the diffusive flux of CO_2 through the soil).

5 Consider a soil at 25°C with water-saturated aggregates of average radius r, just less than the critical radius for the onset of anaerobic conditions. The partial pressures of O_2 and CO_2 in the inter-aggregate pores are 20 and 1 kPa, respectively. The partial pressure of O_2 at the aggregate centres is zero. Assuming steady-state conditions, calculate the average CO_2 partial pressure at the centre of the aggregates. (Note – the solubility of O_2 and CO_2 in water at 25°C and 1 atmosphere pressure is 0.028 and 0.759 mL/mL, respectively). (Hint – see Table 8.2 for D values for O_2 and CO_2 in water and assume that the diffusive flux of CO_2 from the centre of an aggregate to its surface is equal to the flux of O_2 inwards.)

6 (a) Describe the trophic status of the micro-organisms primarily responsible for nitrification in soil.

   (b) What is the key substrate for this form of nitrification?

   (c) Name two inhibitors of nitrification.

7 (a) What type of micro-organisms carry out denitrification in soil?

   (b) Write the reaction for the reduction of NO_3^- to N_2 gas in soil.

   (c) What should happen to the soil pH at the microsite where denitrification takes place?

   (d) Give the sequence in which inorganic ions and compounds other than O_2 are reduced as the reducing power intensifies in an anaerobic soil.

Chapter 9

Processes in Profile Development

9.1 The soil profile

Soil horizons

In Chapter 1, the course of profile development from bare rock to a Brown Forest Soil (Inceptisol (ST) or Tenosol (ASC)) is outlined. Except in peaty soils, the downward penetration of the weathering front greatly exceeds the upward accumulation of litter on the surface, and the mature profile may have the vertical sequence previously illustrated in Fig. 1.4, that is:

• Litter layer, L;
• A horizon (mainly eluvial);
• B horizon (mainly illuvial);
• C horizon (weathering parent material).

Unweathered rock beneath the C horizon that cannot be dug with a spade, even when moist, is called bedrock R. Included in the C horizon can be sediments, unconsolidated bedrock and saprolite, the last being material that shows inherited features of the bedrock even though the constituent minerals have been altered by weathering.

A more comprehensive profile description requires an enlarged glossary of descriptive terms – the horizon notation – intended to convey briefly the maximum information about the soil to an observer. Unfortunately, as with much of the descriptive terminology in soil science (Box 1.1), systems of horizon notation differ from country to country. The system adopted by the Soil Survey of England and Wales (Avery, 1980) is consistent with that proposed for international use (International Society of Soil Science, 1967), and

has largely been adopted in simplified form in the *World Reference Base for Soil Resources* (FAO, 1998) (Table 9.1). The system used in Australia (McDonald *et al.*, 1998) shows some variations that are summarized in Box 9.1. The system used in the USA for Soil Taxonomy, summarized in Box 9.2, uses the concept of diagnostic horizons, which are not necessarily consistent with the A, B and C horizon notation. In all systems, there are two main kinds of horizon, organic and mineral, which are distinguished by their organic matter content.

Organic horizons

An organic horizon must have:

• Greater than 30% organic matter (18% C) when the mineral fraction has 50%* or more clay; or

• greater than 20% organic matter (12% C) when the mineral fraction has no clay; or

• more than a proportionate amount of organic C (between 12 and 18%) if the clay content is intermediate.

Organic horizons that are derived from a superficial litter layer L, and are rarely saturated with water for more than 30 consecutive days, are designated as:

• An F horizon when the organic matter is only partly decomposed; and

• an H horizon when the organic matter is low in minerals and well decomposed.

---

* 60% or more clay for Soil Taxonomy.

Table 9.1 Soil horizon notation. After Avery, 1980.

Horizon	Brief description
O – organic	
Of	Composed mainly of fibrous peat
Om	Composed mainly of semi-fibrous peat
Oh	Humified peat; uncultivated
Op	Humified peat; mixed by cultivation
A – at or near the surface, with well-mixed humified organic matter (by natural processes or cultivation)	
Ah	Uncultivated with > 1% organic matter
Ap	Mixed by cultivation; normally an abrupt lower boundary
Ahg or Apg	Partially gleyed due to intermittent waterlogging; rusty mottles along root channels
E – eluvial, subsurface horizon of low organic content, having lost material to lower horizons.	
Ea	Lacking mottles, bleached particles due to the removal of organic and sesquioxidic coatings
Eb	Brownish colour due to uniformly disseminated iron oxides: often overlies a Bt horizon
Eg	Partial gleying with mottles; often overlies a relatively impermeable Bg or Btg horizon, and sometimes a thin iron-pan
B – subsurface horizon without rock structure, showing characteristic structure, texture and colour due to illuviation of material from above and/or weathering of the parent material	
Bf	Thin iron-pan (< 1 cm); often enriched with organic matter and aluminium
Bg	Dominant gley colours on ped faces, rusty mottles within peds; blocky or prismatic structure
Bh	Translocated organic matter (> 0.6%) with some Fe and Al in coatings on sand grains and small peds
Bs	Enriched with sesquioxides, usually by illuviation; orange to red colour
Bt	Accumulation of translocated clay as shown by clay coatings on ped faces
Bw	Shows alterations to parent material by leaching, weathering and structural reorganization under well-aerated conditions
C – unconsolidated or weakly consolidated mineral material that retains rock structure	
Ck	Contains > 1% secondary $CaCO_3$, as concretions or coatings (also Ak and Bk)
Cg	Dominantly grey to green to blue colours due to waterlogging, typically in sands or recent alluvium
Cgy	As for Cg but contains secondary $CaSO_4$ as gypsum crystals
Cm	Continuously cemented, other than by a thin iron-pan
Cr	Weakly consolidated but dense enough to prevent root ingress except along cracks
Cu	Unconsolidated and lacking evidence of gleying, cementation, etc.
Cx	With fragipan properties, as in Pleistocene deposits: dense but uncemented; firm when dry but brittle when moist
CG – intensely gleyed, bluey-grey to green colours that fade on exposure to air and predominate throughout a structureless soil matrix	

An organic horizon that remains saturated with water for at least 30 consecutive days in most years is called a peaty or O horizon. Wetness favours an increasing thickness of the O horizon, which qualifies as peat when there is:
• More than 40 cm of O horizon material in the upper 80 cm of the profile, excluding litter L and living moss; or
• more than 30 cm of O horizon material resting directly on bedrock R, or very stony material.

Mineral horizons

A mineral horizon is low in organic matter and overlies consolidated or unconsolidated rock. It may be designated A, E, B, C or G, as indicated in Table 9.1, depending on its position in the profile, its mode of formation, colour*, structure,

* Standard colour descriptions are obtained from Munsell colour charts.

Box 9.1 The Australian system of soil horizon notation.

O horizons – Dominated by organic materials in varying stages of decomposition that have accumulated on the mineral soil surface; subdivided into O1 and O2, depending on the degree of decomposition and humification (*cf* F and H horizons).

P horizons – Organic horizons that have accumulated under water or conditions of excessive wetness, commonly called peat (no thickness criteria).

A horizons – Surface mineral horizons, subdivided into:
• A1 – Mineral horizon at or near the surface with some accumulation of humified organic matter;
• A2 – Mineral horizon having, either alone or in combination, less organic matter, sesquioxides or clay than contiguous horizons;
• A3 – Transitional between A and B, dominated by the properties of an overlying A1 or A2.

B horizons – Subsurface mineral horizons characterized by one or more of the following: an accumulation of silicate clay, iron, aluminium or organic matter; a structure and/or consistence different from that of A horizons above or any other horizon immediately below, and stronger colours.
• B1 – Transitional between A and B, dominated by the properties of an underlying B2;
• B2 – Dominant feature is one or more of the following:
  ◦ an illuvial, residual or other accumulation of silicate clay or iron, aluminium or humus;
  ◦ maximum expression of pedological organization as evidenced by a different structure or consistence or stronger colours than horizons above and below;
• B3 – Transitional between B and C, dominated by properties of an overlying B2.

C horizons – Below the A–B soil consisting of consolidated or unconsolidated material, usually partly weathered but little affected by pedogenesis.

degree of weathering and other properties that can be recognized in the field. Three kinds of complex horizon are also recognized:
• A horizon with distinctive features of two horizons, such as a strongly gleyed B horizon, is designated B/G;
• a relatively homogeneous horizon with subordinate features of two contiguous horizons, for instance A and B, is designated AB; and
• a transitional horizon that may comprise discrete parts ($\geq 10\%$ by volume) of two contiguous horizons is called A & B, or E & B for example.

Specific attributes of each major horizon are indicated by lower-case suffixes attached to the

Box 9.2 Soil horizon description in Soil Taxonomy.

Soil horizons relevant to Soil Taxonomy are grouped into:
• Seven diagnostic horizons that have formed at the soil surface; and
• a larger number of diagnostic horizons that have formed below the soil surface, or directly below a litter layer.

A horizon developed at the soil surface is called an epipedon, recognized by a change in the structure of the parent material, and a darkening by organic matter or eluviation of material. Such a horizon may be covered by a layer < 50 cm thick of alluvial or aeolian deposits and still retain its identity; but if the deposit is thicker, the soil becomes a buried soil. To avoid changes in the classification of a soil due to disturbance by ploughing, for example, the properties of an epipedon, except for structure, are determined after mixing the surface soil to 18 cm depth. An epipedon is not necessarily the same as an A horizon. It may include part or all of an underlying B horizon if staining by organic matter extends from the surface into the B horizon.

Subsurface diagnostic horizons are generally regarded as B horizons, although they can be revealed at the soil surface if the profile is truncated by erosion. A detailed description of the diagnostic surface and subsurface horizons used in Soil Taxonomy is given by Soil Survey Staff (1996).

horizon symbol (Table 9.1). Again, there is not complete consistency between national systems so the source books should be consulted for details.

Soil layers

Unless the record of rock lithology or stratigraphy suggests the contrary, it is assumed that the parent material originally extended unchanged to the soil surface. As indicated in Section 5.2, this assumption is often invalid for soils on very old land surfaces subjected to many cycles of weathering, erosion and deposition that have resulted in the layering of parent materials. An example of layering is the K cycle of soil formation on weathered and resorted materials in southeast Australia (Fig. 9.1).

Layers of different lithology are identified by numbers prefixed to the horizon symbols, the uppermost layer being 1 (prefix usually omitted) and those below labelled sequentially 2, 3, etc. Where an old soil has formed on a lower layer and subsequently been buried, its horizons are prefixed by the letter 'b' (or suffixed in the Australian system), as previously illustrated in Fig. 5.9. Buried horizons can be found in very old cleared areas where humans have shifted topsoil for farming or construction purposes.

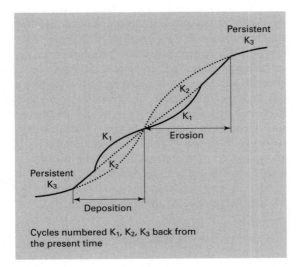

Cycles numbered K_1, K_2, K_3 back from the present time

Fig. 9.1 K cycles of soil formation on a hill slope (after Butler, 1959).

9.2 Pedogenic processes

In this section, the more important processes of profile development (pedogenesis) are discussed. The magnitude of their effect is governed by the interaction of factors defining the initial state of soil formation (parent material, relief), the environment (climate, organisms) and the extent of reaction (time), as discussed in Chapter 5.

Pedogenic processes are considered here as operating in one-dimension (vertically up or down) in a soil profile. But in the landscape, profile development can be much more complex than this simple model suggests, with vertical processes such as leaching, illuviation and mixing by organisms (bioturbation), interacting with lateral processes such as rain wash (including raindrop splash and slope wash) and soil creep, especially on slopes, and deposition of wind-blown material. Examples of pedogenesis involving complex three-dimensional processes are given in Paton *et al.* (1995).

Congruent and incongruent dissolution

Dissolution and hydrolysis

Rock minerals become unstable when exposed to a changed environment. Chemical reactions occur that may be relatively subtle (e.g. oxidation of Fe^{2+} to Fe^{3+}) or drastic, such as the complete dissolution of the mineral in water. Minerals such as halite (NaCl) or gypsum ($CaSO_4.2H_2O$) dissolve to release their constituent elements in the same molar ratio as they occur in the solid, that is

$$NaCl(solid) \leftrightarrow Na^+(liquid) + Cl^-(liquid) \qquad (9.1)$$

and

$$CaSO_4.2H_2O(solid) \leftrightarrow Ca^{2+}(liquid) + SO_4^{2-}(liquid) + 2H_2O. \qquad (9.2)$$

This reaction is called congruent dissolution, as opposed to incongruent dissolution, where the molar ratio of elements in solution differs from that in the solid. Because of this inequality in element release, a compound different from the original mineral gradually forms in the solid phase.

Chemical weathering of primary silicate minerals is a good example of incongruent dissolution by surface hydrolysis (literally the splitting of water molecules). Water molecules at the mineral surface dissociate into H^+ and OH^- and the small, mobile H^+ ions (actually H_3O^+) penetrate the crystal lattice, creating a charge imbalance that causes cations such as Ca^{2+}, Mg^{2+}, K^+ and Na^+ to diffuse out. The potash feldspar orthoclase hydrolyses to produce a weak acid (silicic acid), a strong base (KOH), and a residue of the clay mineral illite

$$3KAl^{IV}Si_3O_8 + 14H_2O \leftrightarrow K(AlSi_3)^{IV}Al_2^{VI}O_{10}(OH)_2$$
$$+ 6Si(OH)_4 + 2KOH. \qquad (9.3)$$

Hydrolysis of the calcic feldspar anorthite, on the other hand, leaves a residue of vermiculite-type clay, which is a weak acid, and the strong base $Ca(OH)_2$

$$3CaAl_2^{IV}Si_2O_8 + 6H_2O \leftrightarrow$$
$$2H^+2[(AlSi_3)^{IV}Al_2^{IV}O_{10}(OH)_2] + 3Ca(OH)_2. \qquad (9.4)$$

Reactions (9.3) and (9.4) are specific examples of the general reaction

$$[M^+silicate] + H_2O \leftrightarrow H^+[silicate \ secondary$$
$$mineral]^- + M^+OH^-, \qquad (9.5)$$

in which M^+ is an alkali (e.g. Na^+) or alkaline earth (Ca^{2+}) cation, or sometimes a transition element (Fe, Mn). Owing to the formation of the strong base MOH, nearly all the primary silicates have an alkaline reaction – the abrasion pH – when placed in CO_2-free distilled water (pH $\cong$ 6). However, Reaction (9.5) proceeds more rapidly to the right when more H^+ ions are present to neutralize the OH^- ions of the strong base. The main sources of this extra acidity in natural, well-drained soils are:
• CO_2 dissolved in the soil water, at partial pressures up to at least 1 kPa (Section 8.2); and
• dissociating carboxyl and phenolic groups of humified organic matter (Section 3.4).

Hydrolysis, neutralization and cation exchange reactions involving a weathering primary silicate, organic matter and the formation of clay mineral residues are illustrated in Fig. 9.2. The reactions leading to the formation of an acid clay (an Al^{3+}-clay) are described in Section 7.2 (see Fig. 7.9).

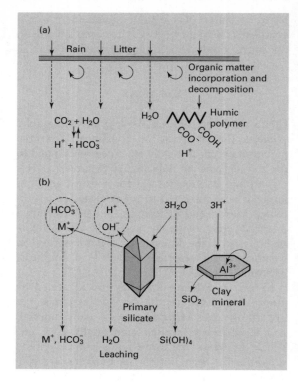

Fig. 9.2 (a) Sources of H^+ ions in soil. (b) Hydrolysis of a primary silicate, weathering products and leaching.

Continuing slow hydrolysis of the adsorbed Al^{3+} ions produces more H^+ ions that are neutralized by further clay mineral decomposition, so the resultant pH rises above that of a pure Al^{3+}-clay or organic acid. Ultimately, clay minerals are completely destroyed to yield gibbsite, or a goethite and hematite mixture in soils with much ferromagnesian minerals. The variation in mineralogy and pH with depth in a highly weathered soil formed *in situ* on basalt illustrates this sequence of events (Table 9.2). Kaolinite is most abundant deep in the profile, where the pH is lowest, but its content decreases towards the soil surface (a zone of more intense weathering) as gibbsite increases and the pH rises steadily.

Oxidation

Weathering of iron-bearing rock minerals in water containing dissolved O_2 and CO_2 is another example of incongruent dissolution. Hydrolysis

Table 9.2 Mineralogical changes in soil formed on weathering basalt. After Black and Waring, 1976.

	Depth (cm)	pH (in water)	Kaolinite (%)	Other clay fraction minerals
Soil surface	0–10			
	10–20	6.5	15–25	H, Q, G
Intensity of	60–90	6.2	30–40	H, Q, G
weathering	120–150	6.1	40–50	H, G
decreases	180–210	6.0	45–55	H
with depth	270–300	5.7	55–65	H
	570–600	4.9	65–75	H
Parent rock	> 600			

H, hematite; Q, quartz; G, gibbsite.

Table 9.3 The relative mobilities of rock constituents. After Loughnan, 1969.

Constituent elements or compounds	Relative mobility*
Al_2O_3	0.02
Fe_2O_3	0.04
SiO_2	0.20
K	1.25
Mg	1.30
Na	2.40
Ca	3.00
SO_4	57

* Expressed relative to Cl taken as 100.

of the mineral releases Fe, present mainly as Fe^{2+}, which then oxidizes to the Fe^{3+} form and precipitates as an insoluble oxyhydroxide – usually either ferrihydrite ($Fe_2O_3.2FeOOH.nH_2O$), which is a necessary precursor of hematite (α-Fe_2O_3), or the stable mineral goethite (α-FeOOH) (Section 2.4). The precipitate may form a coating over the mineral surface that slows the subsequent rate of hydrolysis. Accordingly, in an anaerobic environment where ferric oxide precipitation is inhibited, the rate of weathering of iron-bearing minerals is expected to be faster than under aerobic conditions.

Note that the oxidation of Fe^{2+} to Fe^{3+} according to

$$Fe^{2+} + 2H_2O + {}^1\!/_2O_2 \leftrightarrow Fe(OH)_3 + H^+ \qquad (9.6)$$

is an acidifying reaction. The localized production of H^+ ions will generally accelerate the rate of mineral weathering, as discussed under 'hydrolysis'. Some rock minerals alter directly from one form to another by *in situ* oxidation of Fe^{2+} ions in the lattice – the weathering of biotite mica to vermiculite is one example.

Leaching

Differential mobilities

The mobility of an element during weathering reflects its solubility in water and the effect of pH

on that solubility. Relative mobilities, represented by the Polynov series in Table 9.3, have been established from a comparison of the composition of river waters with that of igneous rocks in the catchments from which they drain. The low mobility of Al and Fe can be explained in terms of the formation and strong adsorption of hydroxy-Al and hydroxy-Fe ions at pH > 4 and 3, respectively, and the precipitation of the insoluble hydroxides. The least mobile element is titanium (Ti), in the form of the insoluble oxide TiO_2, which is used as a reference material to estimate the relative gains or losses of other elements in the profile (Section 2.4).

Percolation depth

Leaching of an element depends not only on its mobility, but also on the rate of water percolation through the soil. In arid areas, even the most soluble constituents, mainly NaCl and to a lesser extent, the chlorides, sulphates and bicarbonates of Ca and Mg, tend to be retained and give rise to saline soils (Section 9.6). As the climate becomes wetter, losses of these salts and silica increase and the soils become more highly leached, except in the low-lying parts of the landscape where drainage water and salts accumulate and modify the course of soil formation. The effect of leaching is well illustrated by the steadily increasing depth to a $CaCO_3$ accumulation layer in soils formed on calcareous loess in the USA, along a line drawn from semi-arid Colorado to humid Missouri (Fig. 9.3).

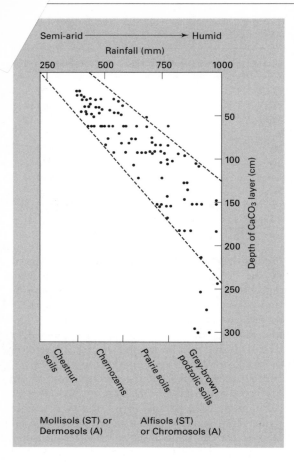

Fig. 9.3 Depth of CaCO₃ accumulation with increasing rainfall (after Jenny, 1941).

Cheluviation

Leaching of immobile Al^{3+} and Fe^{3+} cations can be enhanced by the formation of soluble organo-metal complexes or 'chelates' (Section 3.4) – hence the term cheluviation. Aluminium forms an electrostatic-type complex with the carboxyl groups of humified organic matter in which the average charge per mole of complexed Al is about +1; that is

$$R - COO - [Al(OH)_2(H_2O)_4]^+,$$

where R is the stem of the organic molecule. Ferric iron, on the other hand, is likely to be reduced to Fe^{2+} at the expense of plant-derived reducing agents, with the concurrent formation of a soluble ferrous-organic complex

$$\text{Organic reducing agent } (e^-) + Fe(OH)_3 + 3H^+ \leftrightarrow Fe^{2+}\text{-organic ligand} + 3H_2O. \tag{9.7}$$

The organic compound supplies the electrons that drive this reaction to the right. The stability of the Fe^{2+}-organic ligand complex is favoured by anaerobic conditions (a low redox potential – see Section 8.4); but because H^+ ions are also involved on the left-hand side of Reaction (9.7), the complexes will form at higher redox potentials (more oxidizing conditions), the lower the local pH. Thus, the most active complexing agents are strongly reducing polyphenols that are leached from the canopies of conifers and heath plants, and their freshly fallen litter, and organic acids of the fulvic acid (FA) fraction leached from litter. Polyphenols not involved in cheluviation are oxidatively polymerized and contribute to the characteristic F and H horizons of mor humus that accumulate under coniferous forest and heath vegetation (Section 3.2).

Podzolization

Podzolization involves the cheluviation of Al and Fe from minerals in the upper soil zone and their subsequent deposition, with organic matter, deeper in the profile. Aluminium may also move during podzolization in a colloidal form as part of short-range order hydroxy-aluminosilicates (proto-imogolite, a precursor of imogolite – Section 2.4). The removal of Fe oxides and organic coatings renders the sand grains bleached in appearance, which is a feature of an Ea horizon. As the Fe^{2+}-organic complex is leached more deeply, it becomes less stable through a combination of factors:
• Microbial decomposition of the organic ligands;
• complexing of more metal ions, which increases the cation : ligand ratio in the complex;
• a rise in pH, which favours the oxidation of Fe^{2+} at a given redox potential; and
• the flocculating effect of a higher concentration of salts near the weathering parent material.

The translocated Fe precipitates as ferrihydrite, giving a uniform orange-red colour to the B horizon. The ferrihydrite adsorbs fresh Fe^{2+}-organic complexes, in which the Fe subsequently oxidizes,

Fig. 9.4 An Fe-humus Podzol (Spodosol (ST) or Podosol (ASC)) developed on a well-drained parent material (see also Plate 9.4).

> **Box 9.3** Characteristics of a spodic horizon.
>
> The combination of Bs and Bh horizons generally satisfies the criteria for a spodic horizon, which is the diagnostic horizon of the soil order Spodosol (ST) or Podosol (ASC). A spodic horizon is an illuvial layer > 2.5 cm thick containing 85% or more of 'active' amorphous materials, primarily organic matter and Al, with of without Fe. 'Active' here means materials that have a high pH-dependent charge, a large specific surface area, and a large capacity for water retention. In cultivated soils, a spodic horizon is not part of an Ap horizon, and in uncultivated soils it usually lies below an albic (bleached) horizon, or may be directly beneath a thin O horizon.

and also acts as a coarse filter for humus and Al-organic complexes eluviated from the A horizon. The horizon of Fe and Al oxide accumulation is called a Bs horizon in the British soil profile notation (Table 9.1), and is often overlain by a zone of translocated organic matter accumulation, called a Bh horizon (Fig. 9.4). The end-product of podzolization was called a Podzol in the Great Soil Group terminology of Table 5.4, the exact form of which is determined by the interplay of vegetation, drainage and parent material. Podzolization is favoured by permeable, siliceous parent material, a P : E ratio > 1 for most of the year, and a slow rate of organic decomposition because of the type of litter or acidity of the soil, or both. In Soil Taxonomy, the Podzol is called a Spodosol, characterized by the presence of a spodic horizon (Box 9.3).

Lessivage

Duchaufour (1977) used the term 'lessivage' to describe the mechanical eluviation by water of clay particles from the A horizon without chemical alteration; alternatively, the term clay translocation can be used. The clay is washed progressively downwards in the profile to be deposited in oriented films (argillans) on ped faces and pore walls, gradually forming an horizon of clay accumulation (Bt), or an argillic horizon. Water flow through macropores (old root and worm channels and cracks between peds) facilitates lessivage when the physicochemical conditions are favourable (see below).

Lessivage is unlikely to occur in calcareous soils, or neutral soils that retain a high proportion of exchangeable Ca^{2+} ions, because the clay remains flocculated. Under acid conditions, if exchangeable Al^{3+} and hydroxy-Al ions are predominant on clay surfaces, the clay should also remain flocculated (Section 7.4). However, when adsorbed Al and the Fe of hydroxide films stabilizing microaggregates are complexed by soluble organic compounds, the clay particles are more likely to deflocculate and be translocated. Similarly, when Na^+ exceeds 10–15% of the exchangeable cations, clay deflocculation and translocation are likely, giving rise to solonized soils (Section 9.6).

Table 9.4 Free oxide ratios in the clay fractions of a typical Podzol and Grey-brown Podzolic Soil. After Muir, 1961.

Podzol (Spodosol (ST) or Podosol (ASC))			Grey-brown Podzolic Soil (Alfisol (ST) or Chromosol (ASC))		
Horizon	$\dfrac{SiO_2}{Al_2O_3}$	$\dfrac{Al_2O_3}{Fe_2O_3}$	Horizon	$\dfrac{SiO_2}{Al_2O_3}$	$\dfrac{Al_2O_3}{Fe_2O_3}$
Ea	3.22	5.72	Eb	3.15	3.98
Bh, Bs	1.85	0.84	Bt1	3.05	3.66
Bs	1.83	2.40	Bt2	3.70	3.12
C	2.12	3.71	C	3.21	3.10

The formation of a soil with lessivage (Duchaufour's Sol Lessivé) predisposes it to further pedogenic processes, the course of which depends on drainage in the profile. When the soil is freely drained, with increasing lessivage and acidification in the upper profile, podzolization ensues and the soil may pass through successive stages in the maturity sequence

Lessived Brown Soil → Acid Lessived Soil → Podzolized Lessived Soil → Podzol.

The first three stages correspond to the podzolic or podzolized soils of the Great Soil Group terminology (see Table 5.4). Essentially, the difference between a Podzol and a Sol Lessivé is that clay is destroyed rather than translocated in a Podzol, and the intensity of cheluviation of Fe and to a lesser extent Al is greater in a Podzol. The latter effect is illustrated by the free oxide ratios in different horizons of a Podzol and Sol Lessivé given in Table 9.4. In the latter soil, the ratios $SiO_2 : Al_2O_3$ and $Al_2O_3 : Fe_2O_3$ remain reasonably constant down the profile; but in the Podzol these ratios are high in the Ea horizon and much lower in the Bh, Bs horizon.

A Sol Lessivé is called a 'duplex soil' in Australia (Box 9.4). When the drainage of such a soil is impeded, waterlogging and consequent gleying processes may supervene at variable depths.

Gleying

Collectively, soils that are influenced by waterlogging are called hydromorphic soils (Section 9.5). The most important feature of such soils is gleying (Section 8.4). True gley soils have uniformly

Box 9.4 Soils with an A–B texture contrast.

Lessivage generally leads to a soil in which the clay content increases markedly from the A to the B horizon. Often the A–B boundary is sharp as in the Australian Red Brown Earths (Fig. B9.4.1) and other 'podzolic' soils. Such soils, with a pronounced texture change from the A to

Fig. B9.4.1 Red Brown Earth showing a sharp texture contrast between the A and B horizons (see also Plate B9.4.1).

Box 9.4 *continued*

B horizon (e.g. from sandy loam to clay loam or clay), are widespread in Australia where the generic term 'duplex soil' is used: they also occur in other parts of the world, such as the Piedmont region of eastern USA. In the Australian Soil Classification (Isbell, 2002), duplex soils fall into one of three orders:

• *Chromosols* – soils of strong texture contrast and pH > 5.5 (1 : 5 in water) in the major part of the upper 20 cm of the B2 horizon;

• *Kurosols* – soils of strong texture contrast and pH < 5.5 (1 : 5 in water) in the major part of the upper 20 cm of the B2 horizon;

• *Sodosols* – soils of strong texture contrast and an exchangeable Na^+ content $\geq$ 6% in the fine earth fraction of the major part of the upper 20 cm of the B2 horizon.

The distinction between Chromosols and Kurosols broadly reflects the effect of parent material, climate and the stage reached in the lessivage sequence. In the case of Sodosols, the physicochemistry of the clay environment is overriding.

In Soil Taxonomy (Soil Survey Staff, 1996), soils with marked lessivage are classed as Alfisols, or Ultisols if the base saturation in the subsoil is < 35%.

blue-grey to blue-green colours and are depleted of iron due to its dissolution and removal as Fe^{2+} ions. The reduction of Fe is primarily biological and requires both organic matter and micro-organisms capable of respiring anaerobically. Most of the iron exists as Fe^{2+}-organic complexes in solution or as a mixed precipitate of ferric and ferrous hydroxides, $Fe_3(OH)_8$, which is responsible for the characteristic gley colour (Reaction 8.21). Mottling is a secondary effect in gleys resulting from the re-oxidation of Fe^{2+} in better aerated zones, especially around plant roots and in larger pores. The orange-red mottles have a much higher Fe content than the surrounding blue-grey matrix. Thus, a CG or BG horizon (Table 9.1) frequently has a coarse prismatic structure in which the ped interiors are of uniform gley colour and the exterior surfaces, pores and

root channels have rusty mottling. This pattern is typical of a Groundwater Gley (Entisol (ST) or Hydrosol (ASC)), formed when a watertable resides more or less permanently in the subsoil.

By contrast, in soils with impeded drainage in the upper part of the profile (e.g. many duplex soils), gleying is confined to the A, E or top of the B horizon and is subject to seasonal weather conditions. During summer the soil is usually drier and more aerobic – Fe^{2+} is then oxidized and rusty mottles develop throughout the peds, the surfaces of which may be darkened by organic matter in the A horizon. When the soil wets again (usually in winter), the outer surfaces of the peds develop gley colours. This pattern is typical of a Pseudogley. In colder and wetter climates, the A horizon can remain wet for most of the year even when the mineral soil profile is well drained – this allows a thick O horizon or even peat to develop, and blue-grey gley colours to persist in the mineral soil. This is typical of a Stagnogley (Histosol (ST) or Hydrosol (ASC)). Pseudogleys and Stagnogleys fall into the category of Surface-water Gleys. Profile features of Groundwater and Surfacewater Gleys are illustrated in Fig. 9.5.

The following sections provide illustrative examples of all the foregoing pedogenic processes in action in temperate and tropical soils.

9.3 Freely drained soils of humid temperate regions

The soils over substantial areas of Europe and North America have formed on mixed parent materials of glacial or periglacial origin for a similar period of time. On well-drained sites of slight to moderate relief, two soil sequences are common: one on calcareous, the other on siliceous, parent materials*.

Soils on calcareous, clayey parent material

A general sequence for soil formation with the passage of time, and increased leaching, is shown

---

* Since the examples given are from humid temperate regions in the Northern Hemisphere, no attempt is made in this section to allocate the soils to Australian soil orders.

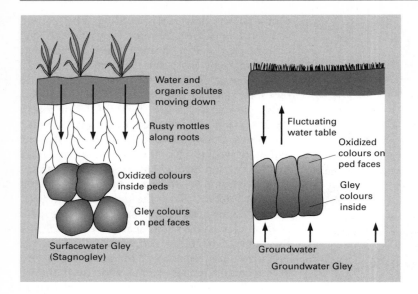

Fig. 9.5 Profile features of a Groundwater Gley and Surfacewater Gley (after Crompton, 1952).

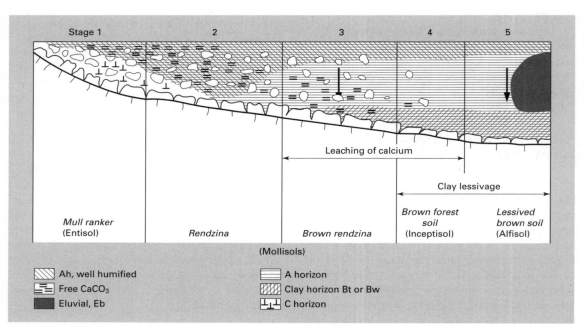

Fig. 9.6 Soil maturity sequence on calcareous clayey parent material under a humid temperature climate (after Duchaufour, 1982).

in Fig. 9.6. The plant succession advances from grassland to deciduous forest. For example, in eastern Canada under deciduous forest, the soil may change from a Brown Forest Soil (Inceptisol) to Grey-brown Podzolic Soil (Alfisol); or to a Grey Wooded Soil (Alfisol) under cooler conditions where conifers flourish. The major difference is that the Grey-brown Podzolic Soil has a mull-like Ah horizon and the Grey Wooded Soil a mor humus layer. Both soil types degenerate, with further leaching, into Brown Podzolic Soils (Ultisols), in which the whole profile becomes acidic and podzolization more obvious. The latter corresponds to the intermediate stages (4 and 5) of Duchaufour's (1982) maturity sequence on siliceous parent material (Fig. 9.7).

Soils on permeable, siliceous parent material

The early stages of this soil sequence, illustrated in Fig. 9.7, feature lessivage under a deciduous forest cover, leading to the formation of a Lessived Brown Soil (Alfisol) and Acid Lessived Soil (Alfisol). With further time and leaching, the podzolizing and lessivage influence strengthens and a Podzolized Lessived Soil forms (Ultisol). Finally, an Iron Podzol or Humus-iron Podzol (Spodosol) forms in the upper part of the deep permeable Eb horizon of the lessived soils. Mackney (1961)

described the terminal stages (5 to 7) of this sequence under oakwoods on heterogeneous acid sands and gravels of the Bunter Beds in the English Midlands. On less acid parent material, however, continued degradation of a Podzolized Lessived Soil (Alfisol) to a Humus-iron Podzol (Spodosol) is normally associated with a change in the vegetation from climax deciduous forest to coniferous forest or heathland.

Soils on mixed siliceous and calcareous parent material

A sequence of soils occurs under beechwoods in the Chiltern Hills of southern England, where the parent material, consisting of siliceous glacial drift mixed with soliflucted material from the underlying Chalk rock, has become acidic after prolonged weathering and leaching (Avery, 1958). As shown

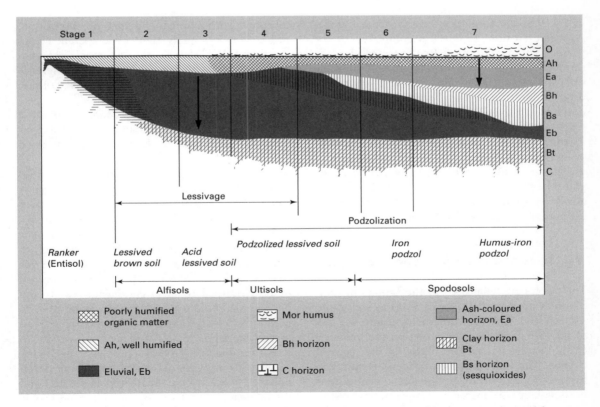

Fig. 9.7 Soil maturity sequence on permeable, siliceous parent material under a humid temperate climate (after Duchaufour, 1982).

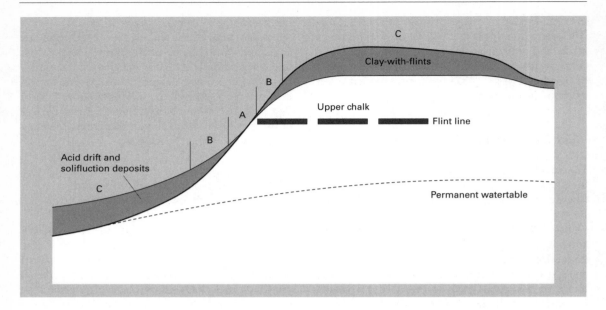

Fig. 9.8 Parent material variations with relief in an old Chalk landscape of the Chiltern Hills (after Avery, 1958).

in Fig. 9.8, soil formation on the steepest slopes A, where erosion is active, is dominated by the Chalk rock and a typical Rendzina (Mollisol) is formed (*cf* stage 2, Fig. 9.6). On the intermediate slopes B, mantled by thin drift and solifluction deposits, immature Brown Forest Soils (Inceptisols) develop. On the deeper drift and acid clay-with-flints of the plateau and gentle slopes C, Lessived Brown Soils and Acid Lessived Soils (Alfisols) occur. Some gleying occurs in the Bt horizons of the lessived soils in the valleys, whereas on the plateau, mor humus accumulates and occasionally Micro-podzols (Spodosols) are found in the Eb horizons of the lessived soils.

9.4 Soils of the tropics and subtropics

The tropics are delineated by the Tropics of Cancer and Capricorn (latitude 23.5° north and south, respectively). The subtropics extend into the low to mid 30° latitudes, depending on the influence of the sea or a large land mass on the regional climate. The following broad generalizations can be made to account for differences in pedogenesis in these lands compared to temperate regions:
• Tropical and subtropical lands are generally hot so that chemical weathering and biochemical

reactions, such as organic matter decomposition, proceed more rapidly than in temperate climes (Section 5.3). Tropical soils have an 'iso' temperature regime, which means that the difference between mean summer and winter soil temperatures is ≤ 5°C.
• Fluctuations in soil water content are more extreme, except for the humid equatorial regions where rainfall is generally high and well-distributed throughout the year.
• Many land surfaces in the tropics and subtropics are pre-Pleistocene in age (> 2 million yr BP) and some date back to the early Tertiary (40–60 million yr BP). Although there has been little recent folding apart from in the orogenic belts along the Andes in South America and in South and South-East Asia, the ancient stable 'shields' and tablelands have been gently upwarped or downwarped to form continental swells and basins on a grand scale. Partial alteration of crystalline rocks (igneous and metamorphic) can extend to depths of 50–100 m and give rise to deep saprolite ('rotten rock') profiles. Thus, many of the soils are very old and deep, having developed on these deeply weathered materials, and the influence of surface organic matter on profile development is much less than in the temperate regions.

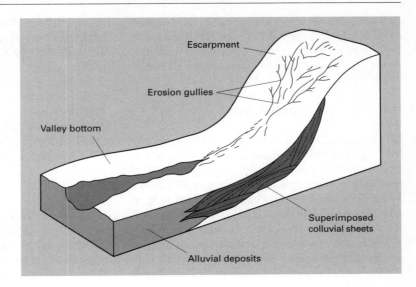

Fig. 9.9 Unconsolidated colluvial and alluvial deposits in an old deeply weathered material in a subtropical landscape (after Thomas, 1994).

• There has been no rejuvenation of parent materials through deposits from Pleistocene ice sheets. However, despite the stability of old tablelands and basins, there have been periods of active erosion and reworking of weathered regoliths on slopes and in valleys. This has produced extensive areas of complex colluvial deposits on slopes, and alluvial deposits in valley bottoms (Fig. 9.9). The former are often underlain by a stone line or stone layer which separates the reworked, unconsolidated material – the pedi-sediment – from the underlying saprolite. The soil formed on the pedi-sediment may bear little relationship to the underlying saprolite.

• As indicated in Chapter 5, age and relief are the more important factors influencing pedogenesis in the tropics and subtropics, although trends associated with differences in parent material are also discernible.

The French pedologist Duchaufour (1982) considered there are three basic kinds of weathering in hot climates that lead to distinctive soil types. In the humid tropics all three may occur together, the change from one to another being determined by local parent material and relief, so that it is possible to treat the processes as stages in the same weathering sequence:

fersiallitization → ferrugination →
ferrallitization.

Fersiallitization

During the fersiallitic process, primary minerals weather to 2 : 1 clays that are partly inherited and partly neo-formed (Section 2.3); Fe^{2+} is also liberated, but is rapidly oxidized and precipitated as ferric hydroxide. The formation of hematite, in particular, as a coating on clays gives many of these soils a distinctive red colour – the process of rubification. Such soils are readily formed in humid Mediterranean-type climates on a wide range of parent materials. Lessivage leading to the development of a red Bt horizon is a characteristic feature, probably aided by the preferential movement of water down cracks as the soil rewets at the end of dry summers. These soils are generally of high base saturation (> 35%) and may be classed in Soil Taxonomy as Mollisols if lacking strong texture contrast between the A and B horizons, or Alfisols (with a strong texture contrast). In the Australian system, the corresponding soil orders would be Dermosols or Chromosols. The Red Brown Earth profile in Fig. B9.4.1 is a good example of a Chromosol (ASC).

In the humid tropics, fersiallitization of relatively young basic rocks may produce Eutrophic Brown Soils (Inceptisols (ST) or Tenosols (ASC)), which, if freely drained, go on to form ferruginous soils (see below). On poorly drained sites, however, such soils develop into mature Vertic

Eutrophic Brown Soils (Vertisols (ST) or Vertosols (ASC)), which have > 35% clay (predominantly 2 : 1 type minerals) and a pronounced tendency to shrink or swell with changing water content.

Ferrugination

In ferrugination, weathering is more complete (only resistant primary minerals such as orthoclase, muscovite and quartz remain), and 1 : 1 clays are synthesized. The Fe oxides formed may or may not be rubified, and 2 : 1 clays are translocated to form a Bt horizon. Depending on the climate, base saturation is variable as in the following examples:
• Under the marked wet and dry seasonality of the savanna regions, base saturation is usually > 35% and Tropical Ferruginous Soils (Alfisols (ST) or Chromosols (ASC)) form.
• In humid climates, however, ferrugination leads to the formation of deep (> 3 m) and more leached Ferrisols (Ultisols (ST) or Chromosols and Kurosols (ASC)). In the humid tropics and on old landscapes, Ferrisols may persist at higher altitudes (where they are humus-rich and very acidic) and on the steeper slopes: otherwise pedogenesis proceeds to the next stage of ferrallitization (see below).

Ferrallitization

For a range of parent materials, on reasonably level but well-drained sites, soil formation in the humid tropics eventually proceeds to the end member of the series:

Tropical Fersiallitic Soil → Tropical Ferruginous Soil → Ferrisol → Ferrallitic Soil.

Virtually all the primary minerals are decomposed leaving quartz, kaolinite (neo-formed), and Fe and Al oxides. Cation exchange capacities < 16 cmol charge (+) per kg clay and the absence of a Bt horizon place most Ferrallitic Soils in the order Oxisol (ST) or Ferrosol (ASC). They are generally acidic (pH 4–5) and of low chemical fertility, but have good physical properties.

Parent material and drainage exert their influence on the degree of ferrallitization in the following way. On acidic parent materials with imperfect drainage, sufficient SiO_2 is retained to allow the neo-formation of kaolinite lower in the profile, which can be up to 50 m deep overall (Ferrallitic Soil with Kaolinite). However, basic rocks weathering on well-drained sites produce shallower soils containing less kaolinite and an abundance of gibbsite, goethite and hematite (Table 9.2). The latter soils are true Ferrallites, which are also known as Krasnozems in Australia.

Laterite

Common in Ferrallitic Soils is an accumulation of a thick zone of sesquioxidic nodules, the genesis of which is described below. While the natural vegetation remains intact, the soil is stable and the sesquioxidic nodules usually soft. However, if the vegetative cover is disturbed or completely removed, the organic-rich A horizon of the old Ferrallitic Soil can be stripped by erosion to expose the sesquioxidic zone, which becomes indurated by dehydration at high temperatures. The result is a pedological structure called laterite. The profile of a typical Ferrallitic Soil with Kaolinite that degenerates to a deep laterite is illustrated in Fig. 9.10a. A typical profile before erosion comprises:
1 An uppermost concretionary zone of Fe oxides and gibbsite, which may be massive if cemented, but loosely nodular or pisolitic* if uncemented;
2 an intermediate mottled zone – orange to red Fe oxide deposits in a pale-grey matrix of kaolinite; and
3 lowermost, a pallid zone of pale-grey to white kaolinite and bleached quartz grains, sometimes weakly cemented by secondary SiO_2, and usually increasing in salt concentration towards the weathering zone.

Ancient laterites that were much eroded and dissected during the Pleistocene often survive as prominent escarpments and crusts in the landscapes of tropical Africa, Australia and South-East Asia. An example of such a laterite escarpment in the Kimberley region of north-west Australia is shown in Fig. 9.11. Truncation by erosion gives rise to a new weathering surface (Fig. 9.10b) on which distinctive contemporary soils have formed:

---

* Resembling small dried peas in size and shape.

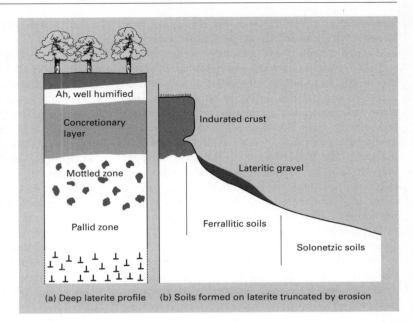

Fig. 9.10 (a) A deep laterite profile (Oxisol (ST) or Ferrosol (ASC)). (b) Soil catena on a truncated laterite (Oxisol (ST) or Ferrosol (ASC)).

Ah, well humified

Concretionary layer

Mottled zone

Pallid zone

Indurated crust

Lateritic gravel

Ferrallitic soils

Solonetzic soils

(a) Deep laterite profile

(b) Soils formed on laterite truncated by erosion

Fig. 9.11 A highly dissected laterite profile in the Kimberley region of northwest Australia (see also Plate 9.11).

impoverished Ferrallitic Soils on the upper two laterite zones, and Solonetz-type soils (Section 9.6) in the pallid zone where salts have accumulated.

The formation of laterite profiles in a landscape is complex. In some cases the zones of sesquioxide accumulation are the residuum of prolonged intense weathering. The pallid zone appears to form through the reduction and solution of Fe under anaerobic conditions in a zone of groundwater influence. The reduced Fe (Fe^{2+}) moves in solution both laterally and vertically, the vertical movement following seasonal, or longer-term, fluctuations in the height of the watertable. Re-oxidation and precipitation of insoluble $Fe(OH)_3$ can contribute to the development of the overlying mottled and concretionary zones.

Fluctuating aerobic–anaerobic conditions associated with variations in regional drainage produce other distinctive pedological features in tropical and subtropical landscapes. For example, organic-complexed Fe^{2+} (Section 9.2), dissolving in the zone of a fluctuating watertable, can move vertically and laterally with the water until the Fe^{2+} is oxidized and re-precipitated at a point in the landscape where better aeration supervenes. Better aeration may occur during the dry season of a seasonal subtropical climate, or when the drainage water encounters a different parent material. The result is the formation an ironstone layer, or plinthite, typical of a Groundwater Laterite (Oxisol (ST) or Ferrosol (ASC)) (Fig. 9.12). The plinthite hardens and becomes impenetrable when exposed at the soil surface or when the regional watertable falls due to dissection by rivers or long-term climate change, and the soil profile becomes drier.

9.5 Hydromorphic soils

Soils showing hydromorphic features (Section 9.2) are found in catenas in many parts of the world: the Vertic Eutrophic Brown Soils and Groundwater Laterites are two examples already cited. Hydromorphism is especially prevalent in humid temperate regions, such as north-west Europe, where Groundwater Gleys and Surfacewater Gleys can be found on most parent materials. Two soil sequences found in the north and west of England and Wales serve to illustrate the point:

1 At altitudes > 300 m, where the annual rainfall is > 1250 mm, the high P : E ratio and low temperatures favour blanket peat formation (see below), even on well-drained parent materials. Proceeding downslope, a typical catena consists of:

Blanket Peat (Histosol)* → Thin Iron-pan Soil (Entisol) → Brown Podzolic Soil (Ultisol) or Podzol (Spodosol).

This sequence is illustrated in Fig. 9.13. Gley colours predominate under the peat and to the base of the wet soil zone as the peat attenuates to a thin O horizon downslope. Approximately in the midslope area, a distinctive iron-pan – shiny black above and rusty-brown beneath – may be found, as Fe^{2+} in solution moves down into the

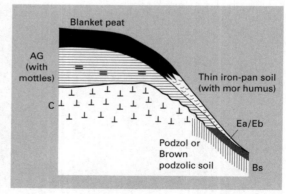

Fig. 9.13 Catenary sequence on cool wet uplands in the west of England and Wales. Symbols as in Fig. 9.6 and 9.7.

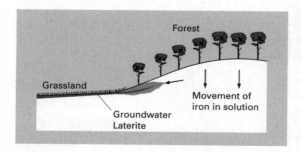

Fig. 9.12 Groundwater Laterite formation in a humid tropical environment.

* See footnote for Section 9.3.

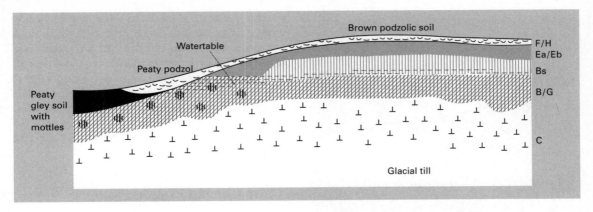

Fig. 9.14 Catenary sequence on cool, imperfectly drained lowlands in the west of England and Wales. Symbols as in Fig. 9.6 and 9.7.

better aerated lower horizons and is oxidized. This soil is an example of a Surfacewater Gley (Stagnogley type).

2 At lower altitudes under an annual rainfall of 750–1000 mm, a catenary sequence similar to that of Fig. 9.14 forms on gentle slopes, often overlying relatively impermeable glacial till. Moving downslope, the sequence consists of

Brown Podzolic Soil (Ultisol) → Peaty Podzol (Spodosol) → Peaty Gley Soil (Histosol).

The better drained upslope soils show rusty mottling along root channels in an otherwise weakly gleyed Ea or Eb horizon underlying a thin Ah. However, as the watertable rises downslope, the whole profile becomes gleyed, apart from weak rusty mottling near the surface, and a Peaty Gley Soil (Histosol) forms. The soil at the bottom of the slope is an example of a Groundwater Gley.

Peat formation

A prerequisite for peat formation is very slow decomposition of organic matter due to a lack of O_2 under continuously wet conditions. This condition is achieved by various combinations of relief and climate (especially a high P : E ratio), together with the permeability and base status of the underlying rocks. The two main types of peat are described as follows.

Basin or topogenous peats

Irrespective of climate, basin peats form where drainage water collects. If the water is neutral to alkaline, even though plant decomposition rates are substantial, the abundant growth of sedges, grasses and trees produces an accumulation of well-humified remains, identified as fen peat. The ash content of this peat is high (10–50%) due to the accession of colloidal mineral particles in the drainage waters. Large areas of fen peat are found, for example, in Florida, the Fens of England and the Okavango Swamp of Botswana, the latter two being areas of relatively low rainfall.

Blanket bog (upland or climatic peat)

A P : E ratio > 1 for all or part of the year and cool temperatures predispose to the formation of blanket peats, described as ombrogenous deposits, over much of Ireland, the Highlands of Scotland, and the Pennines in England, especially on acidic rocks (e.g. Fig. 9.13). The vegetation of hardy grasses, Ericas and *Sphagnum* moss provides an acid litter, and because the only source of water is rain, which is dilute in bases, plant decomposition is extremely slow and humification imperfect. The ash content of this peat is usually < 5%.

The formation of basin and fen peats began earlier in the Holocene period (post-Pleistocene) than did the blanket peats, many of which originated during the warm, wet Atlantic phase lasting

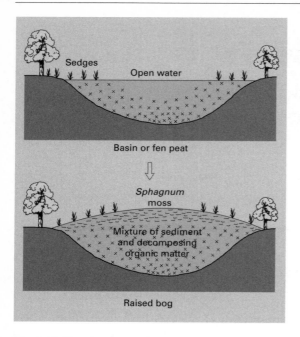

Fig. 9.15 Transformation of a basin or fen peat to a raised bog.

from *c.* 7500 BP to 4000 BP. Where present-day climate is favourable (P : E ratio > 1), basin and fen peats are often transformed to raised bog (Fig. 9.15). The build-up of sediment and peat in a wet hollow eventually raises the living vegetation above the water level so that continued plant growth comes to depend more heavily on rain. Slowly, acid-tolerant and hardy species such as heather, cottonwood and *Sphagnum* invade and displace the species typical of fen peat.

Acid Sulphate Soils

In tidal swamps and brackish sediments, in the presence of abundant organic matter, ferric oxides from the sediments and SO_4^{2-} from the tide-water are reduced according to Reactions (8.19) and (8.20), respectively. The end products are Fe^{2+} ions (solution), H_2S (gas) and insoluble FeS_2 (Reaction 8.22). While these 'sulphidic' sediments remain submerged, or at least saturated through the influence of a high watertable, they are relatively harmless; but if exposed through falling water levels the FeS_2 will oxidize and an Acid Sulphate Soil forms (Entisol (ST) or Hydrosol (ASC)) (Box 9.5).

Box 9.5 Genesis of an Acid Sulphate Soil.

A schematic representation of the build up of sulphidic sediments in a tidal swamp, and their coverage with recent peat and alluvium, is shown in Fig. B9.5.1. Such sediments are widespread in West Africa, South-East Asia, northern South America and northern Australia. Dent and Pons (1995) described three stages in the development of an Acid Sulphate Soil on such sediments:

• *Unripe sulphidic clay* – Under tidal water or peat or alluvium, up to 15% FeS_2, a smell of H_2S, high soluble salts and pH 5–7;
• *raw Acid Sulphate Soil* – Drained but subject to periodic flooding, lower soluble salts, pH 3–5 in the upper oxidized zone and 5–7 in the subsoil, rusty mottles along roots and yellow mottles of jarosite ($KFe_3(SO_4)_2(OH)_6$). The reactions occurring in the presence of O_2, namely

$$FeS_2 + 7/2O_2 + H_2O \leftrightarrow Fe^{2+} + 2SO_4^{2-} + 2H^+ \quad (B9.5.1)$$

and

$$Fe^{2+} + 1/4O_2 + 5/2H_2O \leftrightarrow Fe(OH)_3 + 2H^+, \quad (B9.5.2)$$

are very acidifying and cause the pH to fall as low as 2. Such a low pH is lethal to most plant and animal life;
• *ripe Acid Sulphate Soils* – Drained; gleyed at depth with H_2S and FeS_2 present, pH 5–7 and high salt concentration; overlain by a layer with jarosite and goethite mottles, pH 3–4, coarse prismatic structure; overlain by a layer with many dark brown mottles, prismatic to angular blocky structure, pH 3–4; topped by an Ah horizon of reddish brown clay, low in salts, pH *c.* 4.

Acid Sulphate Soils require large amounts of lime to counteract their acidity. Watertables should be kept as high as possible to minimize FeS_2 oxidation and allow the pH to rise through natural reducing reactions (Section 8.4).

Box 9.5 *continued*

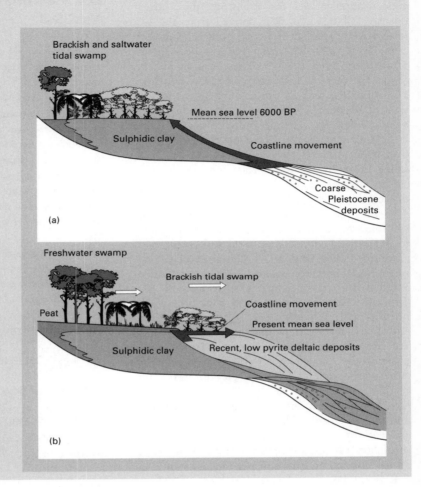

Fig. B9.5.1 (a) Schematic
cross-section of a tropical
delta showing older sulphidic
sediments accumulated under
a steadily rising sea level.
(b) Schematic cross-section
of a tropical delta before
forest clearance showing
accumulation of newer
sediment of low pyrite content
and peat during a period of a
relatively stable sea level (after
Dent and Pons, 1995).

Worldwide, there are some 12–14 million ha of
coastal plains and tidal swamps where the surface soil is already very acid, or will become so
if drained. An area at least as large of sulphidic
sediments, covered during the last 10,000 years
by peat and alluvial deposits, is also at risk of
acidification if cleared and drained.

9.6 Salt-affected soils

Sources of salt

'Salt' is a generic term covering all the soluble
salts in soil, mainly comprising Na, K, Ca and Mg

chlorides and sulphates, and Na bicarbonates
and carbonates. Salts derived from past cycles of
geological weathering have accumulated in oceans,
lakes and groundwater, and act as a source of
salt in soil in the following ways:

• *Inheritance* – sediments laid down under water
naturally entrain salts (connate salts), which are
inherited by the soils that ultimately form on these
sediments;

• *weathering* – during soil formation, salts are
continually released by weathering of residual
primary minerals and secondary silicates;

• *atmospheric deposition* – wind blowing over
the sea whips up spray to form aerosols of salt
crystals, which may subsequently be deposited

on land as dry deposition; or they may act as condensation nuclei for raindrops that subsequently fall on land. Salt derived from the sea in this way is called cyclic salt and can amount annually to 100–200 kg/ha for distances up to 50–150 km from the sea. These are significant inputs for very old soils containing few weatherable minerals, such as those found in the southwest of Western Australia;

• *human activity* – changes in hydrology, through large scale clearing of deep-rooted vegetation or over-application of irrigation water, can result in saline groundwater rising too close to the soil surface. Examples of these processes are discussed in Chapter 13.

A soil is saline if the salt concentration in the soil solution exceeds *c.* 2500 mg/L, corresponding to an electrical conductivity of the soil's saturation extract > 4 dS/m (Box 13.3). Naturally occurring saline soils, common in arid regions, are called White Alkali Soils or Solonchaks (Aridisols (ST))*. Typically, they have a uniform dark-brown Ah horizon with salt efflorescences on the surface and interflorescences within the pores. The salts are brought up by capillary rise from saline groundwater. Below the Ah horizon the soil is structureless and frequently gleyed. Although there is exchange of Na^+ for Ca^{2+} and Mg^{2+} on the clay surfaces, the clay remains flocculated and the structure stable provided that a high salt concentration is maintained. Frequently, sufficient secondary $CaCO_3$ precipitates at depth to form an Ak horizon and the soil pH ranges from 7 to *c.* 8, depending on the partial pressure of CO_2 in the soil air.

Solonchak–Solonetz–Solod soil sequence

Climatic change with rainfall increasing, or a lowering of the regional groundwater, can have serious consequences for a Solonchak, especially if the exchangeable Na^+ content exceeds 10–15% of the CEC (Section 13.2). Dilution of the soil solution as salts are slowly leached causes the diffuse double layers at clay surfaces to expand and increased swelling pressure to be exerted

* Not well accommodated in the Australian Soil Classification; could be a Rudosol or Hydrosol.

Box 9.6 Formation of a Black Alkali Soil.

The increase in pH of a Solonchak soil as it is leached by rain water is due to a combination of effects:

• While Ca^{2+} and Mg^{2+} remain the dominant exchangeable cations, the soil pH is controlled by the activity ratio $(H^+)/\sqrt{(Ca^{2+} + Mg^{2+})}$ (see the ratio law, Section 7.2)) as the soil solution becomes more dilute. Thus, if the soil solution concentration decreases from 10^{-2} to 10^{-4} M, for the activity ratio to remain constant, H^+ ion activity in solution must decrease by a factor of 10: that is, the pH would rise by 1 unit from, say, 7 to 8.

• As the soil is leached and the soil solution becomes more and more dilute, the exchangeable Na^+ percentage increases; but the weakly adsorbed Na^+ ions are gradually displaced by H^+ ions produced by the hydrolysis of water at the clay surface, according to the reaction

$$Na^+\text{-clay} + HOH \leftrightarrow H^+\text{-clay} + NaOH. \qquad (B9.6.1)$$

The pH rises because NaOH is a strong base. The pH may rise as high as 10–10.5, the actual value being dependent on reversible reactions involving NaOH, CO_2, $NaHCO_3$ and Na_2CO_3, that is

$$NaOH + CO_2 \leftrightarrow NaHCO_3 \qquad (B9.6.2)$$

and

$$NaOH + NaHCO_3 \leftrightarrow Na_2CO_3 + H_2O. \qquad (B9.6.3)$$

The soil is now at the 'black alkali' stage because the high pH causes organic matter to disperse, further weakening the cohesion of soil peds and causing a black staining on mineral surfaces.

---

within clay domains (Section 7.4). The soil pH also rises and the soil may degenerate to a Black Alkali Soil (Box 9.6).

Owing to swelling pressure and dispersion of organic matter at high pH, the Na-clay becomes

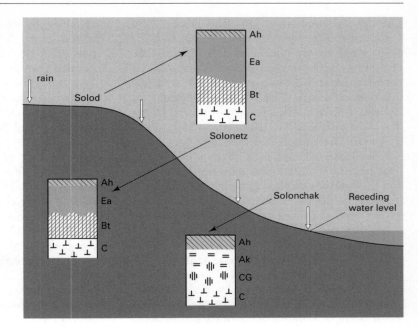

Fig. 9.16 Solonchak–Solonetz–Solod maturity sequence resulting from a declining influence of groundwater (after Kubiena, 1953). Symbols as in Fig. 9.6 and 9.7.

unstable and deflocculates. Clay particles are illuviated into the lower part of the profile where they are flocculated by the higher salt concentration there. The Bt horizon that develops is very poorly drained, often with signs of gleying, and has massive columnar peds capped with organic coatings (Fig. 4.4). The overall process of clay deflocculation and illuviation is called solonization, and the soil formed is a Solonetz (Alfisol (ST) or Sodosol (ASC)).

Continued leaching of a Solonetz leads to the complete removal of soluble salts and most of the basic exchangeable cations from the A horizon, which becomes acid, and the soil becomes a Solodized Solonetz (Alfisol (ST) or Sodosol (ASC)). The H-clay slowly decomposes to release Al^{3+}, Mg^{2+} and SiO_2 (Fig. 9.2). Thus, as salt leaching and acidification progresses down the profile into the Bt horizon, the H,Na-clay becomes an Al, Mg-clay: the massiveness of that horizon decreases and permeability improves, facilitating further leaching. This final step in the degradation of a Solonchak is called solodization, and the resultant soil with its sandy, bleached E horizon and disintegrating Bt horizon is called a Solod or Soloth (Sodic Alfisol (ST) or Kurosol (ASC)). Profiles typical of this maturity sequence are illustrated in Fig. 9.16.

9.7 Summary

A soil profile develops as the weathering front penetrates more deeply into the parent material and organic residues accumulate at the surface. The principal pedogenic processes operating are mineral dissolution (congruent and incongruent), hydrolysis, oxidation-reduction, leaching, cheluviation, clay decomposition, lessivage (clay translocation), gleying and salinization, variations in the intensity of which with depth give rise to one or more soil horizons. Superimposed lithological discontinuities are recognized as soil layers, numbered from the top down.

Plant residues on the surface form a litter layer L, which may give rise to a superficial organic horizon comprising F and H subdivisions, depending on the degree of decomposition and humification. An organic horizon that remains wet for more than 30 consecutive days in most years is called a peaty or O horizon. An O horizon > 30 cm thick on bedrock R, or > 40 cm thick in other situations, is called peat. A mineral horizon is designated A, B and C from the surface down, with C being weathering parent material. This notation is expanded with suffixes to denote particular pedological features (e. g. iron oxide accumulation) or processes (e.g. lessivage, gleying).

Clay translocation which produces profiles with a marked A–B texture contrast ('duplex soils') is predisposed by the cheluviation of sesquioxidic cementing agents, or the deflocculation of clay particles due to high exchangeable Na^+ and low salt concentrations. The resulting 'podzolic' or 'solodic' soils are widespread in Australia, falling into three orders of the Australian Soil Classification – the Chromosols, Kurosols and Sodosols.

Pedogenesis in temperate climates is often dominated by the influence of surface organic matter, and the soluble complexing agents released from it. Many of the parent materials are also young, having been deposited when the Pleistocene ice sheets retreated. With the passage of time, a soil passes through a maturity sequence, the stages of which differ between calcareous and siliceous parent materials, for example, and are also influenced by the vegetation type and drainage. Examples are given in Figs 9.6–9.8. Furthermore, subtle changes in the influence of parent material, climate and drainage can produce a comparable range of profile variations in soils of the same age, as illustrated in Figs 9.13 and 9.14.

In tropical and subtropical regions, temperatures are higher and more constant than in temperate regions. Chemical weathering and organic matter decomposition proceed more rapidly, and extremes of water content are common. Compared with recently glaciated and periglacial areas of the Northern Hemisphere, much of the tropical and subtropical land surface is very old and flat so that soils have been forming on the same parent material for millions of years. However, on slopes and in valley bottoms in these old landscapes, soils have formed on deep colluvial and alluvial deposits of reworked material overlying deeply weathered rock (saprolite). The junction between these frequently disparate materials is often marked by a stone line. The main trends in pedogenesis can be differentiated by the extent of primary mineral decomposition, the leaching of bases and silica, and the neo-formation of clay-fraction minerals (kaolinite and the sesquioxides).

In the humid tropics, the most severe weathering environment, the three processes:

fersiallitization → ferrugination → ferrallitization

can occur as a sequence in time, or side by side, with the change from one to another being determined by local parent material and relief. The end stage – ferrallitization – produces characteristic laterite formations, the soils on which are called Oxisols (ST) or Ferrosols (ASC).

Hydromorphic processes (including gleying) can affect soil development under any climate. In cold wet climates, Surfacewater Gleys, including Pseudogleys and Stagnogleys (Entisols (ST) or Hydrosols (ASC)) occur on many parent materials, even ones that are well-drained. Continuously wet and cold climates produce Peaty Gleys and Blanket Peats (Histosols (ST) or Organosols (ASC)). A watertable within the soil profile for all or part of a year gives rise to a Groundwater Gley (Histosol (ST) or Entisol (ST), depending on the organic matter content, or Hydrosol (ASC)). Potentially problematic variants of the Groundwater Gleys found in tidal marshes and forming on drained sediments in river deltas are the Acid Sulphate Soils (Entisols (ST) or Hydrosols (ASC)).

Salinization due to inherited salts (connate salt), cyclic salt or the rise of saline groundwater produces Solonchak soils (Aridisols (ST)) which, on being leached undergo solonization (Solonetz phase – Alfisols (ST) or Sodosols (ASC)), and finally solodization (Solod or Soloth phase – Sodic Alfisols (ST) or Kurosols (ASC)).

References

Avery B. W. (1958) A sequence of beechwood soils on the Chiltern Hills, England. *Journal of Soil Science* **9**, 210–24.

Avery B. W. (1980) *Soil Classification for England and Wales.* Soil Survey Technical Monograph No. 14. Lawes Agricultural Trust, Harpenden.

Black A. S. & Waring S. A. (1976) Nitrate leaching and adsorption in a krasnozem from Redland Bay, Qld. II. Soil factors influencing adsorption. *Australian Journal of Soil Research* **14**, 181–8.

Butler B. E. (1959) *Periodic Phenomena in Landscapes as a Basis for Soil Studies.* CSIRO Soil Publication No. 14. CSIRO Publishing, Melbourne.

Crompton E. (1952) Some morphological features associated with poor soil drainage. *Journal of Soil Science* **3**, 277–89.

Dent D. L. & Pons L. J. (1995) A world perspective on acid sulphate soils. *Geoderma* **67**, 263–76.

Duchaufour P. (1977) *Pédologie. I. Pédogenese et Classification.* Masson, Paris.

Duchaufour P. (1982) *Pedology Pedogenesis and Classification* (Translated by T. R. Paton). Allen & Unwin, London.

FAO (1998) *World Reference Base for Soil Resources.* World Resources Report No. 84. FAO, Rome.

International Society of Soil Science (1967) Proposal for a Uniform System of Soil Horizon Designations. *Bulletin of the International Society of Soil Science* 31, 4–7.

Isbell R. F. (2002) *The Australian Soil Classification*, revised edn. Australian Soil and Land Survey Handbooks Series Volume 4. CSIRO Publishing, Melbourne.

Jenny H. (1941) *Factors of Soil Formation.* McGraw-Hill, New York.

Kubiena W. L. (1953) *The Soils of Europe.* Murby, London.

Loughnan F. C. (1969) *Chemical Weathering of the Silicate Minerals.* Elsevier, New York.

Mackney D. (1961) A podzol development sequence in oakwoods and heath in central England. *Journal of Soil Science* 12, 23–40.

McDonald R. C., Isbell R. F., Speight J. G., Walker J. & Hopkins M. S. (1998) *Australian Soil and Land Survey Field Handbook*, 2nd edn. (Reprinted). Australian Collaborative Land Evaluation Program, Canberra.

Muir A. (1961) The podzol and podzolic soils. *Advances in Agronomy* 13, 1–56.

Paton T. R., Humphreys G. S. & Mitchell P. B. (1995) *Soils. A New Global View.* Yale University Press, New Haven.

Soil Survey Staff (1996) *Keys to Soil Taxonomy*, 7th edn. United States Department of Agriculture Natural Resources Conservation Service, Washington DC.

Thomas M. F. (1994) The Quaternary legacy in the tropics: a fundamental property of the land resource, in *Soil Science and Sustainable Land Management in the Tropics* (Eds J. K. Syers & D. L. Rimmer). CAB International, Wallingford, pp. 73–87.

Further reading

Avery B. W. (1990) *Soils of the British Isles.* CAB International, Wallingford.

Birkeland P. W. (1999) *Soils and Geomorphology*, 3rd edn. Oxford University Press, New York.

Duchaufour P. (1995) *Pédologie: Sol, Végétation, Environment.* 4th edn. Masson, Paris.

Juo A. S. R. & Franzluebbers K. (2003) *Tropical Soils. Properties and Management for Sustainable Agriculture.* Oxford University Press, New York.

Young A. & Young R. (2001) *Soils in the Australian Landscape.* Oxford University Press, Melbourne.

Example questions and problems

1 (a) What is the difference between a soil horizon and a soil layer?
  (b) Define an epipedon.
  (c) What are the requirements for an O horizon to be recognized as peat?

2 Identify any four processes, operating vertically or laterally, by which a soil profile develops in the landscape.

3 (a) When a marine sediment containing 2.5% (by weight) of finely-divided iron pyrites (FeS_2) is reclaimed by drainage, the upper 0.5 m becomes aerobic. (i) Write equations for the reaction of FeS_2 with O_2, and (ii) describe what happens to the sediment pH in the top 0.5 m.
  (b) Once settled after drainage, the bulk density of the sediment is 1.5 Mg/m^3. Calculate how much lime (as pure $CaCO_3$) needs to be mixed into the top 25 cm of each cubic metre to neutralize the acidity produced on complete oxidation of the FeS_2 present.

(Molar mass of Fe, S, Ca, C and O = 56, 32, 40, 12 and 16 g, respectively).

4 (a) Define the characteristic feature of a duplex soil.
  (b) Give an example of a soil order representing duplex soils in (i) the Australian Soil Classification and (ii) Soil Taxonomy.

5 (a) What is the threshold clay content (% by weight) that defines a 'clay soil'?
  (b) Name the dominant type of clay mineral in a Vertisol (ST) or Vertosol (ASC)?

6 (a) Describe the factors predisposing to the formation of blanket peat.
  (b) (i) What is plinthite, and (ii) what happens to plinthite on exposure to air?

7 (a) What is the origin of cyclic salt?
  (b) Write an equation for the hydrolysis of a Na-clay.
  (c) Give an estimate of the pH attained in a Black Alkali Soil.

Chapter 10

Nutrient Cycling

10.1 Nutrients for plant growth

The essential elements

There are 17 elements without which green plants cannot grow normally and reproduce. On the basis of their concentration in plants, these essential elements are subdivided into:

• The macronutrients C, H, O, N, P, S, Ca, Mg, K and Cl which occur at concentrations > 1000 mg/kg (plant dry matter basis); and

• the micronutrients Fe, Mn, Zn, Cu, B, Ni and Mo which are generally < 100 mg/kg.

Carbon, H and O are only of passing interest, because they are supplied as CO_2 and H_2O, which are abundant in the atmosphere and hydrosphere; likewise Cl, which is abundant and very mobile as the Cl^- ion. Of the others, with the exception of N which comprises 78% of the atmosphere, the major sources of the essential elements are weathering minerals in the soil and parent material. *The nutrient supplying power of a soil is a measure of its fertility*, the full expression of which depends on the influence of the plants grown, the environment, and soil management (Part 3).

Plants absorb other elements, some in considerable quantities such as Na, which is essential for halophytes, e.g. saltbush (*Atriplex vesicaria*), and beneficial for many other species because it can partially substitute for K (Marschner, 1995). Sodium is also essential for animals and humans. Silicon (Si) is essential for certain wetland grasses, and beneficial for a number of species of economic importance such as paddy rice, sugar cane and sugar beet. Cobalt (Co) is essential for symbiotic N_2 'fixation' in legumes, and non-legumes such as alder (Section 10.2), and is also essential for ruminant animals. Selenium (Se) and iodine (I), although of no benefit to plants other than a few algae, are vital for the health of animals and humans. Deficiency of Se causes white muscle disease in animals, and I deficiency causes goitre in humans. Vanadium (Va) can substitute for Mo to some extent in the N_2 fixation process of bacteria. Claims have been made in China that the rare earth elements lanthanum (La) and cerium (Ce) are beneficial for plant growth, but this has not been substantiated elsewhere.

Nutrient cycling

The flow of nutrients in the biosphere is continuous between three main compartments:

• An *inorganic store*, chiefly the soil;

• a *biomass store*, comprising living organisms above and below ground; and

• an inanimate *organic store*, on and in the soil, formed by the residues and excreta of living organisms.

The general nutrient cycle is illustrated in Fig. 10.1. The relative size of the nutrient stores differs for the various elements, as do the importance of the pathways for inputs and outputs and the partitioning of an element between 'soil' and 'non-soil'. Transformations within each store often involve complex chemical and biochemical processes.

There are also differences in the way elements are distributed within the soil. Elements that are released at depth by mineral weathering are

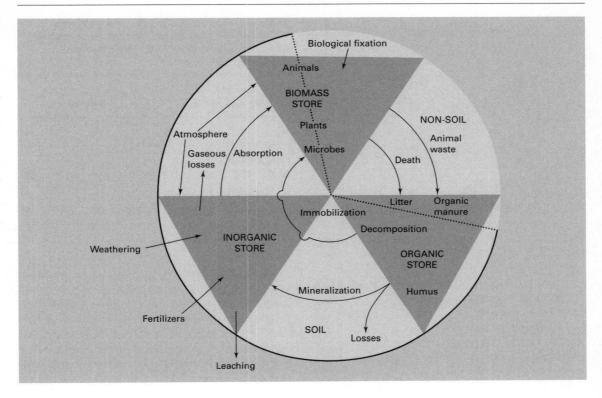

Fig. 10.1 Fundamentals of a nutrient cycle.

absorbed by plant roots and transported to the shoots, to be deposited finally on the soil surface in litter or animal excreta. Accessions from the atmosphere add to the surface store. Nitrogen always accumulates in the organic-rich A horizon, with the content declining gradually with increasing depth; P is similar, but the decline with depth is more abrupt because phosphate ions are strongly adsorbed and retained near the surface (Fig. 10.2). Sulphur, like N, accumulates as organic-S in the surface of temperate-region soils, but not necessarily in tropical soils, due to the greater mineralization of organic-S and leaching of SO_4^{2-} ions. Many of the latter soils show a bimodal S distribution, with organic-S in the surface and SO_4^{2-} ions adsorbed in the subsoil at sites of positive charge (Fig. 10.2b). The distribution of most cations – Fe^{3+}, Ca^{2+}, Mg^{2+}, Mn^{2+}, Cu^{2+}, Zn^{2+}, Ni^{2+} and K^+, for example – usually correlates with clay accumulation. But the 'heavy metals' Cu, Zn, Fe and Mn are also associated with soil organic matter.

The inorganic store

For elements such as P, K, Ca and Mg, plant roots tap only a small fraction of the soil's inorganic store during a single growth season. This fraction is withdrawn from an 'available pool', which is made up of:
• Ions in the soil solution; and
• exchangeable ions adsorbed by clay minerals and organic matter.
Less readily available ions are held in sparingly soluble compounds such as gypsum, calcite and inorganic phosphate compounds, or held in non-exchangeable forms in clay lattices (K^+ and NH_4^+).
   Inputs to the available pool occur by:
1 Weathering of soil and rock minerals (Sections 5.2 and 9.2);
2 precipitation and dry deposition (Section 10.2);
3 mineralization of organic matter (Section 3.1 and following sections); and
4 the application of organic manures, organic and inorganic fertilizers (Section 11.1 and Chapter 12).

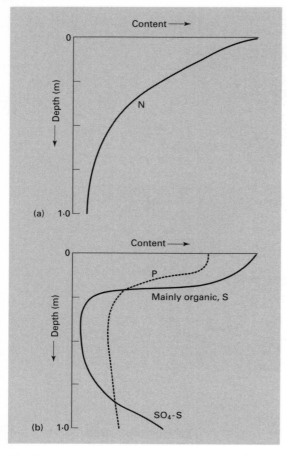

Fig. 10.2 (a) and (b) Distribution of N, P and S in a soil profile.

Table 10.1 The distribution of selected elements (kg/ha) between plant biomass and soil for contrasting vegetation types in West Africa. After Nye and Greenland, 1960.

Plant biomass (including roots and litter)				Soil (0–30 cm)			
N	K	Ca	Mg	N	K	Ca	Mg
Mature forest (40 yr old)							
2040	910	2670	350	4590	650	2580	370
Imperata cylindrica grassland							
46	105	13	24	1790	190	2910	380

Although small (0.5–2 t C/ha), the soil microbial biomass has been identified as the 'eye of the needle' through which all C returned to the soil must eventually pass. Virtually the same is true of N – 98% or more of the soil N is in organic combination, and the metabolism of this N by soil micro-organisms is intimately associated with that of C (Section 3.1).

The organic store

The cycle shown in Fig. 10.1 is completed by the decomposition of plant litter and animal excreta by the soil mesofauna and micro-organisms, a topic discussed in Chapter 3. Specific aspects of mineralization, as they affect the availability of individual nutrients to plants, are discussed in the following sections.

10.2 The pathway of nitrogen

Mineralization and immobilization

Depending on the soil type and environmental conditions, the quantity of soil N in the root zone (*c*. 1 m deep) ranges from 1 to 10 t/ha, much of which occurs in the top 15–20 cm. Fertile agricultural soils in Britain, for example, have an average N content of 4000 kg/ha to 15 cm depth (Anon., 1983). Organic N is transformed during microbial decomposition of organic matter according to the general reaction

The biomass store

Storage of nutrients in the plant biomass depends on the mass of vegetation produced. As Table 10.1 shows, the store of K, Ca and Mg (less so N) in mature forest under high rainfall in the tropics is comparable to that in the top 30 cm of soil; but the store of these elements in relatively unproductive grassland is much less than in the topsoil. K is a notable exception because it cycles rapidly through the soil-plant system. All biomass is important in nutrient cycling, irrespective of the size of the store, because the remnants of living organisms and their excreta are the substrate on which soil micro-organisms feed.

Organic N (proteins, nucleic acids) $\rightarrow NH_4^+ + OH^-$.

$$(10.1)$$

This step, referred to as ammonification, is obviously an alkalizing reaction. Subject to any limitations imposed by soil temperature, water content, pH and aeration, NH_4^+ is oxidized to NO_3^- according to the reaction

$$NH_4^+ + 2O_2 \rightarrow NO_3^- + H_2O + 2H^+, \qquad (10.2)$$

which is called nitrification. A detailed description of nitrification is given in Section 8.3. Reactions (10.1) and (10.2) together describe the process of mineralization. Note that two H^+ ions are produced per mole of NH_4^+ oxidized, so the net effect is to produce one mole of H^+ per mole of NO_3^- formed. Thus, mineralization of organic N to NO_3^- is an acidifying process that leads to accelerated acidification in soils prone to NO_3^- leaching (Section 11.3).

NH_4^+ and NO_3^- comprise the pool of mineral N on which plants feed. Ammonium is held as an exchangeable cation that is readily displaced into the soil solution and, with NO_3^-, moves to the roots in the water absorbed by the transpiring plant – a process of mass flow. Both ions can also move from regions of higher to lower concentration by diffusion (mass flow and diffusion are discussed under 'phosphate availability'). Thus, all the mineral N (except for non-exchangeable NH_4^+ in micaceous clay crystals) is available to the plant.

The balance between mineralization and immobilization of N during decomposition is dependent on the C : N ratio of the substrate (see Box 3.4). However, the net mineralization rate can be described by the equation

$$\frac{dN}{dt} = -kN, \qquad (10.3)$$

where N is the amount of organic N per unit soil volume and k is called the decay coefficient, which has the units of reciprocal time ($1/t$). Over an interval of 1 year or more, the amount of N mineralized (N_{min}) is given by kN, the value of which is highly variable, ranging between 10 and 300 kg/ha (Box 10.1). Whether this is enough to meet the needs of plants depends on their uptake of N, as discussed below.

Box 10.1 Variation in net mineralization in soils.

The decay coefficient k in Equation 10.3 is an averaged parameter. There are large differences in net mineralization rate between fresh residues (from crops or pastures), animal excreta (dung and urine), and resident soil organic matter (SOM). For example, fresh residues of C : N ratio < 25 can have an effective $k \approx 0.3$ (1/yr) in the year after their addition to soil, compared to a k value of 0.01–0.03 (1/yr) for SOM. Thus, N_{min} depends on:
• The amount of residues and organic manures returned to the soil, relative to SOM;
• the C : N ratio of the residues and manure, their placement in the soil and their distribution (which affects their accessibility to micro-organisms);
• soil management (soil pH and whether or not the soil is cultivated or drained); and
• environmental factors (especially temperature and effective rainfall).

The amount of N in crop residues in the UK ranges from as low as 20 (linseed) to 150 kg/ha (oilseed rape), excluding any N returned from roots fragments and rhizosphere deposition (Section 3.1), which is very difficult to measure. N returned in residues under highly productive grassland can be much larger (300–400 kg/ha/yr), and contributes to the large 'flushes' of net mineralization observed if the grassland is ploughed out. For example, in the UK, N_{min} in the year after ploughing a pasture can range from 60 (1 year old sward, no N applied) to 300 kg/ha (old sward, 300 kg N/ha/yr fertilizer input). In south-eastern Australia, mineral N contents of 100–200 kg N/ha to 1.2 m depth have been measured at the end of summer under annual ryegrass-subterranean clover pastures. Both species die out during the hot, dry summers and add to the pool of plant residues, dung and urine that is mineralized.

Table 10.2 N contents of selected agricultural crops.

	N content (kg/ha)
Spring malting barley, 4.5 t grain/ha	80
Wheat, 6 t/ha (grain and straw)	120
Potatoes, 30 t/ha (fresh weight)	200
Grass, 10 t/ha (dry matter)	250
Winter oilseed rape, 3.2 t/ha	250
Maize, 13 t/ha (grain and stover)	360
Sugar cane, 120 t/ha (fresh weight)	450

Plant uptake

In natural ecosystems, plant growth is generally slower than in agricultural systems and the annual uptake of N is relatively small: Cole *et al.* (1967) quote figures of 25–78 kg N/ha/yr for a coniferous forest in North America. One of the highest figures is 248 kg N/ha/yr estimated by Nye and Greenland (1960) for mature secondary rainforest in tropical Ghana. Cultivated crops are more demanding, so N contents are generally high (80–450 kg N/ha per crop, as shown in Table 10.2) and the mineralizing capacity of the soil is often insufficient to achieve optimum growth. In this case, the magnitude of gains and losses in the N cycle, as shown in Fig. 10.3, are of great significance.

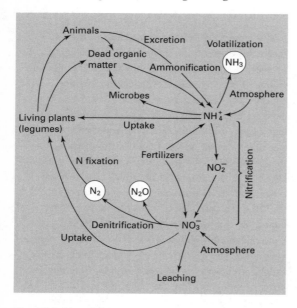

Fig. 10.3 N cycling and transformations in the soil-plant ecosystem.

Gains of soil N

Atmospheric inputs

Apart from N released as N_2 and N_2O during denitrification (Section 8.4), N enters the atmosphere as the oxides NO and NO_2 that are formed during the combustion of fossil fuels (the primary source), biomass burning and by lightning discharges. Nitrogen is also released as NH_3 gas by volatilization from manures, rotting vegetation and soil, especially calcareous soil following the surface application of ammonium fertilizers (Section 12.2). Dust blown into the air may carry both NH_4^+ and NO_3^- ions, as well as organic N compounds.

Nitrogen in the atmosphere returns to the soil-plant system in several forms. Oxides of N react with hydroxyl free radicals to form nitric acid (HNO_3), which contributes to 'acid rain' (more than half of the total rainfall acidity in the UK). Some NH_3 is absorbed through the stomata in plant leaves while the remainder is dissolved in rain or forms salts, such as $(NH_4)_2SO_4$ (Equation 10.11), which are dissolved in the rain or deposited on soil and plant surfaces as dry deposition. The total N input from the atmosphere ranges from 5 to 60 kg/ha/yr, depending on how bad the air pollution is. Exceptionally, total deposition may reach 100 kg N/ha/yr in forests close to intensive livestock activities (large feedlots and dairies). Within the total it is difficult to identify precisely the relative contributions of dry deposition and rainfall, but a ratio between 2 and 3 : 1 is reasonable. The average atmospheric N contribution in northwest Europe (excluding biological N_2 fixation) is between 20 and 30 kg/ha/yr, which is small compared with other inputs to the N cycle for intensively cropped agricultural soils, but highly significant for natural ecosystems.

Fertilizers and manures

Under natural vegetation, plant litter and animal manures provide the bulk of the N returning to the soil. In agriculture, however, where much of the crop is removed from the land, the input of N in chemical fertilizers (Chapter 12) or through the growth of legumes is necessary to boost the soil N supply for subsequent crops.

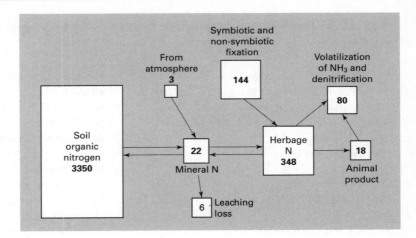

Fig. 10.4 N inputs, outputs and the main soil N components for a soil under a young clover-grass pasture in New Zealand. The quantity of N is given in kg/ha for soil components and as kg/ha/yr for fluxes (after White and Sharpley, 1996).

Biological N_2 fixation

A small minority of micro-organisms is exceptional in being able to reduce molecular N_2 to NH_3 and incorporate it into amino acids for protein synthesis. These micro-organisms live either independently in the soil (as free-living heterotrophs or photosynthetic organisms), or in symbiotic association with plants. The process of N_2 reduction is called nitrogen fixation, and it enables the organisms to grow independently of mineral N.

The annual turnover of biologically fixed N_2 in the biosphere is *c.* 150 Mt, which is small compared with the total of *c.* 15 Gt N in terrestrial vegetation and *c.* 1500 Gt N in SOM. Nevertheless, in natural ecosystems, and pastures based on legumes such as clover, the input of biologically fixed N_2 is a most important component of the N cycle. An example of N inputs and losses for a white clover – ryegrass pasture in New Zealand is given in Fig. 10.4. The biochemistry of N_2 fixation is summarized in Box 10.2.

Free-living N_2 fixers

Regulation of N_2 fixation in aerobic organisms is complicated because the reduced nitrogenase enzyme must be protected from O_2, yet O_2 is required for the cell's respiration and oxidative phosphorylation to produce ATP. It is not surprising, therefore, that the ability to fix N_2 is not widespread among non-photosynthetic, aerobic organisms. The more important of the free-living

Table 10.3 The major free-living N_2 fixing organisms.

Bacteria	Aerobes	*Azotobacter*, *Beijerinckia* spp.
	Micro-aerobic	*Azospirillum brasilense*, *Thiobacillus ferrooxidans*
	Facultative anaerobes	*Klebsiella* spp, *Bacillus* spp.
	Obligate anaerobes	*Clostridium pasteuranium*, *Rhodospirillum*, *Chlorobium* and *Desulphovibrio* spp.
Cyanobacteria (blue-green algae)	Photosynthetic aerobes	*Nostoc*, *Anabaena*, *Cloeocapsa* spp.

or non-symbiotic N_2 fixing organisms are listed in Table 10.3.

The distribution and activity of N_2 fixing organisms in soil is determined by factors such as aeration, pH, supply of organic substrates, mineral N concentration and the availability of micronutrients, especially Cu, Mo and Co. *Azotobacter* and the blue-green algae are restricted to neutral and calcareous soils, but spore-forming anaerobes like *Clostridium* are tolerant of a wide range of pH. At high concentrations of mineral N, the N_2 fixers are out-competed by more aggressive heterotrophic species. Thus, the rhizosphere of plant roots, where mineral N is low and there is an abundance of exudates of high C : N ratio, is a favoured habitat for free-living N_2 fixing organisms. Many of these

Box 10.2 The biochemistry of N_2 fixation in free-living organisms and symbiotic associations.

The splitting of the dinitrogen molecule ($N \equiv N$) and the reduction of each part to NH_3 are catalysed by the nitrogenase enzyme, which consists of two proteins – a Fe protein and a Mo–Fe protein, roughly in the ratio of $2 : 1$.
The reaction may be written

$$N_2 + 8H^+ + 8e^- + n\text{Mg.ATP} \rightarrow$$
$$2NH_3 + n\text{Mg.ADP} + n\text{PO}_4^{3-}, \qquad (\text{B}10.2.1)$$

for which the prerequisites are:
• A reductant of low redox potential to supply electrons, e^- (usually ferredoxin);
• an ATP-generating system; and
• a low partial pressure of O_2 at the site of nitrogenase activity.

On average, the amount of N fixed in symbiotic associations, especially involving legumes, is an order of magnitude greater than that fixed by free-living organisms. This reflects the large amount of

energy available from carbohydrate metabolism in a higher plant, and the favourable environment for N_2 reduction created within a root nodule (see Fig. 10.5). Within a nodule, the site of fixation is the bacteroid surface to which electrons are transported by a reduced co-enzyme and donated to $N \equiv N$ bound to nitrogenase. Synthesis of the nitrogenase is induced in the bacteria under favourable conditions within the nodule where the O_2 partial pressure is regulated on a microscale. Control of O_2 partial pressure is achieved through a barrier to the diffusion of dissolved O_2 in the nodule's cortex (Marschner, 1995), and the presence of the pigment leghaemoglobin (which gives a characteristic pink colour to nodular tissue that is fixing N_2). Because the pigment has a very high affinity for O_2, it is able to transfer O_2 to the respiring bacteroids while keeping the free O_2 concentration in the membrane-bound sacs very low.

organisms and associated nitrogenase activity have been identified in the rhizosphere of variety of crops, including wheat, maize, pasture grasses and sugar cane. This has been called associative N_2 fixation.

Nitrogenase activity in soil is detected by the acetylene reduction technique. When nitrogen is excluded from the nitrogenase site by an excess of acetylene (C_2H_2), this gas is reduced to ethylene (C_2H_4) according to the reaction

$$C_2H_2 + 2H^+ + 2e^- \rightarrow C_2H_4, \qquad (10.4)$$

and the C_2H_4 detected by gas chromatography. Although the identification of functional nitrogenase activity does not necessarily mean that N_2 is being fixed, an estimate of N fixed can be made if a molar ratio of $1 : 3$ for N_2 fixed to C_2H_2 reduced is assumed.

Because of the high energy consumption of N_2 fixation (c. 16 mol ATP per mol N_2 reduced), photosynthetic organisms tend to make the greatest contribution to non-symbiotic N_2 fixation: for

example, up to 50 kg N/ha/yr is attributed to blue-green algae and photosynthetic bacteria in paddy rice culture. Estimates for other farming systems lie in the range 0–10 kg N/ha/yr. There is some evidence that contributions from free-living organisms may be higher in temperate deciduous woodlands (20–30 kg N/ha/yr), and exceptionally up to 100 kg N/ha/yr in some tropical forests.

Symbiotic N_2 fixation

Symbiosis denotes the cohabitation of two unrelated organisms that benefit mutually from the close association. In the case of N_2 fixation, the invasion of roots of the host plant by a microorganism (the endophyte) culminates in the formation of a nodule in which carbohydrate is supplied to the endophyte, and amino acids formed from the reduced N are made available to the host.

There are three main symbiotic associations:
• Nodulating species of the family Fabacae (Leguminosae) in which the endophyte is a

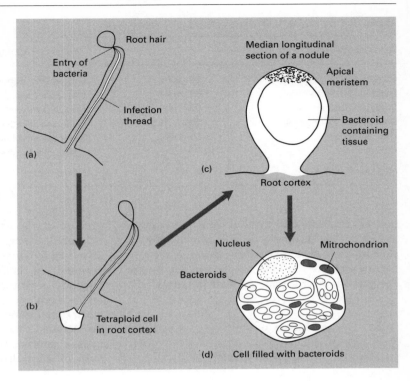

Fig. 10.5 Sequence in nodule formation on a legume root. (a) Bacteria invade a root hair through an invagination of the cell wall – the infection thread. (b) Tetraploid cell penetrated – this cell and adjacent cells are induced to divide and differentiate to form a nodule. (c) Bacteria released, divide once or twice and enlarge to form bacteroids that respire but cannot reproduce. (d) Groups of bacteroids enclosed within a membrane and N_2 fixation begins (after Nutman, 1965).

bacterium of the genera *Rhizobium* and *Bradyrhizobium**;
• non-legumes, including the genera *Alnus* (alder), *Myrica* (bog-myrtle), *Elaeagnus* and *Casuarina*, in which the endophyte is usually an actinomycete (*Frankia*); and
• lichens, which are an association of a fungus and blue-green alga, and the aquatic fern *Azolla* which lives in association with the blue-green alga *Nostoc*.

Judged by their numbers and widespread distribution, legumes are by far the most important symbiotic N_2 fixers. The *Rhizobium* and *Bradyrhizobium* bacteria, when not infecting a legume root, live as heterotrophs in the soil. The success of root infection, nodule initiation and subsequent N_2 fixation is determined by one or more factors that operate at any one of several points in the complex sequence of events, shown diagrammatically in Fig. 10.5. These factors include the following:

* The genus *Bradyrhizobium* includes slow-growing rhizobial strains isolated from soil that also produce alkali, as opposed to acid, when grown in pure culture.

• Low pH and low Ca that inhibit root hair infection and nodule initiation;
• competition between effective and non-effective bacterial strains at the time of root infection. A non-effective strain produces many nodules that do not fix N_2, yet their presence inhibits nodulation by an effective strain;
• because of the large energy demand of fixation, a plant will preferentially absorb mineral N, especially NH_4^+, when it is readily available, rather than fix N_2;
• inadequate photosynthate supply and water stress reduce a nodulated root's fixation capacity;
• nodulated legumes need more P, Mo and Cu than non-nodulated legumes, and have a unique requirement for Co;
• nodulation and N_2 fixation are generally stimulated by simultaneous infection of the roots by arbuscular mycorrhizas (Section 10.3), due mainly to the enhanced supply of P to the plant.

Because of these environmental and nutritional constraints and inherent differences in host-bacterial strain performance, the quantity of N fixed is highly variable; but estimates for

Table 10.4 Estimation of N₂ fixation by legumes (kg N/ha/yr).

Temperate species		Tropical and subtropical species	
Clovers (*Trifolium* spp.)	55–600	Grazed grass-legume pastures:	
Lucerne (*Medicago sativa*)	55–400	*Stylosanthes* spp.	10–30
Soybeans (*Glycine max*)	90–200	*Macroptilium atropurpureum*	44–129
Broad beans (*Vicia faba*)	200	Grain and forage legumes:	
Peas (*Pisum* spp.)	50–100	Beans (*Phaseolus vulgaris*)	64
		Pigeon pea (*Cajanus cajan*)	97–152
Median value	*c.* 200	Median value	*c.* 100

several temperate and tropical legumes are given in Table 10.4. Legume N is made available by the sloughing-off of nodules, through the excreta of grazing animals returned to the soil, or by the decomposition of legume residues in the soil.

Losses of soil N

Crop removal

Much of the crop N is removed in harvested material, except under grazing conditions when approximately 85% of the N is returned in animal excreta. The fate of cereal crop residues is important, because the straw from a 4 t/ha wheat crop contains 20–25 kg N. The straw may be baled and used as animal bedding, eventually forming farmyard manure, or burnt, when most of the N is volatilized, or ploughed back into the soil.

N leaching loss

Nitrogen in the NO_3^- form is very vulnerable to leaching. Nitrate is concentrated mainly in the surface 20–25 cm of soil where it is produced as an end product of mineralization of organic N, or from fertilizers. Nitrate concentrations fluctuate markedly due to the interaction of temperature, pH, water content, aeration and species effects on nitrification. The following effects should be noted:
1 *Temperature* – in temperate regions, low winter temperatures (< 5°C) retard nitrification, which then attains a peak in spring and early summer when NO_3^- concentrations of 40–60 mg N/kg may be found in surface soils under bare fallow. Higher values may be recorded in warmer, drier soils

because of the capillary rise of soil water and minimal leaching.
2 *Soil water content* – in warm climates of little seasonal temperature variation, soil mineral N fluctuates with the cycle of wet and dry seasons, as has been observed in East Africa. Following partial sterilization of the soil during the long dry season, at the onset of the rains the surviving micro-organisms multiply rapidly and produce a flush of mineralization from easily decomposed organic substrates (Fig. 10.6). High NO_3^- concentrations are the result, but these decrease with time as the available organic N is consumed and NO_3^- is removed by plant uptake and leaching.

During the cold and wet winters of temperate climates, plant N uptake is small and much of the soil NO_3^- is leached. This effect has been observed in well-drained soils in northwest Europe, New Zealand and southern Australia (annual rainfall > 600 mm). However, wetness in a poorly

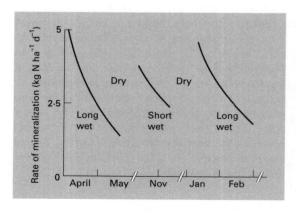

Fig. 10.6 Seasonal changes in mineral N in an East African soil (after Birch, 1958).

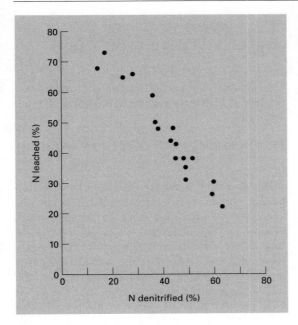

Fig. 10.7 Relationship between leaching and denitrification for grazed grass and white clover-based pastures in a humid temperate environment (after Garret *et al.*, 1992).

drained soil predisposes to denitrification (when the air-filled porosity is < 15% at the field capacity). Thus, there can be an inverse relationship between the proportion of soil NO_3^- lost by leaching and that lost by denitrification, depending on the drainage status and temperature of the soil, as shown in Fig. 10.7.

3 *Natural grassland and forest communities* – the mineral N found in grassland and forest soils is predominantly NH_4^+, which suggests that nitrifica-

tion may be suppressed, possibly due to the secretion of inhibitory compounds by the roots of certain species. Normally, the input of N in rain and from biological N_2 fixation roughly balances the loss by leaching in natural plant communities; but clear-felling a forest accelerates the rate of nitrification and, in the absence of plants to absorb the NO_3^-, leads to large N losses in the drainage (Section 10.4).

4 *Soil physical properties* – soil texture and structure influence NO_3^- leaching. Sandy soils, which generally have high hydraulic conductivities and low water-holding capacities, lose NO_3^- rapidly by leaching when rainfall exceeds evaporation. The dissolved NO_3^- moves downwards by mass flow at the same velocity as the percolating water, a process sometimes called 'miscible displacement'. As discussed in Section 6.3, for fertilizer-derived NO_3^- initially at the soil surface, the distance z travelled in time t into the soil is given by

$$z(t) = \frac{J_w t}{\bar{\theta}}, \tag{10.5}$$

where J_w is the water flux and $\bar{\theta}$ the average water content of the wet soil zone. The term $J_w/\bar{\theta}$ defines the average pore water velocity during leaching. Once steady state conditions are established during rainfall or irrigation, $J_w t$ is approximately equal to the amount of rain or irrigation water (mm) that has infiltrated. The contrast between NO_3^- leaching in a sandy soil and a poorly structured clay soil is illustrated in Table 10.5. This type of leaching is satisfactorily described by the convection-dispersion equation (see Box 6.7).

Table 10.5 Average pore water velocities and fertilizer leaching depths for soils of different texture and structure during rainfall or irrigation. After White, 2003.

Soil type	Rainfall or irrigation intensity (mm/h)	Average volumetric water content θ during infiltration (m³/m³)	Average pore water velocity (mm/h)	Depth z of the fertilizer front after 5 h (cm)
Sandy soil	10	0.25*	40	20
Clay soil with a massive structure**	10	0.45*	22	11
Well-structured clay soil with macroporosity	10	0.1†	100	50

* Representative values of θ at field capacity.

** Note that the infiltration rate of this soil may fall below the rainfall intensity so that ponding could occur. The depth of leaching would then be less.

† Macropores create a preferred pathway for water and solute movement, and much of the soil matrix is by-passed.

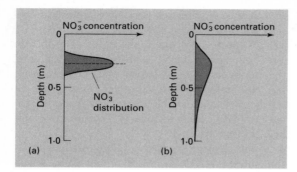

Fig. 10.8 Effects of soil structure on depth of leaching of NO_3^- originally concentrated at the surface, following 100 mm of rain. (a) A poorly structured clay loam of average volumetric water content $\theta = 0.4$ m³/m³. (b) A well-structured clay loam in which the effective volume for NO_3^- transport is only 0.1 m³/m³.

However, in well-structured clay and clay loam soils some of the percolating water moves rapidly down macropores by a process called preferential flow, with the water movement through the small pores inside aggregates being very much slower. In this case, the effective volume for NO_3^- transport can be much smaller than the average volumetric water content $\bar{\theta}$, and the leading front of leached NO_3^- is deeper than would be expected by a process of miscible displacement (Table 10.5). Whereas a 'slug' of fertilizer NO_3^- moves downwards as a symmetrical bulge by miscible displacement, as shown in Fig. 10.8a, the NO_3^- concentration profile will be more stretched out in a soil in which preferential flow occurs (Fig. 10.8b).

Gaseous N losses

Volatilization. Animal excreta containing urea (or uric acid in the case of poultry) are susceptible to volatile losses of N. For example, heavily fertilized grassland (receiving up to 400 kg N/ha/yr) that is intensively grazed can lose much more N (up to 50% of that applied) by volatilization of NH_3 than comparable grassland cut for hay only. Catalysed by the enzyme urease, which is widespread in soil, urea is hydrolysed to ammonium carbonate which then hydrolyses to ammonium hydroxide according to the reactions

$$(NH_2)_2CO + 2H_2O \rightarrow (NH_4)_2CO_3 \qquad (10.6)$$

and

$$(NH_4)_2CO_3 + 2H_2O \leftrightarrow 2NH_4OH + H_2CO_3. \qquad (10.7)$$

The alkaline solution of NH_4OH is unstable, particularly at high temperatures, and releases NH_3 gas according to

$$NH_4OH \leftrightarrow NH_4^+ + OH^- \leftrightarrow NH_3 + H^+ + OH^-. \qquad (10.8)$$

Whether NH_4^+ ions are produced by ammonification as shown in Equation 10.1, or by the hydrolysis of urea in a urine patch, the overall process of NH_3 volatilization is pH neutral, in contrast to nitrification which is potentially acidifying. However, the soil in the immediate vicinity of hydrolysing urea will initially increase in pH, and subsequently fall back to its original pH as NH_3 escapes. Loss of NH_3 by this process also occurs from ammonium and urea fertilizers under certain conditions (Section 12.2).

Denitrification. There are two possible mechanisms of nitrogen loss by denitrification:
• *Chemo-denitrification* – when it occurs, is mainly a result of NO_2^- accumulation following heavy applications of ammonium or urea fertilizers. This process is therefore discussed in Section 12.2.
• *Biological denitrification* is far more important than chemo-denitrification. It occurs mainly in the top 10–20 cm of soil when the air-filled porosity is < 15%, and culminates in the release of N_2O and N_2, as discussed in Section 8.4. Quantification of the N loss is difficult because, although N_2O concentrations in the soil air can be measured accurately by gas chromatography, N appearing as N_2 gas can only be distinguished from background N_2 gas if the original N substrate is labelled with ^{15}N. Short-term rates of denitrification can be measured in the field by injecting sufficient C_2H_2 into the soil air (10% by volume) to block the reduction of N_2O to N_2, and then determining the flux of N_2O emitted from a limited area of surface. Denitrification rates are highly variable both temporally and spatially and may reach 0.1–0.2 kg N/ha/day for short periods, depending on organic substrate availability, temperature, pH and the distribution of anoxic zones in the soil. As shown in Figure 10.7, the proportion of

available NO_3^- lost by denitrification can range from 15 to 65%. Deep leaching of NO_3^- to a poorly aerated subsoil during winter does not necessarily predispose to denitrification because there may be insufficient organic substrate at depth for rapid microbial growth.

10.3 Phosphorus and sulphur

Plant requirements

Crop analyses indicate that comparable amounts of P and S are removed in harvested products each year, except for the Brassicas (kale, mustard and rape), which have a higher requirement for S than P (Table 10.6). Nevertheless, the two elements differ markedly in the balance of inputs and outputs by various pathways in the soil-plant cycle, and to a lesser extent in the transformations they undergo in the soil.

Table 10.6 Amounts of P and S in crop products (kg/ha).

	P	S
Barley, 4 t/ha (grain and straw)	12	15
Potatoes, 30 t/ha	25	20
Grass, 10 t/ha (dry matter)	30	15
Maize, 13 t/ha (grain and stover)	50	50
Kale, 50 t/ha (fresh weight)	25	100

The soil reserves of P and S

Atmospheric inputs

Phosphorus concentrations in rain are very low, and more P is deposited by dry deposition (mainly dust), particularly during dry weather. The average rate of deposition in the UK is 0.3 kg P/ha/yr. Amounts of S deposited from the atmosphere can be much higher, ranging from 1 to > 20 kg/ha/yr. The major source is sulphur dioxide (SO_2) from the burning of fossil fuels, plus small contributions near coasts in the form of methyl sulphides and H_2S released from marine sediments. Several pathways are possible in the deposition of atmospheric S on soil and vegetation:

- Direct absorption of SO_2 through the stomata of leaves;
- oxidation of SO_2 to SO_3 according to the reaction

$$SO_2 + {}^1\!/_2 O_2 \leftrightarrow SO_3. \tag{10.9}$$

Reaction (10.9) is normally slow, but is accelerated in the presence of ozone (O_3) or hydrogen peroxide (H_2O_2);

- sulphuric acid also forms according to the reaction

$$SO_3 + H_2O \leftrightarrow H_2SO_4 \tag{10.10}$$

and comes down in rain (hence the term acid rain), or is neutralized by reacting with NH_3 gas according to the reaction

$$2NH_3 + H_2SO_4 \leftrightarrow (NH_4)_2SO_4; \tag{10.11}$$

- some $(NH_4)_2SO_4$ is dissolved in rain but most is deposited directly on vegetation, which together with SO_2 absorption by plants, comprises dry deposition. In industrial areas where total S inputs are > 20 kg/ha/yr, 80% of the input is dry deposition; in other areas where inputs are lower (5–20 kg S/ha/yr), the ratio of wet to dry deposition is approximately 1 : 1 (McGrath *et al.*, 2002). Trends in S emission and deposition are outlined in Box 10.3.

Forms of sulphur

Soil S is derived originally from sulphidic minerals in rocks that are oxidized to sulphate on weathering. Clay and shales of marine origin are frequently rich in sulphides, and the profiles of recently formed soils on marine muds may have SO_4-S contents up to 50,000 kg/ha. Normally, soil S is predominantly in an organic form and ranges between 200 and 2000 kg/ha.

Organic S is released by microbial decomposition, depending on the C : S ratio of the residues, which lies between 50 and 150 for well-humified organic matter. Net mineralization of organic S is low on average, with decay coefficients between 0.01 and 0.03 1/yr. Sulphur appears to be stabilized in humus in much the same way as N, and indeed the N : S ratio (between 6 and 10) is much

Box 10.3 Changes in S emissions and deposition in industrialized countries and their implications.

In response to concern over the adverse effect of acid deposition on natural ecosystems, particularly in Europe and North America, governments have required industry to reduce S emissions, especially those from fossil fuels. Taking the UK as an example, total S emissions fell by 50% between 1970 and 1993 and the relative contribution of S to the total acid deposition also fell substantially. The distribution of atmospheric inputs of S over the UK in 1999 is shown in Fig. B10.3.1a. As a result of much lower inputs over the last 30 years, the area of soils deficient in S for crop and pasture growth

has increased. Model predictions of S depositions, based on planned reductions in S emissions, suggest that inputs by 2010 are likely to fall to < 8 kg S/ha/yr over nearly all of the UK (Fig. B10.3.1b), so the area of S-deficient soils is likely to continue to expand. This trend has implications for the use of S fertilizer (Section 12.4).

Sulphur emissions and deposition are higher in parts of central Europe, and in rapidly industrializing countries such as China and India (but the latter are largely unquantified). Depositions are generally less in southern hemisphere countries where population

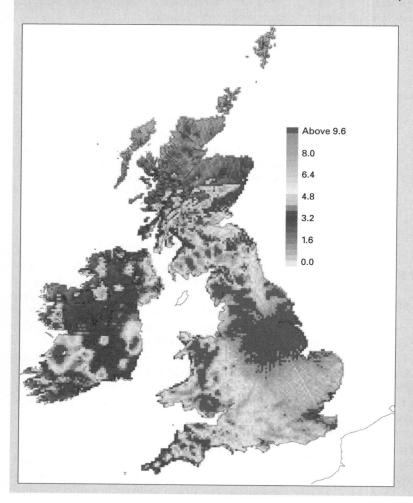

Above 9.6
8.0
6.4
4.8
3.2
1.6
0.0

Fig. B10.3.1 (a) Total S deposition (kg/ha/yr) for the UK in 1999 predicted by the Hull acid rain model (HARM). Figure published with permission of the National Expert Group on Transboundary Air Pollution (2001) (see also Plate B10.3.1a).

Box 10.3 *continued*

densities are lower and industrial areas less extensive.

Reductions in S deposition from the atmosphere in mature industrialized countries have also slowed the rate of acidification of soil and natural waters

(Section 11.3). A disturbing trend, however, is that N oxide emissions (mainly from vehicles) have risen by about 20% over the same period that S emissions have declined, and these now account for more than half the acid deposition in Europe.

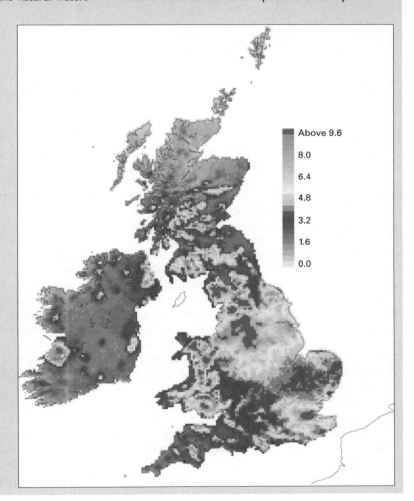

Fig. B10.3.1 (b) Total S deposition (kg/ha/yr) for the UK in 2010 predicted by the Hull acid rain model (HARM). Figure published with permission of the National Expert Group on Transboundary Air Pollution (2001) (see also Plate B10.3.1b).

less variable than the C : S ratio. Sulphate (SO_4^{2-}) ions released are only weakly adsorbed at positively charged sites, in competition with $H_2PO_4^-$ and organic anions, and so are prone to leaching. SO_4-S adsorbed on surfaces and in solution comprises the labile pool of S that is available to plants (see below).

Forms of phosphorus

Soil P contents range from 500 to 2500 kg/ha, of which 15–70% may exist in the strongly adsorbed or insoluble inorganic forms; the remainder occurs as organic P that is the major source of P for soil micro-organisms and mesofauna. Phosphate

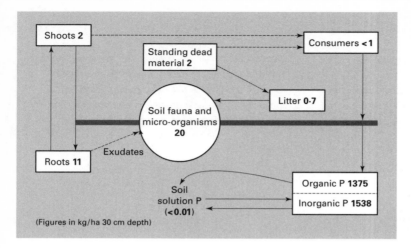

Fig. 10.9 P cycle in a native prairie grassland in Western Canada (after Halm et al., 1972).

in nucleic acids and nucleotides is rapidly mineralized, but when the C : P ratio rises above *c.* 100, P is immobilized by the micro-organisms, especially bacteria, which have a relatively high P requirement (1.5–2.5% P by dry weight compared to 0.05–0.5% for plants). Bacterial phosphate residues comprise mainly the insoluble Ca, Fe and Al salts of inositol hexaphosphate, which are called phytates. Thus, in closed ecosystems with insignificant P inputs, the soil organisms are highly competitive with higher plants for P, as illustrated by the P cycle in a natural grassland (Fig. 10.9). In this case, the soil biomass (mesofauna and micro-organisms) is by far the largest pool of P in the cycle, apart from the relatively inert inorganic and organic pools in the solid phase. Note also the very low amount of P in the soil solution.

Not only do micro-organisms mineralize organic P, but some groups secrete organic acids, such as α-ketogluconic acid, which attack insoluble Ca phosphates and release the phosphate. Some species of the micro-organisms *Aspergillus, Arthrobacter, Pseudomonas* and *Achromobacter*, which are abundant in the rhizosphere of plants, have this ability. However, most of the P is absorbed by the micro-organisms themselves in the intensely competitive environment of the rhizosphere, and little is immediately available to the plants.

Phosphate availability

Orthophosphate ions ($H_2PO_4^-$ and HPO_4^{2-}) released by mineralization are rapidly adsorbed by soil clays and sesquioxides where their desorbability steadily declines with time. Phosphate also occurs in poorly defined, insoluble Fe, Al and Ca compounds (see below), which slowly revert to more stable forms with time. The combined effect of these processes that make P less available to plants is called 'phosphate fixation'. In terms of its availability to plants, phosphate in soil is often divided into two broad components:
• P on surfaces that can be readily desorbed, plus phosphate ions in solution – the *labile pool*; and
• P occluded in sesquioxide coatings on surfaces, or held in insoluble compounds or organic combinations – the *non-labile pool*.

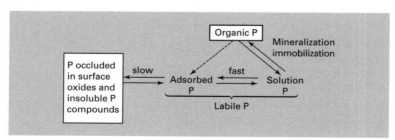

Fig. 10.10 Phosphate transformations in soil.

Transformations between the different forms of
P in soil are shown in Fig. 10.10. More informa-
tion on labile and non-labile P in soil is given in
Box 10.4.

Effect of pH change

Soil pH, clay and sesquioxide content, and exchangeable Al^{3+} all influence P availability. In
highly weathered soils where the sorption of P by
sesquioxides is prevalent, or in well-fertilized acid
soils where $FePO_4$ and $AlPO_4$-type compounds
may exist and dissolve incongruently to release
more P than Fe or Al, P solubility increases with
pH rise (Fig. 10.11). By contrast, in P-fertilized
alkaline and calcareous soils, insoluble Ca phosphates comprise the bulk of the non-labile P, and
the solubility of these compounds increases as the
pH falls. Overall, the availability of P in these
soils is greatest between pH 6 and 7, as shown in
Fig. 10.11.

However, acid soils of low P status containing
more clay minerals than sesquioxides show a
contrary trend in P availability with pH change.
Between pH 4.5 and 7, exchangeable Al^{3+} hydrolyses to form hydroxyl-Al ions that adsorb P
because of their residual positive charge. These
ions also polymerize readily so that a hydroxyl-
Al complex with incorporated $H_2PO_4^-$ ions forms
on the surface of clay minerals, resulting in a
minimum in P concentration over the pH range
5–6.5. Above pH 7, Al begins to dissolve as
the aluminate anion and P is released into the soil
solution.

Box 10.4 Forms of labile and non-labile inorganic P in soil.

Labile P should be immediately available to plants
growing in the soil. Soil P that can be desorbed
in a solution of 0.01 M $CaCl_2$, which approximates
the ionic strength of the soil solution of many
agricultural soils, is classed as labile. Other estimates
of labile P have been made by extracting a soil
sample in 0.5 M $NaHCO_3$, buffered at pH 8.5, for
30 minutes, or shaking a sample with an anion-
exchange resin. By raising the pH and supplying the
competitive anion HCO_3^-, $NaHCO_3$ solution
desorbs some of the surface 'ligand-exchange' P
(Section 7.3). A resin, on the other hand, supplies
not only a source of competitive anion (e.g.
HCO_3^-), but also provides a large sink for the
desorbed phosphate ions. Thus, estimates of soil
labile P vary according to the strength of bonding
of the adsorbed P and the desorption or extraction
method that is used (White, 1980).

Non-labile P is not immediately available to
plants, but may become so if the labile pool is
depleted, or there is a significant change in the soil's
physicochemical environment. Phosphate adsorbed
by hydroxy-Al polymers on clay surfaces, and that
adsorbed by ligand exchange on discrete sesquioxide
particles, can become occluded in these materials.
The mechanism is unclear, but may involve solid-
state diffusion and/or highly localized dissolution
and re-precipitation of the oxyhydroxides due to
changing physicochemical conditions. Because P ions
may be complexed on or within these oxides and
hydroxides, they are often referred to as 'sorbed P'.
Hydroxy-Al compounds containing P are most
stable in the pH range 4.5–7.0; Fe oxyhydroxides
containing P can dissolve and re-precipitate under
fluctuating redox conditions (Section 8.4).

Phosphate also forms insoluble salts with Fe,
Al and Ca. The crystalline minerals strengite
($FePO_4.2H_2O$) and variscite ($AlPO_4.2H_2O$) are only
stable below c. pH 1.4 and 3.1, respectively, and are
unlikely to exist in soil. However, in less acid soils,
slightly more soluble compounds with an Al : P or
Fe : P mole ratio c. 1 (amorphous $AlPO_4$ and $FePO_4$)
can form, especially in the vicinity of a dissolving
granule of superphosphate (Section 12.3). Above
pH 6.5, phosphate forms insoluble salts with Ca,
such as octacalcium phosphate $Ca_4H(PO_4)_3.5H_2O$,
which reverts to the more stable hydroxyapatite
$Ca_{10}(PO_4)_6(OH)_2$, and phosphate may also be
present as an impurity in calcite crystals. In the
latter case, the release of P depends on calcite
dissolution, which is primarily controlled by the
partial pressure of CO_2 in the soil air (Section 11.3).

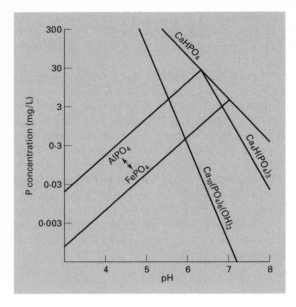

Fig. 10.11 P solubility-pH relations at 25°C for some Ca, Fe and Al phosphates (after Lindsay, 1979. From *Soils for Fine Wines* by Robert E. White, copyright © 2003 by Oxford University Press, Inc. Used by permission of Oxford University Press, Inc.).

Mycorrhizas

Because of low P concentrations in soil solutions and competition from micro-organisms in the rhizosphere, many plants have evolved mechanisms to enhance their absorption of P when it is in short supply. One of the most important of these is the mycorrhiza, a symbiotic association of fungus that lives with a root. The great majority of higher plants are mycorrhizal, and most host plants benefit nutritionally from the symbiosis. From this viewpoint, there are two main types of mycorrhiza.

1 *Ectomycorrhizas* – these are primarily associated with woody perennials: members of the Pinaceae (pines), Fagaceae (oak, beech), Betulaceae (birch, alder), *Eucalyptus* and poplar. Many fungal genera have been identified in ectomycorrhizas, mainly members of the Basidomycetes and Ascomycetes. The fungal mycelium forms a sheath some 0.03 mm thick around the root cylinder, which may be up to 40% of the colonized root weight, and also grows through the intercellular spaces of the root cortex, forming a distinctive network of hyphae called a Hartig net. The major benefit appears to be for P nutrition because of:

• The extra absorbing surface provided by the external hyphae, and the greater longevity of infected roots, which continue to absorb P after non-mycorrhizal roots have ceased active uptake; and

• the possibility that the fungus hydrolyses organic phosphates and makes inorganic P available to the host.

2 *Arbuscular mycorrhizas* (AM). Previously called vesicular-arbuscular mycorrhizas, the fungus does

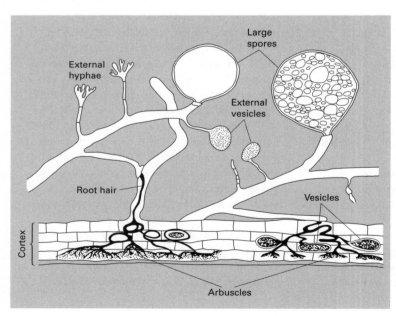

Fig. 10.12 Diagram of an arbuscular-mycorrhiza infecting a plant root (after Nicholson, 1967).

Box 10.5 Nutrient depletion zones around absorbing plant roots.

A depletion zone develops around a plant root when the rate of intake of a nutrient exceeds the rate at which the labile pool of that nutrient can be replenished at the root surface. An example is given in Fig. 10.13. For most nutrients, replenishment from the non-labile pool is too slow to meet the demand of a rapidly absorbing root, so movement of labile nutrients from the surrounding soil into the depletion zone must occur. Ions move in solution by mass flow and diffusion, and potentially by surface diffusion in the solid phase. However, it has been found that, although a high proportion of labile P ions such as $H_2PO_4^-$ is adsorbed, the contribution of surface diffusion to ion transport near roots is negligible compared to the flux in solution (Tinker and Nye, 2000). For ions such as Ca^{2+}, Mg^{2+} and SO_4^{2-}, the concentration in the soil solution is normally high enough for mass flow alone to meet root demand, and no depletion zone develops (in some cases a surplus of the ion occurs at the root surface). However, for ions such as $H_2PO_4^-$, K^+, Zn^{2+}, Mn^{2+} and Cu^{2+} that are strongly adsorbed and at low concentrations in solution, mass flow is insufficient to meet root demand and movement

to the root is primarily by diffusion. The radial diffusive flux J_s (see Box 6.7) is given by

$$J_s = -D_e \frac{dC}{dx}, \qquad (B10.5.1)$$

where C is the ion concentration in the soil solution, x is the radial distance from the root surface and D_e, the effective ion diffusion coefficient, is given by

$$D_e = D\theta f \frac{dC}{dC'}. \qquad (B10.5.2)$$

In this equation, D is the ion diffusion coefficient in bulk water, θ is the volumetric water content, f is an impedance factor (range 0–1) to account for the tortuous pathway for diffusion, and dC'/dC is the soil's buffer power for the nutrient. Buffer power is related to the slope of the Q/I relation (Section 7.2), except that C' is the amount of labile nutrient per unit volume, not per gram of soil. Note that D_e decreases when θ and f are small, and dC'/dC is large. A depletion zone around a root will extend a distance $\sqrt{D_e t}$ from the surface, where t is the time from when the root begins to absorb the nutrient.

not form a sheath but grows in the root cortex, forming two characteristic structures – the vesicles, which are temporary storage bodies in the intercellular spaces, and the arbuscules, which are branched structures that grow intracellularly and release solutes into the cell (Fig. 10.12). The endophyte is a Phycomycete of the family Endogonaceae. The fungal spores are present in most soils, but only germinate when a suitable host root is near: if the young mycelium does not penetrate a host root, it dies. However, once established in the root (and the host–fungal relationship is not very strain-specific), the hyphae ramify widely outside the host tissue.

AM fungi are found in nearly all higher plants and in many conifers and ferns. The incidence of infection is especially high in infertile soils, where N and P uptake can be enhanced. However, the mycorrhizal state confers most advantage for P uptake, first because the external hyphae can increase by several-fold the effective length of the

absorbing roots, and second because the hyphae provide a pathway for rapid transport across the 'depletion zone' that develops around the root in the case of immobile elements such as P (Fig. 10.13). The benefit is most notable in plants such as onion and citrus, which have very few or no root hairs (Box 10.5). AM fungi are of considerable importance in the cycling of P in natural ecosystems and for the establishment of pioneer species in harsh habitats, such as on sand dunes and mine spoil. The incidence of infection generally declines in cultivated crops, however, because P fertilizer has a more rapid effect on growth than does a mycorrhizal infection. AMs also enhance plant uptake of Cu and Zn, and the suppression of AMs by P fertilizer may explain the appearance of Cu and Zn deficiencies in some P-fertilized crops, especially on high pH soils (Section 10.5). Members of the Cruciferae (brassicas) and Chenopodiaceae (buckwheat) are non-mycorrhizal and have evolved with other mechanisms, such as

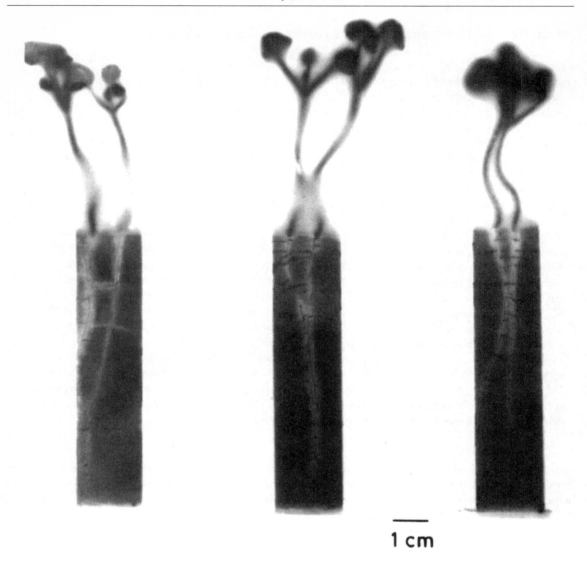

Fig. 10.13 Autoradiograph of roots of rape plants in soil labelled with ^{33}P showing a zone of depletion around the main roots and an accumulation of P within the root axes and laterals (after Bhat and Nye, 1973).

an ability to acidify their rhizospheres, to enhance P uptake from P-deficient soils.

The AM fungus, which is an obligate symbiont, benefits from C compounds supplied by the host. This supply amounts to 6–10% of the photosynthetically fixed C, but there is not necessarily a growth-depressing effect of this C diversion on the host because of plant adaptations that have developed to counteract the demand (Tinker and Nye, 2000).

P and S losses by leaching and in surface runoff

With the exception of very sandy soils, and other situations where soil P has accumulated in large amounts through the repeated application of P fertilizer and animal manures (Section 12.3), P losses by leaching are usually < 1 kg/ha/yr. However, losses up to 3 kg/ha/yr have been measured in drain flow from under-drained soils, and losses

Table 10.7 P and S losses (kg/ha/yr) in drainage from neutral and calcareous soils in southern England. After Williams, 1975 and others.

	PO_4-P	SO_4-S
Woburn – neutral, sandy soil over Oxford Clay	0.05	100
Saxmundham – sandy, calcareous boulder clay	0.12	86
Wytham – brown calcareous soil over Oxford Clay	0.13	78

Table 10.8(a) Annual return (= uptake) of K, Ca and Mg (kg/ha/yr) by a 40 yr-old forest in West Africa. After Nye and Greenland, 1960.

	K	Ca	Mg
Litter fall	68	206	45
Timber fall	5	82	8
Leaching (throughfall – rainfall)	220	29	18
Total	293	317	71

Table 10.8(b) Annual return (= uptake) of K, Ca and Mg (kg/ha/yr) by a 58 yr-old *Eucalyptus obliqua* forest. After Attiwill and Leeper, 1990.

	K	Ca	Mg
Litter fall	7	24	10
Leaching (throughfall – rainfall)	13	6	2
Total	20	30	12

10 times greater than this may occur from arable soils, when runoff causes erosion and removes small soil particles that have been enriched with P from fertilizer or manure applications (Catt *et al.*, 1998).

Sulphate forms sparingly soluble gypsum in arid region soils, but is readily leached from soils of humid regions. For example, where fertilizers containing SO_4-S are used (single superphosphate, ammonium sulphate), leaching losses of SO_4-S are equivalent to 30–80% of the fertilizer S in the year following application. An exception occurs with soils high in sesquioxides and of low pH where SO_4^{2-} ions are retained by adsorption at depth, as shown in Fig. 10.2b. Since SO_4^{2-} ions are not strongly adsorbed in the surface soil layer because of the influence of negatively charged organic matter, SO_4-S losses in runoff are usually small. Table 10.7 shows data from three sites on neutral and calcareous soils in southern England that illustrate the contrast between leaching losses of S and P.

10.4 Potassium, calcium and magnesium

Plant requirements

Annually, agricultural crops remove between 5 and 25 kg Mg/ha, and exceptionally up to 50–60 kg/ha for high yields of maize or oil palm. Ca requirements range from 10 to 100 kg/ha and the K requirement is even higher at 75–300 kg/ha.

There are few estimates of the rate of cation uptake by perennial vegetation, especially forests, because it is difficult to measure the amount and composition of the annual growth increment. By assuming that in a mature forest the rate of nutrient return balances the rate of uptake from the soil, Nye and Greenland (1960) obtained the estimates of cation uptake shown in Table 10.8a. Considerably lower values were reported by Attiwill and Leeper (1990) for a mature Eucalyptus forest in southern Australia (Table 10.8b), and even lower values for K and Ca by Cole *et al.* (1967) for a 36 year-old Douglas fir plantation in Washington, USA. Note that considerable quantities of these nutrients, especially K, are leached from the tree canopies (calculated as the difference between nutrients in direct rainfall and in throughfall – see Section 6.1).

Soil reserves

The pool of exchangeable cations Ca^{2+}, Mg^{2+} and K^+ provides the immediate source of these elements for plants. Insoluble reserves may occur as calcium and magnesium carbonates, potash feldspars and interlayer K^+ in micaceous clays. Total reserves are therefore very variable, reflecting the conditions of soil formation, but exchangeable Ca^{2+} values usually lie between 1000 and 5000 kg/ha, Mg^{2+} between 500 and 2000 kg/ha and $K^+ \sim 1000$ kg/ha.

Atmospheric inputs

Amounts of K and Ca in all forms of precipitation and dry deposition are comparable and lie between 1 and 20 kg/ha/yr. Magnesium can be much higher, especially on sea coasts (because ocean water is rich in Mg), amounting to as much as 150 kg/ha/yr. Thus, Mg inputs roughly balance combined losses through plant removal and leaching, but not so for K and Ca. A striking feature of K input to the soil is the magnitude of its leaching from plant canopies, amply demonstrated by the figures for the tropical tree species and *Eucalyptus obliqua* in Table 10.8. This process is one of the main reasons for the rapid cycling of K in natural ecosystems in which plants and their residues can accumulate 40–60% of the total labile K in the root zone.

Mineral weathering

Apart from recent glacial or volcanic parent materials, the reserve of weatherable primary minerals is usually small in the top 30 cm of soil where 80–90% of the plant roots are located. Nevertheless, the slow release of nutrients from the 'geologic reserve' at greater depth is often crucial for the stability of natural plant communities for which elemental losses by leaching and erosion occasionally exceed atmospheric inputs. For example, data from West Africa suggested that 58, 64 and 14 kg/ha of K, Ca and Mg, respectively, were pumped up annually from the subsoil by deep tree roots.

Leaching losses of K, Ca and Mg

Cation leaching is accelerated in cultivated soils, where nitrification occurs, for two reasons:
1 There is a net production of one mol of H$^+$ per mol of NO$_3^-$ produced during ammonification and nitrification (Reactions 10.1 and 10.2). These H$^+$ ions displace exchangeable Ca^{2+}, Mg^{2+} and K$^+$ from negatively charged clay and organic matter;
2 conversion of NH$_4^+$ to NO$_3^-$ means that the moles of anion charge in solution increase relative to the moles of cation charge. However, the extra anions are balanced by cations, especially Ca^{2+}, displaced from the exchange surfaces. Thus, Ca^{2+} and NO$_3^-$ predominate in the water draining through soil.

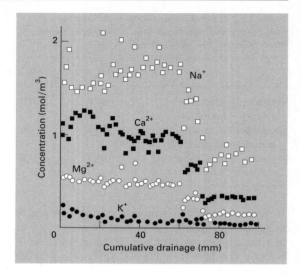

Fig. 10.14 Concentrations of Ca^{2+}, Mg^{2+}, Na$^+$ and K$^+$ in drainage from a pasture near the coast in New Zealand (after Heng *et al.*, 1991).

For agricultural soils in Britain, for example, Ca leaching losses range from 90 to 400 kg/ha/yr. There is a direct correlation between Ca losses and pH, with higher losses (> 200 kg/ha) usually occurring from soils that have been limed or that contain free CaCO$_3$. Although Ca^{2+} is normally the most abundant cation in soil drainage water, for acid soils close to the sea Na$^+$ will often be the dominant cation, as shown by results for a silt loam under pasture in New Zealand (Fig. 10.14).

Suppression of nitrification in soils under natural grassland or forest will reduce cation loss by leaching. For example, the net leaching loss of K, Ca and Mg from an acid podzolic soil under an undisturbed deciduous forest in New Hampshire, USA, was 0.6, 9.2 and 2.2 kg/ha/yr, respectively (Bormann and Likens, 1970). These net losses were balanced by weathering of the parent material. Nitrification appeared to be of minor significance in the soil under the forest, and NO$_3^-$ losses in the drainage balanced inputs in precipitation. However, after the forest was clear-felled (and the timber was neither removed nor burnt), nitrification increased sharply and net N leaching loss rose to 120 kg N/ha/yr, while net K, Ca, Mg losses rose to 13, 92 and 15 kg/ha/yr, respectively.

10.5 Trace elements

Definitions and sources

Those elements whose total concentration in the soil is normally < 1000 mg/kg are called trace elements. They fall into three categories:

1 The micronutrients Cu, Zn, Mn, B, Ni and Mo, which are essential to plants at normal concentrations in the plant (ranging from 0.1 mg/kg for Mo to 100 mg/kg for Mn), but become toxic at higher concentrations. Iron (Fe) is the one micronutrient that is not strictly a trace element;

2 Elements such as Se, I and Co which are not essential for plants, but are essential for animals;

3 Elements such as Li, Be, As, Hg, Cd and Pb, which are not required by plants or animals and are toxic to either group at concentrations greater than a few mg/kg in the organism.

With industrialization and the generation of wastes from industry and large cities, as well as the input of chemicals in intensive agriculture, reports of undesirably high concentrations of many of these elements in soils, plants, animals or water are becoming more frequent. We speak of contamination and pollution of the environment (Box 10.6).

Trace elements in soil are normally derived from the parent material and natural deposition from the atmosphere (e.g. from volcanoes, dust and sea spray). During the 20th century, however, anthropogenic sources increased markedly, especially from mining, industrial processing, motor vehicles, agricultural chemicals and urban waste. These sources have increased trace element loadings in soil-plant systems via the atmosphere and direct solid or liquid applications (e.g. spreading animal waste, sewage sludge or in landfills).

Influence of parent material

Trace elements are largely bound in mineral lattices, to be released by weathering, so that the type of parent material determines the natural abundance of the individual elements in soil. Iron, Cu, Mn, Zn, Co, Mo, Ni and Cr occur in ferromagnesian minerals, common in ultrabasic and basic igneous rocks, and Se in black shales and rock phosphates (Haygarth, 1994). Fe, Mn, and Mo also occur as insoluble oxides (Co is often

> **Box 10.6** Soil contamination and pollution.
>
> The term 'heavy metals' is frequently used for metals and metalloids involved in contaminating or polluting the environment. By implication, these elements are considered toxic to either plants or animals if concentrations exceed a certain threshold. This terminology is vague and the terms 'micronutrient' and 'trace element' used here are to be preferred. Threshold concentrations and any adverse effects should be defined for each element.
>
> The terms 'contamination' and 'pollution' tend to be used interchangeably. Sometimes 'contamination' is restricted to the presence of a substance not normally found in soil, which may or may not have a harmful effect. Commonly, 'pollution' is used as a generic term to cover the introduction by humans of any substance having an adverse effect on the environment. It has a pejorative connotation. The topic of contamination is discussed further under **Soil quality** in Chapter 15.

co-precipitated in MnO_2), and Zn, Fe, Pb and Cu as equally insoluble sulphides in sedimentary rocks. Boron (B) occurs as the resistant mineral tourmaline in acid igneous rocks and to some extent may substitute for Al in the tetrahedral sheet of 2 : 1 clay minerals. Typical trace element contents of soils formed on parent rocks of different basicity are presented in Table 10.9.

Atmospheric inputs

Trace elements enter the air as gases, aerosols and particulates and return to the soil-plant system mainly by dry deposition. Metals that are used extensively in industry – Cd, Zn, Cu, Ni and Pb – show the greatest enrichment in the air of industrialized regions, relative to metals like Fe or Ti that are naturally abundant. The amounts deposited are therefore likely to be greater in industrial regions of the northern hemisphere than in remote rural areas. Table 10.10 demonstrates this contrast, where estimated annual rates

Table 10.9 Trace element contents (kg/ha) of soils on parent materials of different rock type. After Mitchell, 1970 and others.

Trace element	←Increasing basicity			
	Serpentine	Andesite	Granite	Sandstone
Co	160	16	< 4	< 4
Ni	1600	20	20	30
Cr	6000	120	10	60
Mo	2	< 2	< 2	< 2
Cu	40	20	< 20	< 20
Mn	6000	1600	1400	400
Pb	2	12	36	24

Table 10.10 Estimated rates of trace element deposition (g/ha/yr) from the atmosphere over Europe and North America. After Sposito and Page, 1984.

Element	Europe		North America	
	Minimum	Median	Minimum	Median
Mo	< 0.3	–	0.2	–
Co	0.3	–	0.2	4.7
Cd	0.8	3	< 0.2	6.5
Ni	6.3	39	< 16	142
Mn	14	68	0.1	236
Cu	13	536	7.9	441
Zn	20	19	–	788
Pb	87	189	71	4257

of deposition of several trace elements are given. The minimum values are representative of rural areas, whereas the median values are biased towards urban areas because of the greater number of observations there.

Soil reactions and availability to plants

Free ions and complexes in solution

The reactions involved in the turnover of a trace element in the soil-plant system are summarized in Fig. 10.15. The chemistry of an individual element determines the extent to which it participates in any one of these reactions, and hence its avail-

ability to plants. The rate of any reaction depends not only on the concentration of reactants and products, but also on steriochemical factors (for complex formation) and environmental conditions. Ni and Cr are two elements for which the supply in the environment is rarely inadequate to meet plant demand.

Most trace elements – Fe, Cu, Mn, Zn, Co, Cd, Ni and Pb – occur as cations. Others such as Cr can occur as an anion (chromate ions $Cr_2O_7^{2-}$ and CrO_4^{2-}) or the cation Cr^{3+}, depending on its oxidation state. Of the remainder, Mo, As and Se occur as oxyanions – MoO_4^{2-} (molybdate), $H_2AsO_4^-$ and $HAsO_4^{2-}$ (arsenate), SeO_4^{2-} (selenate) and SeO_3^{2-} (selenite). Boron occurs as H_3BO_3 (boric acid),

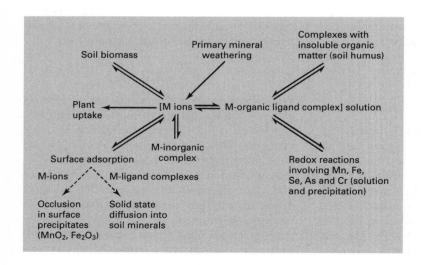

Fig. 10.15 Trace element (M) equilibria in soil.

which dissociates to H^+ and $B(OH)_4^-$ (borate anion) at pH > 8, and I occurs as I^-. The cations form complexes of varying stability with inorganic and organic ligands. The most abundant ligand is H_2O so the cations are normally hydrated. The principles of complex formation are described as follows

$$aM^{n+} + bL^{l-} \leftrightarrow \{M_aL_b\}^q, \qquad (10.12)$$

where charge balance requires that

$$an - bl = q, \qquad (10.13)$$

and q, the valency of the complex, may be +, 0 or −. The stability of the complex is given by the *conditional stability constant* ^cK_s, defined by the expression

$$^cK_s = \frac{[M_aL_b^q]}{[M^{n+}]^a[L^{l-}]^b}. \qquad (10.14)$$

The constant is called conditional because its value usually depends on the solution composition and temperature. Metal-inorganic ligand complexes may form, depending on the pH and concentration of suitable ligands. Some examples are $CdCl^+$, $MnSO_4^0$, $NiSO_4^0$, and $ZnHPO_4^0$, all of which modify the metal's adsorption characteristics and its mobility in the soil. Computer programs have been developed to enable 'speciation' of elements in soil solutions (e.g. Sposito and Coves, 1995). However, many metals form much more stable complexes with organic ligands. These complexes can be bidentate, tridentate or even quadridentate 'chelates' (Section 3.4); the greater the number of bonds, the greater the stability. The complexes exist in solution as anions or cations of low charge : mass ratio, but they can also be adsorbed, as discussed below.

The cations Fe, Cu, Mn, Ni and Zn also hydrolyse at pH values between 3 and 10 and eventually precipitate as insoluble hydroxides. This is a special case of Reaction (10.12); for example,

$$Fe^{3+}(soln) + 3OH^-(soln) \leftrightarrow Fe(OH)_3^0(solid). \qquad (10.15)$$

Reaction (10.15) is a reversible precipitation-dissolution reaction for which the thermodynamic solubility constant K_{sp} is defined by

$$K_{sp} = \frac{(Fe^{3+})(OH^-)^3}{(Fe(OH)_3)}. \qquad (10.16)$$

Taking the activity of pure solid $Fe(OH)_3$ as 1, Equation 10.16 becomes

$$K_{sp} = (Fe^{3+})(OH^-)^3. \qquad (10.17)$$

Whether the insoluble hydroxide or a soluble organic complex (FeL_b^q) is thermodynamically more stable at a given pH depends on which chemical form maintains the lower activity of free Fe^{3+} ions in solution. This fact underlies the choice of organic compounds (chelating agents) to form soluble complexes with micronutrients, such as Fe, that become unavailable to plants at high pH due to the precipitation of the insoluble hydroxide (Box 10.7).

Trace elements in the solid phase

The concentration of individual trace elements in the soil solution is normally very low (< 10^{-7} M). The surface adsorbed component shown in Fig. 10.15 accounts for about 10% of the total of any trace element in soil. The remainder is held in insoluble precipitates, primary and secondary minerals, or is complexed with soil organic matter. Metal cations are adsorbed as follows:
• Non-specifically, by electrostatic attraction to negatively charged surfaces, in competition with the macronutrient cations Ca^{2+}, Mg^{2+} and K^+; or
• through the formation of outer-sphere (OS) complexes (Section 7.1); or
• specifically through the formation of inner-sphere (IS) complexes on oxyhydroxide surfaces (Section 7.1).

Inner-sphere complexes are more stable than OS complexes because no water molecules separate the cation from the surface and a degree of covalent bonding is involved. Hydrated Cu^{2+} ions, for example, form OS complexes with the cavities in the silica sheet of 2 : 1 clay surfaces from which they can be displaced by Ca^{2+}. On the other hand, Cu^{2+} ions forming IS complexes with carboxyl groups on organic matter are not exchangeable to Ca^{2+}. Although stable complexes are formed between trace metal ions and organic ligands both in solution and on surfaces, because the organic C in solution (up to 50 mg/L) is very

Box 10.7 Organic complexes and micronutrient availability.

Deficiency of Fe is manifest in plants as lime-induced chlorosis; that is, interveinal yellowing of leaves due to the lack of chlorophyll. The deficiency of micronutrients such as Fe and Mn can be corrected by supplying these elements in a soluble complex. For example, ethylenediamine tetraacetic acid (EDTA) is a synthetic chelating agent that forms stable complexes with most trace elements. Fig. B10.7.1 shows the effect of pH on the balance between the stability of $Fe(OH)_3$ and the complex $FeEDTA^-$. The change from the chelate to the hydroxide being the more stable phase occurs c. pH 9 normally, but in calcareous soils this change occurs c. pH 8 because of increasing stability of the $CaEDTA^{2-}$ complex at high pH. Thus, chelating agents, such as EDDHA (ethylene diamine di(o-hydroxyphenylacetic acid)), which have a higher affinity for Fe than for Ca, should be used on calcareous soils. However, there is evidence that plants take up more Fe from complexes of lower stability, indicating that the complex may have to be broken down before the Fe can be absorbed into the cell. Complexing agents such as EDTA and DTPA (diethylene triamine pentaacetic acid) are used to extract soil samples to measure the bioavailability of trace elements (Box 10.8).

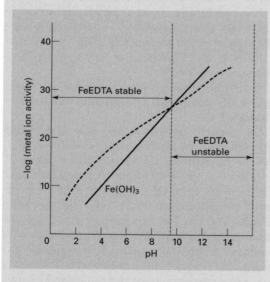

Fig. B10.7.1 pH stability range for $FeEDTA^-$ in soil (after Martell, 1957).

much less than the C in the solid phase, organic complex formation tends to immobilize metals rather than to solubilize them. Sewage sludge, particularly that derived from industrial wastes, often contains high concentrations of metals such as Cu, Zn, and Ni in fairly labile organic combination, which may pose problems when large applications are made to agricultural soils (Section 11.1).

The adsorption of metal cations increases as the pH rises, with maximum adsorption occurring at or below the pK_1 value for cation hydrolysis. The effect of pH is primarily due to the increase in negative charge of variable charge surfaces, but the presence of even low concentrations of the hydrolysed cations also contributes to increased adsorption. These factors explain why the availability of most trace elements to plants decreases with pH rise.

Anions such as MoO_4^{2-} are most strongly adsorbed at low pH on sesquioxide surfaces by a ligand-exchange mechanism analogous to that described for phosphate (Section 7.3). B is adsorbed by the same mechanism on oxide surfaces as $B(OH)_4^-$, but as this species only appears at pH > 8, B solubility decreases at high pH. Mo is therefore the only micronutrient to increase in availability as the soil pH rises. The two arsenate anions are also adsorbed by ligand exchange, but in the case of Se, SeO_4^{2-} (the more stable form, selenate) is only weakly adsorbed whereas SeO_3^{2-} (selenite) is adsorbed by ligand exchange. Thus, selenate can be quite mobile in soil and is the form taken up by plants. Both As and Se anions can be methylated in biological reactions in soil and released as gases (e.g. $(CH_3)_2HAsO_2$ and $(CH_3)_2Se$).

The adsorbed phase of a trace element comprises the labile component of the soil reserve, the size of which can determine the bioavailability of the element. Measuring trace element bioavailability is important for assessing not only the deficiency or otherwise of micronutrients for crop growth, but also the potential hazard of these elements when the soil loading is high (Box 10.8).

The effect of poor drainage on trace element solubility

Manganese and Fe in insoluble compounds are mobilized in waterlogged soils as Mn^{2+} and Fe^{2+}

Box 10.8 Bioavailability of trace elements.

In Section 7.2, the concept of a quantity/intensity (Q/I) relationship is introduced, which controls the availability of a nutrient ion, such as K^+, to plants. This concept of nutrient availability, or bioavailability, is developed further in Section 11.2. With respect to micronutrients and trace elements, the concentration of the free ionic species in solution at an absorbing surface (for example, a root or algal cell) controls the rate of uptake by the organism. But because the free ion concentration is normally very low, the extent to which that concentration can be maintained by dissociation of a complex, or desorption from the solid phase, becomes the controlling factor in determining bioavailability. Thus, Tiller (1996) and others have argued that the Q factor is paramount in determining trace element bioavailability and that it should be measured as the total potentially labile element (per unit mass of soil). Tiller et al. found that acids were not suited to measuring Q because of their partial neutralization in calcareous soils and their dissolution of soil minerals. They preferred to extract soil in 0.05 or 0.1 M EDTA for periods up to 7 days to completely remove elements with a high affinity for soil surfaces and present at heavy loadings.

Unfortunately, much of the work on trace element chemistry in soils has been empirical and led to the use of a bewildering range of complexing, extracting and dissolving reagents (Beckett, 1989). The more rational Q/I approach has not often been applied, so that many conflicting results on bioavailability have been obtained: so much so that legislation on soil contamination by heavy metals is framed in terms of total soil content, not bioavailable content. In many cases, therefore, critical values for unacceptable contamination have been set quite low and the risk associated with higher total soil contents overestimated (Chaney and Oliver, 1996). Where some differentiation of the ease of release of an element from the solid phase is needed to refine the estimate of bioavailability, the following fractionation sequence is suggested:

1 Weakly bound forms desorbed in dilute $CaCl_2$ or NH_4NO_3;
2 strongly bound forms extracted in 0.05 or 0.1 M EDTA (prolonged if necessary); and
3 the residue dissolved in strong acids or measured by X-ray fluorescence.

Table 10.11 The effect of drainage condition on trace element availability in soil.

Soil	Extractable* element content (mg/kg soil)			
	Co	Ni	Mo	Cu
Freely drained	1.3	1.3	0.06	2.6
Poorly drained	1.9	3.4	0.19	6.6

* Extracted in CH_3COOH or EDTA.

ions, as discussed in Section 8.4. In acid, waterlogged soils, chromate ions can be reduced to Cr^{3+}, but this forms very insoluble oxides and hydroxides. In addition, the solubility of Co, Ni, Mo and Cu (elements not reduced at the redox potentials attained in soils) may be increased under waterlogged conditions due to the accelerated weathering of ferromagnesian minerals, as shown in Table 10.11. The dissolution of Mn oxides can also release co-precipitated Co into the soil solution.

Crop uptake and leaching losses

Estimates of micronutrients in harvested products vary widely, but the range (in g/ha/yr) is approximately as follows:

Mn	34–500
Mo	< 0.2–500
B	27–160
Zn	31–80
Cu	6–80.

Comparing these values with the minimum inputs from the atmosphere in Table 10.10 suggests that soils in rural areas may suffer a net depletion of these elements by crops. On the other hand, these elements occur as trace contaminants in fertilizers and liming materials (e.g. basic slag) and unknown amounts are released through mineral weathering each year.

Because of their precipitation in insoluble compounds and strong retention by mineral and organic surfaces, micronutrient losses by leaching are very small, except for B in acid to neutral soils, and Fe and Mn in some gleyed soils (Section 9.2). Nevertheless, without significant atmospheric inputs, cumulative losses from very old soils eventually lead to widespread micronutrient deficiencies, as occur in the coastal areas of eastern and southern Australia. Leaching of Se from naturally-rich sediments in the San Joaquin Valley of California has caused serious pollution of surface and underground water.

10.6 Summary

The macronutrient elements C, H, O, N, P, S, Ca, Mg, K and Cl are essential for plant growth and reproduction and occur in concentrations > 1000 mg/kg dry matter. The micronutrient elements Fe, Mn, Zn, Cu, B, Ni and Mo occur at concentrations < 100 mg/kg dry matter. Except for Fe, micronutrients are included in the trace elements – elements normally occurring at a concentration < 1000 mg/kg soil. Also included are elements such as Cr, Se, I and Co that are essential only for animals, and Li, Be, As, Hg, Cd and Pb that are not required by plants or animals, and are toxic at concentrations greater than a few mg/kg in the organism.

Gains and losses of nutrients in natural ecosystems, comprising the soil–plants–animals–atmosphere, are roughly in balance so that continued growth depends on the cycling of nutrients between the biomass, organic and inorganic stores. Removals from agricultural systems, especially of crop products, generally exceed natural inputs unless these are augmented by fertilizers and organic manures, and in the case of some trace elements, by accessions from the atmosphere.

Nutrients in the biomass store pass to the organic store by excretion from, or on the death of, living organisms. The greater part of the organic and inorganic stores is in the soil. Additions to the inorganic store occur through mineralization of organic residues, weathering and atmospheric inputs (rainfall and dry deposition), and from fertilizers in agricultural systems. The part of the inorganic store that supplies nutrients to plants and other organisms is called the available or labile pool, consisting of ions in solution and ions adsorbed by clays, sesquioxides and organic matter. Nutrients and potentially toxic elements held in insoluble precipitates or strongly adsorbed complexes (organic or inorganic) are considered non-labile. The bioavailability of a nutrient or potentially toxic element is best characterized by a quantity/intensity (Q/I) graph, which describes the relationship between the amount of labile nutrient in the solid phase and its concentration in the soil solution. However, many empirical extraction methods have been developed that give composite estimates of both the Q and I factors.

The ability of a minority of plants in symbiotic association and free-living micro-organisms to reduce N_2 gas to NH_3 within a cell (nitrogen fixation) provides a unique input of N to the biomass and hence organic stores. Nitrogen is transformed from the organic to inorganic store by ammonification followed by nitrification, according to

$$Organic\text{-}N \rightarrow NH_4^+ + OH^-$$

and

$$NH_4^+ + 2O_2 \rightarrow NO_3^- + H_2O + 2H^+.$$

However, losses occur through volatilization of NH_3 (from NH_4^+) at high pH, especially following the hydrolysis and decomposition of urea in animal excreta, and through NO_3^- leaching and NO_3^- reduction (denitrification) to N_2O and N_2 under anoxic conditions. Note that the ammonification-nitrification process is potentially acidifying. Substantial amounts of N may be returned to soil-plant systems from the atmosphere, with dry deposition of NH_3 and $(NH_4)_2SO_4$ being particularly significant in urban and industrialized areas, and around large animal feedlots.

Atmospheric inputs of S are much greater than P, especially in industrialized regions, but are now declining steadily due to the reduction in S emissions from fossil fuel combustion. Biological transformations of P and S in soil are similar, but P cycling is more conservative than S because the ions $H_2PO_4^-$ and HPO_4^{2-} are strongly adsorbed, or form insoluble precipitates with Fe, Al and Ca. Sulphate ions (SO_4^{2-}) are only weakly adsorbed and do not form insoluble precipitates (except for $CaSO_4$ in arid soils). Sulphate is also reduced to sulphides under intense reducing conditions. Many plants have evolved associations between their roots and fungi called mycorrhizas, which enhance P absorption (and also Cu and Zn) from deficient soils. In contrast to SO_4^{2-}, which is readily leached (except from acid sesquioxidic soils), P losses by leaching are usually small because P concentrations in the soil solution are very low. Exceptions occur where light-textured soils have been regularly treated with P fertilizers, or large amounts of organic manures.

Calcium, Mg, K and Na are held mainly as exchangeable cations, the supply of which buffers the soil solution against depletion. Losses of these cations are accelerated when anions such as NO_3^-, SO_4^{2-} and HCO_3^- are plentiful in the percolating water, and also under acid conditions when H^+ and Al^{3+} ions displace the cations from exchange sites. The micronutrient cations Fe, Mn, Cu, Zn, Ni and Co, and the trace element cations Cr, Hg, Cd and Pb, form complexes with negatively charged surfaces, the complexes formed with organic compounds being more stable than the inorganic ones. An ability to form stable complexes with soluble organic ligands (chelates) makes micronutrient cations more available to plants by preventing their precipitation as insoluble hydroxides at alkaline soil pHs. Molybdenum, B, Se, As and I occur as anions in solution. Molybdate is most strongly adsorbed by sesquioxides at low pH, and becomes more available as the pH increases. Less strongly adsorbed, but following the same trend with pH, are As, Se and I. B exists as undissociated H_3BO_3 at pH values up to 8, above which the acid dissociates and $B(OH)_4^-$ is adsorbed, thereby decreasing B availability.

References

Anon. (1983) *The Nitrogen Cycle of the United Kingdom*. Royal Society Study Group Report. Royal Society, London.

Attiwill P. M. & Leeper G. W. (1990) *Forest Soils and Nutrient Cycles*. Melbourne University Press, Melbourne.

Beckett P. H. T. (1989) The use of extractants in studies on trace metals in soils, sewage sludges and sludge-treated soils. *Advances in Soil Science* 9, 143–76.

Bhat K. K. S. & Nye P. H. (1973) Diffusion of phosphate to plant roots in soil. I. Quantitative radiography of the depletion zone. *Plant and Soil* 38, 161–75.

Birch H. F. (1958) The effect of soil drying on humus decomposition and nitrogen availability. *Plant and Soil* 10, 9–31.

Bormann F. H. & Likens G. E. (1970) Nutrient cycles of an ecosystem. *Scientific American* 223, 92–101.

Catt J. A., Howse K. R., Farina R., Brockie D., Todd A., Chambers B. J., Hodgkinson R., Harris G. L. & Quinton J. N. (1998) Phosphorus losses from arable land in England. *Soil Use and Management* 14, 168–74.

Chaney R. L. & Oliver D. P. (1996) Sources, potential adverse effects and remediation of agricultural soil contaminants, in *Contaminants and the Soil Environment in the Australasia-Pacific Region* (Eds R. Naidu, R. S. Kookana, D. P. Oliver, S. Rogers & M. J. McLaughlin). Kluwer, Dordrecht, pp. 323–59.

Cole D. W., Gessel S. P. & Dice S. F. (1967) Distribution and cycling of nitrogen, phosphorus, potassium and calcium in a second-growth Douglas fir ecosystem, in *Primary Productivity and Mineral Cycling in Natural Ecosystems* (Ed. H. E. Young). Ecological Society of America, New York.

Garrett M. K., Watson C. J., Jordan C., Steen R. W. J. & Smith R. V. (1992) The nitrogen economy of grazed grassland. *Proceedings of the Fertiliser Society* 326, 3–32.

Halm B. J., Stewart J. W. B. & Halstead R. L. (1972) The phosphorus cycle in a native grassland ecosystem, in *Isotopes and Radiation in Soil-plant Relationships including Forestry*. International Atomic Energy Agency, Vienna, pp. 571–86.

Haygarth P. M. (1994) Global importance and global cycling of selenium, in *Selenium in the Environment* (Eds W. T. Frankenberger & S. Benson). Marcel Dekker, New York, pp. 1–27.

Heng L. K., White R. E., Bolan N. S. & Scotter D. R. (1991). Leaching losses of major nutrients from a mole-drained soil under pasture. *New Zealand Journal of Agricultural Research* 34, 325–34.

Lindsay W. L. (1979) *Chemical Equilibria in Soils.* Wiley, New York.

Marschner H. (1995) *Mineral Nutrition of Higher Plants*, 2nd edn. Academic Press, London.

Martell A. E. (1957) The chemistry of metal chelates in plant nutrition. *Soil Science* 84, 13–41.

Mitchell R. L. (1970) Trace elements in soils and factors that affect their availability. *Geological Society of America, Special Paper* 140, 9–16.

National Expert Group on Transboundary Air Pollution (2001) *Transboundary Air Pollution: Acidification, Eutrophication and Ground-level Ozone in the UK.* UK Department of Environment, Food and Rural Affairs, London.

Nicholson T. H. (1967) Vesicular-arbuscular mycorrhiza – a universal plant symbiosis. *Science Progress, Oxford* 55, 561–81.

Nutman P. S. (1965) Symbiotic nitrogen fixation, in *Soil Nitrogen* (Eds W. V. Bartholomew & F. E. Clark). Agronomy No. 10, American Society of Agronomy, Madison, WI.

Nye P. H. & Greenland D. J. (1960) *The Soil under Shifting Cultivation.* Commonwealth Bureau of Soils, Technical Communication No. 51, Harpenden.

Sposito G. & Coves J. (1995) SOILCHEM on the Macintosh, in *Chemical Equilibrium and Reaction Models* (Eds R. H. Loeppert, A. P. Schwab & S. Goldberg). Soil Science Society of America Inc, Madison, WI, pp. 271–87.

Sposito G. & Page A. L. (1984) Cycling of metal ions in the soil environment, in *Metal Ions in Biological Systems*, Volume 18 (Ed. H. Siegel). Marcel Dekker, New York.

Tiller K. G. (1996) Soil contamination issues: past, present and future, a personal perspective, in *Contaminants and the Soil Environment in the Australasia-Pacific Region* (Eds R. Naidu, R. S. Kookana, D. P. Oliver, S. Rogers & M. J. McLaughlin). Kluwer, Dordrecht, pp. 1–27.

Tinker P. B. & Nye P. H. (2000) *Solute Movement in the Rhizosphere.* Oxford University Press, New York.

White R. E. (1980) Retention and release of phosphate by soil and soil constituents, in *Soils and Agriculture* (Ed. P. B. Tinker) Critical Reports on Applied Chemistry, Vol. 2. Blackwell Scientific Publications, Oxford, pp. 71–114.

White R. E. (2003) *Soils for Fine Wines.* Oxford University Press, New York.

White R. E. & Sharpley A. N. (1996) The fate of non-metal contaminants in the soil environment, in *Contaminants and the Soil Environment in the Australasia-Pacific Region* (Eds R. Naidu, R. S. Kookana, D. P. Oliver, S. Rogers & M. J. McLaughlin). Kluwer, Dordrecht, pp. 29–67.

Williams R. J. B. (1975) The chemical composition of water from land drainage at Saxmundham and Woburn (1970–1975). Rothamsted Experimental Station Report for 1975, Part 2, pp. 37–62.

Further reading

Alloway B. J. (Ed.) (1990) *Heavy Metals in Soils.* Blackie, Glasgow.

Harter R. D. & Naidu R. (1995) Role of metal-organic complexation in metal sorption by soils. *Advances in Agronomy* 55, 219–63.

Peoples M. B. & Craswell E. T. (1992) Biological nitrogen fixation: investments, expectations and actual contributions to agriculture. *Plant and Soil* 141, 13–39.

Read D. J., Lewis D. H., Fitter A. H. & Alexander I. J. (Eds) (1992) *Mycorrhizas in Ecosystems.* CAB International, Wallingford.

Woomer P. L. & Swift M. J. (Eds) (1994) *The Biological Management of Tropical Soil Fertility.* Wiley & Sons, Chichester.

Example questions and problems

1. (a) Name the three main nutrient stores in soil.
   (b) Identify three sources from which these stores can be augmented.
   (c) Describe the two pathways by which S from the atmosphere is returned to the land.
   (d) Suppose that the average rate of S emission as SO_2 from an industrial region is 20 kg/ha/yr. If all this SO_2 is oxidized in the atmosphere and subsequently deposited on the land, calculate the expected acid input (in kmol H^+/ha). (The atomic masses of H, O and S are 1, 16 and 32 g, respectively).

2. (a) What is labile phosphate?
   (b) What is 'phosphate fixation'?
   (c) Name the two main groups of fungal–plant root symbiotic associations.
   (d) What is a nutrient depletion zone?
   (e) Identify the main mechanism by which phosphate ions move through soil to a plant root.

3. Composted chicken manure is ploughed into a sandy soil at the rate of 5 t dry matter/ha before a crop of leeks is grown. The C : N ratio of the manure is 15. From previous experimental work, it is known that the decay coefficient k for this manure is 1.5 (1/yr). Assume that the C content of the manure is 50% and that the leeks are expected to take up 60 kg N/ha over 60 days before being harvested. Calculate as follows.
   (a) How much of the composted manure (t C/ha) will decompose in the soil during 60 days? (Use the equation $C(t) = C(0)\exp(-kt)$).
   (b) How much mineral N (kg N/ha) would you expect to be produced from the manure during this period?
   (c) If the amount of soil mineral N at the time of sowing the crop was 10 kg N/ha, and leaching is expected to remove 15 kg N/ha before the crop is harvested, how much more N than that supplied by the manure is required by the crop?

4. A cereal grower on a black cracking clay soil wants to increase his wheat yield to 4 t grain/ha by applying N fertilizer. Such a crop should also produce 4 t of straw and root dry matter. The estimated N content of the grain and other plant material is 1.5% and 0.5%, respectively (dry matter basis). The grower estimates the initial soil mineral N content to be 20 kg N/ha, and N losses during crop growth (120 days) will be 25 kg N/ha. He seeks your advice on the amount of N fertilizer to apply. From a literature search, you find that the rate coefficient k for mineralization of soil organic N in this soil type is 0.03 (1/yr). The soil organic N content is 4000 kg N/ha to 15 cm depth.
   (a) Calculate how much urea fertilizer (46% N) is required per ha to meet the grower's objective.
   (b) What would be your recommendation on how to apply this urea? (Hint – see Section 12.2).

5. Rain falls on wet soil in an orchard at a rate of 6 mm/h for 3 hours. Soluble urea fertilizer was applied to the soil surface 6 days before rain started. The soil has a volumetric water content of 0.45 m^3/m^3 at saturation. Answer the following:
   (a) Write equations for the hydrolysis of urea in the soil,
   (b) Which forms of mineral N (derived from urea) would you expect to find in the soil after 6 days?
   (c) Which form of N would leach? and
   (d) How far into the soil will the leading edge of the fertilizer (and its reaction products) move during the rain period?

6. Urea fertilizer is applied to an acid soil at the rate of 75 kg/ha and becomes evenly distributed through the top 5 cm after rain.
   (a) Assuming that after 5 days 50% of the urea has hydrolysed and no losses of N have occurred, calculate the amount of NH_4-N present in the soil to 5 cm depth (in kg N/ha).
   (b) Suppose that after another 5 days all the urea has hydrolysed and 50% of the NH_4-N formed has been nitrified. Calculate how much NO_3-N is formed.
   (c) Noting all the reactions in the conversion of urea to NH_4^+ to NO_3^-, calculate the net amount of H^+ or OH^- ions formed in 10 days (in kmol/ha). (Urea fertilizer contains 46% N and the atomic masses of N and H are 14 and 1 g, respectively).

7. Give the ionic form in which the following micronutrient elements commonly occur in soil:
   (a) manganese; (b) copper; (c) boron; (d) molybdenum.

Part 3

Soil Management

'And earth is so surely the food of all plants that with the proper share of the elements which each species requires, I do not find but that any common earth will nourish any plant.'

Jethro Tull (1733) in *Horse-hoeing Husbandry*, republished by William Cobbett in 1829

Chapter 11

Maintenance of Soil Productivity

11.1 Traditional methods

Shifting cultivation

Shifting cultivation, also called 'slash-and-burn' agriculture, is widespread in the tropics where it is practised on some 500 million ha of rainforest, open forest and grassland savanna, which is about one-quarter of the potential arable land of these regions. Traditionally, the cycle begins with hand-clearing of the natural vegetation followed by burning of residues *in situ*, which in the case of rainforest releases a large store of available nutrients for the growth of subsequent crops such as maize, cowpeas, cassava and groundnuts. Large trees may be preserved and afford some protection to the newly exposed soil. However, loss of nutrients from the ash by erosion and leaching can be high (Table 11.1), and the encroachment of weeds that compete with the crops is rapid.

After two or three crops the land is abandoned to the regenerating forest and the farmer moves to a new site, where the cycle is repeated. In recent years, clearance of the vegetation and cultivation of crops have become mechanized, a practice that has led to soil structural degradation and increased loss of soil nutrients through erosion (Hulugalle, 1994).

In traditional shifting cultivation very little chemical fertilizer is used. Regrowth of the native vegetation, especially the forest, is essential for the stability of such an agricultural system. During this regrowth period, which has been called the bush fallow (Nye and Greenland, 1960), the soil's fertility is restored by deep-rooting trees or perennial grasses. These plants draw nutrients from deep in the subsoil and weathering parent material, and return large quantities of litter (up 10 t C/ha/yr) to the soil surface. There is also an input of symbiotically and non-symbiotically-fixed N in soils of the humid forest regions. Experience in the savanna regions suggests that a cycle of 2–4 years of cropping and 6–12 years of fallow can maintain soil fertility in the long term, but in the wet forest zones a 1–2-year cropping period and 10–20 years of fallow are preferred.

Traditional shifting cultivation is conservative of natural resources when properly managed, but cannot support dense agrarian populations. As population pressure has increased, fallow periods have been shortened with the result that soil degradation occurring during the cropping period is not reversed during the fallow. For example, in Africa, Oldeman *et al.* (1991) have estimated that

Table 11.1 Rates of soil erosion under shifting cultivation and undisturbed rainforest. After Brady, 1996.

Land use	Erosion (t soil/ha/yr)		
	Minimum	*Median*	*Maximum*
Undisturbed rainforest	0.03	0.30	6.2
Shifting cultivation			
– bush fallow	0.05	0.15	7.4
– cropping period	0.4	2.8	70

22% of the usable agricultural land is severely degraded. This problem is discussed further under 'Recent developments'.

Rotational cropping

In the more closely settled temperate regions of Europe, competition for farming land forced the adoption of intensive cropping systems much sooner than in the tropics. One of the earliest systems was the simple 3-year rotation of autumn cereal–fallow–spring cereal, practised in England for 1500 years. This wholly arable system was largely displaced in the 18th century by rotations, such as the Norfolk four-course, in which grass-clover pastures and root crops were alternated with cereals, thus permitting close integration of livestock and arable farming. Added advantages were the N_2 fixed by legumes and the abundant residues left in the soil after the pasture phase.

However, depressed cereal prices in the 1880s and a growing belief in the value of grass as a 'soil conditioner' led to the temporary pasture or ley phase being extended for 3 or more years. Gradually the arable–ley rotation evolved into the practice of mixed farming. The essence of this system is the rearing of livestock on farm-produced grass and cereals, in the course of which much of the nutrient taken up from the soil is returned in organic manures.

In the winter rainfall zone of southern Australia, arable–ley rotations became the mainstay of agriculture following the widespread use of super-phosphate and introduction of the annual pasture legume *Trifolium subterraneum* (subclover) early in the 20th century. Phosphate was applied, N was fixed by the legume during the pasture phase, and these nutrients cycled through the grazing animals. Soil organic matter slowly increased and soil structure improved. After 2 or more years, the pasture was ploughed in and cereals were sown, nourished mainly by the reserves built up during the pasture phase. However, on many soils in the zone of annual rainfall > 500 mm, after 40–50 years of such agriculture, accelerated soil acidification has been observed (Section 11.3). The area of dryland salinization has also expanded in rainfall zones up to 800 mm per year (Section 13.2).

Table 11.2 Typical concentrations of macronutrients in organic manures. After Chadwick and Chen, 2002 and ADAS, 1982.

	Dry matter (%)	N*	P*	K*
Cattle farmyard manure (FYM)	25	6.0	1.5	6.6
Pig FYM	25	7.0	3.0	4.1
Poultry (layer) manure	30	16	5.7	7.5
Broiler litter	60	30	10.9	14.9
Cattle slurry	6	3.0	0.5	2.9
Pig slurry	6	5.0	1.3	2.5
Sewage sludge				
Liquid digested	4	2.0	1.0	0.1
Dewatered digested	50	15	5	–

* kg/t fresh weight for FYM or kg/m^3 for slurry.

Organic manures

Farmyard manure (FYM) and deep litter (from stall-fed animals), cattle and pig slurry, poultry manure and sewage sludge are examples of organic manures (Table 11.2). Their beneficial effects lie as much in soil structural improvement from the large additions of organic matter, as in the contribution made to a soil's nutrient supply. Organic manures also encourage flourishing populations of small animals, especially earthworms, and microorganisms that may produce phytohormones (indolyl-acetic acid, cytokinins and abscisic acid) in the rhizosphere. The following categories of organic manure are important:
• *FYM* consists of cattle dung and urine mixed with straw. It is very variable in composition and although low in macronutrients, at recommended application rates of 40–50 t/ha, the quantities of micronutrients supplied are roughly equivalent to those removed in a succession of 4–5 crops (Table 11.3);
• *slurry* is a mixture of dung and urine and washing water from animal houses and dairies. Typical compositions of some representative manures and slurries are given in Table 11.2;
• *sewage sludge* – a number of types of sludge may be produced from raw sewage. Liquid raw sludge from primary sedimentation tanks is

Table 11.3 Micronutrients (kg/ha) in a FYM application compared with those removed in crops. After Cooke, 1982.

	Mn	Zn	Cu	Mo
FYM (45 t/ha)	3.36	1.12	0.56	0.01
Total in four arable crops in succession	2.50	1.80	0.30	0.01

sometimes offered to farmers, but, more commonly, sludge produced after secondary treatment of sewage is used. Anaerobic fermentation reduces pathogen numbers and smell, and produces a black liquid digested sludge containing 2–4% dry matter (DM) (Table 11.2). Dewatered sludges (raw or digested) contain 40–50% DM and are often called 'sludge cake'. When a flocculant, such as lime or $FeSO_4$ has been used before dewatering, the sludge has better 'condition' and is of value in improving the structure of reconstituted soil on land rehabilitation sites. However, much of the soluble N and P is lost during the dewatering treatment so that the increase in total N and P concentration through dewatering is not in proportion to the increase in dry matter percentage (Table 11.2). Constraints on the use of sewage sludge on agricultural land are discussed in Box 11.1.

Green manure and compost

Green manure and compost are examples of organic 'manures' derived from plant materials. The practice of ploughing in a quick-growing leafy crop before maturity is called green manuring. A major benefit is the avoidance of NO_3^- leaching in susceptible soils, because soil NO_3^- is taken up by the green manure crop and slowly released to a

Box 11.1 The use of sewage sludge or biosolids on agricultural land.

To emphasize the potential benefits of the nutrients supplied, sewage sludge is often now referred to as 'biosolids'. Biosolids can be applied at rates of 50–80 t DM/ha, provided that the acceptable levels of heavy metals such as Cd, Cu, Zn, Ni, Hg and Pb, which might affect human or animal health, are not exceeded. Accordingly, in European Union (EU) countries and the USA, limits have been set for the concentration of metals in biosolids that are used for land application. Annual loading limits have also been calculated, based on a 10-yr average (Table B11.1.1). In the EU, these limits will be lowered in stages up to 2025. Even though Cd is low in sludges (generally < 50 mg/kg), there is particular concern in Australia and New Zealand because the maximum permissible concentration in food (other than meat and fish products) is set at 0.05 mg Cd/kg fresh weight. One of the reasons for this caution is that superphosphate used in these countries has traditionally had a relatively high Cd concentration, having been made from Pacific Island phosphate rocks that are of high Cd concentration.

Note that the maximum permissible concentrations are expressed in terms of the total element, which is conservative because as observed in Box 10.8 the bioavailable concentration will be much lower. A soil approaching its limiting concentration should be kept at or above pH 6.5 for arable crops or 6.0 for grassland, because the higher the pH the more likely a metal is to be precipitated as an insoluble hydroxide (Section 10.5).

Table B11.1.1 Concentration limits and annual loading limits for heavy metals in biosolids used for land application. After Epstein, 2003.

Element	Maximum concentration (mg/kg dry matter)		Loading limit averaged over 10 years (kg/ha/yr)	
	USA	EU	USA	EU
Cadmium (Cd)	39	20–40	1.9	0.15
Mercury (Hg)	17	16–25	0.85	0.1
Lead (Pb)	300	750–1200	15	15
Nickel (Ni)	420	300–400	21	3
Copper (Cu)	1500	1000–1750	75	12
Zinc (Zn)	2800	2500–4000	140	30

subsequent cash crop as the residues of low C : N ratio decompose. If the green manure crop is a legume, there is an additional boost to the soil N supply from symbiotically-fixed N. The long-term effect of green manuring on soil organic matter is minimal, since the succulent residues are rapidly decomposed and contribute little to soil humus. Similarly, beneficial effects on soil structure through the stimulation of microbial activity are ephemeral.

Compost is made by accelerating the rate of humification of plant residues in well-aerated, moist heaps (note that animal manures may be added to compost heaps to lower the C : N ratio of the mix). Ideally, a compost of C : N ratio 10–12 and acceptable nutrient content can be made in 6–8 weeks. The heap should be turned regularly to keep the water content and temperature as uniform as possible. The heap temperature should exceed 55°C for at least 3 days to kill pathogens and weed seeds.

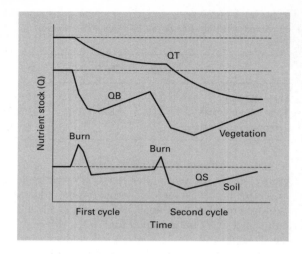

Fig. 11.1 Changes in the nutrient stock in slash-and-burn agriculture as a function of time. QT = total nutrient stock in the ecosystem; QB = nutrient stock in plant biomass; QS = nutrient stock in the soil, and QT = QB plus QS (after Juo and Mann, 1996).

Recent developments

Since the 1950s, demand for food created by a rapidly growing world population and a need to grow cash crops for export earnings have induced a swing away from traditional methods of agriculture in many parts of the world, often with diminished sustainability of the production systems. In the wet tropics, for example, although properly managed shifting cultivation conserves natural resources, it supports relatively few people (less than one-tenth of the world's population). Also, as population pressure on natural resources has increased, migrants who have no knowledge of, nor the incentive to follow traditional practices, have moved into the shifting-cultivation areas. Thus, the rate of forest clearance has increased in Indonesia, Malaysia and the Amazon Basin for example, the bush-fallow period has shortened, and the total stock of nutrients in an ecosystem has declined (Fig. 11.1). The consequences of these changes in the soil and vegetation are serious not only for food production, but also because of greater net inputs of CO_2 and other radiatively active gases to the atmosphere, which contribute to the enhanced 'greenhouse effect' (Box 3.1). More permanent systems of

agriculture are being developed for these tropical areas; however, these will require substantial inputs of nutrients from external sources to sustain high levels of production.

In temperate regions, the intensification of agriculture, especially after World War II, has led to livestock and arable farming becoming increasingly separated, so that animal manures are not plentiful in areas where they are most needed. Inevitably, there has been a switch to cereal monoculture – 'the one-course rotation' – and increasing reliance on chemical fertilizers in place of traditional manuring and fallowing. The need to maximize annual economic returns per ha has meant that green manure crops cannot be grown on a large scale. In many cases high economic returns can only be achieved with the aid of irrigation, which has placed increased pressure on scarce water resources in countries such as China, India, Australia and parts of the USA.

Nevertheless, agriculture was generally very successful over the last four decades of the 20th century in producing enough food to satisfy a growing world population, now > 6 billion (Box 11.2). Production increased annually at about 1% in developed countries and 3% in developing countries, although problem areas remain in Central Africa,

Box 11.2 The Green Revolution.

Beginning in the 1950s, the 'Green Revolution' was the international community's response to the need to increase per capita food production, particularly in developing countries, where in 1961 some 80% of the population of 1.7 billion people had a daily food intake < 2100 calories (FAO, 1995). The programme focused on three related activities:

• Breeding of early maturing, high-yielding varieties (HYVs) of wheat and rice;
• promotion and distribution of 'high input' packages including fertilizers, pesticides and the regulation of water supplies; and
• implementation of these technical innovations in the most favourable agro-climatic regions and suitable rural communities.

The extraordinarily successful outcome of the programme was that food production per head was some 18% higher in 1990 than 1961 (Fig. 11.2),

and in developing countries only 8.5% of the population of c. 4 billion had a daily food intake < 2100 calories. Notwithstanding this success, evidence emerged in the 1990s that yields from the HYVs had in some instances reached a plateau, the causes of which have been variously ascribed to increased incidence of pests and disease, and soil degradation, associated with intensive monoculture. With high inputs of fertilizers and pesticides, there have inevitably been losses to the environment so that the quality and quantity of water in many densely settled and farmed regions have deteriorated, especially in the Ganges Plain of India, the eastern seaboard of China, and the Mekong Delta of South-East Asia. Addressing these problems to develop and implement more sustainable agricultural systems is the challenge for the current generation of soil scientists (Chapter 15).

Central America and parts of South and East Asia. The rate of population growth slowed in the last decade of the 20th century, and is forecast to fall further up to 2015 (Bruinsma, 2003), so that with modest increases in production the food supply per capita is steadily improving (Fig. 11.2). In many developing countries both higher yields per ha and more land under cultivation have contributed to the rise in production; in others

such as China and India, which are developing rapidly as industrialized nations, land under cultivation has actually decreased, but much improved annual yields per ha have more than compensated for this loss. In developed western countries, yields have increased through the introduction of better crop varieties, high inputs of fertilizer and widespread use of pesticides. However, this achievement has not been without

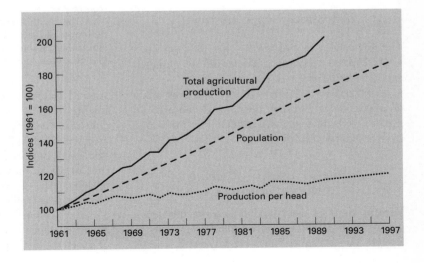

Fig. 11.2 World agricultural production, population and production per head (after FAO, 1995 and Bruinsma, 2003).

cost in terms of accelerated soil acidification (Section 11.3), deterioration of soil structure (Section 11.4), accelerated soil erosion (Section 11.5), eutrophication of water bodies (Section 12.3), soil salinization (Section 13.2) and environmental contamination (Section 15.1). Partly in reaction to these problems of high-intensity 'production' agriculture, the business of low-input 'organic farming' has expanded over the last few decades in many developed western countries, a topic discussed in Section 15.2.

11.2 Productivity and soil fertility

Soil, climate, pests, disease, genetic potential of the crop, and man's management are the main factors governing land productivity, as measured by the yield of crop or animal product per ha. This book is primarily concerned with soil properties – how they interact with soil management and their effect on productivity. Soil fertility connotes primarily the combined effect of chemical and biological properties, and is probably the most important single soil factor affecting productivity. The modification of soil fertility by the proper use of fertilizers is discussed in Chapter 12. Soil physical properties, notably structure, are normally secondary although these become of prime importance when soils are susceptible to structural degradation, or once the fertility problems have been solved.

Soil nutrient supplying power

The assessment of soil fertility has two aspects:
• *Qualitative* – the aim is to identify which nutrients are deficient (or in excess); and
• *quantitative* – the aim is to estimate how much of a particular limiting nutrient is required to achieve optimum growth under the prevailing environmental conditions.

Qualitative aspects – diagnosis of nutrient deficiency

Various signs of disorder indicate the deficiency (or toxicity) of one or more essential elements in a plant. For example, P deficiency typically causes

(a)

(b)

Fig. 11.3 (a) Grapevine leaves showing Fe deficiency (courtesy of Scholefield Robinson Horticultural Services, Netherby, South Australia).
(b) Leaves of *Glycine javanica* showing K deficiency (normal leaf in the lower-centre).
(See also Plate 11.3a and b).

stunted growth and N deficiency an overall yellowing of leaves and stunted growth. Interveinal chlorosis due to Fe deficiency (Fig. 11.3a) associated with high soil pH was discussed in Box 10.7. Potassium deficiency shows as marginal 'spotting' that develops into complete necrosis of the leaf margins (Fig. 11.3b). However, such visual symptoms appear only after the plant has suffered a check in growth due to 'hidden hunger' for the deficient element. Growth and the concentration of the element in a plant's tissues are closely linked,

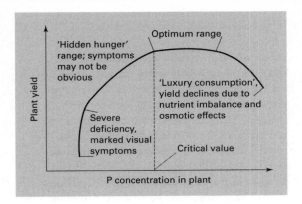

Fig. 11.4 The relationship between yield and plant tissue concentration of P.

as shown in Fig. 11.4, so that plant analysis or tissue testing is useful in diagnosing deficiency (or toxicity).

Note in Fig. 11.4 that the 'critical value' is the element concentration in the plant below which an increase in supply leads to an increase in yield. To be of use in deficiency diagnosis, such values must be determined for individual crops, preferably when no other nutrient limits growth. Under these conditions the critical value should be independent of soil type. However, soil water supply to the plant has a significant influence on nutrient concentrations, with concentrations usually being lower under dry conditions, even when the soil supply of the nutrient is adequate. Thus, soil water availability should be monitored when plant nutrient concentrations are being evaluated against critical values.

Quantitative aspects – soil testing

The quantitative approach involves analysing a soil to measure the amount of nutrient that is available for plant growth. This is called soil testing. Fundamentally, the concept of nutrient 'availability' is more complex than at first appears because:
• Plants take up nutrient ions from the soil solution around the roots in which the total amount of nutrient is usually insufficient to meet plant demand;
• the concentration of a nutrient in the soil solution is buffered by the whole labile pool, which

includes an easily desorbable surface phase (e.g. for Ca, Mg, K, P and S) and also may include a readily mineralizable organic fraction (e.g. for N, S and P); and
• the rate of replenishment of a nutrient at a root surface depends not only on the soil's buffering capacity for that nutrient, but also on the rate of movement to the root surface by mass flow and diffusion (Section 10.2).

This complexity may be resolved by measuring the quantity/intensity (Q/I) relationship for a nutrient (Box 10.8). The concentration of an ion is independent of the size of the system studied and is therefore a measure of its intensity factor (I). The amount of an ion extracted by an anion or cation exchange resin, or a neutral salt solution, corresponds to the easily desorbable labile component and hence measures its quantity factor (Q). The change in I with change in Q depends on the slope of the Q/I graph, which defines the buffering capacity of the soil for that particular nutrient. Further, the effective diffusion coefficient for an ion such as $H_2PO_4^-$, which is mainly adsorbed, depends primarily on the reciprocal of the buffering capacity as discussed in Box 10.5. Fig. 11.5 illustrates the contrasting phosphate buffering capacities (PBC) of two soils. Clearly, the P concentration in solution (I) in soil A, which has a low PBC, will decrease more sharply as P is

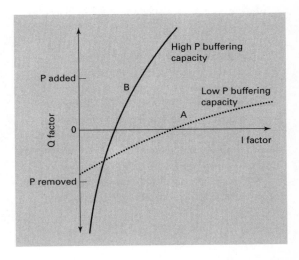

Fig. 11.5 Q/I relations for two soils showing contrasting phosphate buffering capacities.

Box 11.3 Soil testing in practice.

Throughout the world, a great variety of empirical soil tests are used to measure 'available' nutrients, even for a single element such as P. Broadly, these tests employ:
- Neutral salt solutions, e.g. 0.01 M $CaCl_2$ (for P and K), 0.01 M $Ca(H_2PO_4)_2$ (for S), 2 M KCl (for mineral N); or
- anion or cation exchange resins (e.g. for P, S, K, Mg and Ca); or
- acids or alkalis of varying strength (e.g. 0.005 M H_2SO_4 for P or 0.5 M $NaHCO_3$ for P); or
- complexing agents (e.g. 0.005 M DTPA for Cu, Zn, Mn and Fe).

The milder extractants (neutral salt solutions) tend to reflect the I factor of a soil's nutrient supply, whereas the more severe extractants (acids and complexing agents) are dominated by the Q factor. Specific details of the methods are given in books such as Westerman (1990) and Rayment and Higginson (1992). Irrespective of whether the Q or I factor (or both) is measured, a soil test must be calibrated against crop yield, usually on a variety of soils with different rates of the element supplied in fertilizer. The resultant graph of yield response (percentage increase over the control) against the test values for the suite of soils allows a critical value to be established (Fig. B11.3.1).

In many countries a solution of 0.5 M $NaHCO_3$ buffered at pH 8.5 (the 'Olsen test') is the standard extractant for soil P. In the UK, no soil test is considered satisfactory for N, but in other European countries and many states of the USA, measurements of 'residual' soil mineral N (exchangeable NH_4^+ and solution NO_3^-) made in late winter have been found useful for predicting fertilizer N requirements for crops growing in the following spring and summer. The depth of sampling may range from 0.6 to 1.8 m.

Besides the choice of soil test, the representivity of any sample analysed needs to be considered.

Soil is naturally very variable (Chapter 5). The spatial distribution of available nutrients is variable, not only because of the natural variability in soil properties but also because of localized plant removals, biological transformations, fertilizer inputs, and the non-uniform return of dung and urine from animals. Variations in temperature and moisture with time also affect nutrient availability. Hence, to improve the accuracy of a soil test, a sufficient number of samples must be taken to encompass the full variability of the area tested. Analysing a large number of such samples (15 or more) for each test is not cost-effective, so the samples are usually bulked, mixed and then subsampled to provide a representative 'composite' sample for analysis. The sampling depth is normally 0–10 cm, although this is not appropriate for N as indicated above.

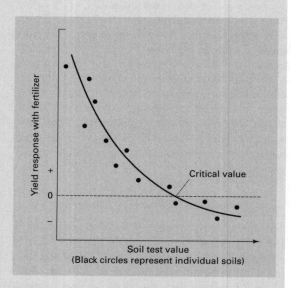

Fig. B11.3.1 Yield response to fertilizer in relation to soil test value (points represent individual soils tested).

absorbed by plants than for soil B, which has a high PBC. Conversely, soil A needs much less P fertilizer than soil B to raise I to a satisfactory value for plant uptake.

For N, not only the total amount of soil NH_4^+ and NO_3^-, but also the amount of readily mineralizable organic N may need to be measured to give Q. In some cases the stable isotope (^{15}N) has been used to measure Q using the principle of isotopic dilution of a trace amount of the isotope in the labile pool. Similarly, the radioactive isotope (^{32}P) has been used to measure the amount of labile soil P using the principle of isotopic exchange. This amount has been called an E value (when isotopic exchange takes place in a shaken soil suspension), or an L value if a test plant is grown and used to measure the extent of isotopic exchange in the soil – Section 12.3).

The implications of these concepts for practical soil testing are discussed in Box 11.3.

Quantitative assessment of fertilizer needs

The amount of fertilizer needed to correct a specific nutrient deficiency can be estimated in the course of calibrating a soil test by means of a fertilizer rate trial, in which the yield at different rates of fertilizer (amounts per ha) are measured.

The response to fertilizer frequently follows a law of diminishing returns, which means that the yield increase for successive equal increments of fertilizer becomes steadily less. If y is the yield and x the amount of fertilizer element (e.g. N) in kg per ha, this law can be expressed as

$$y = A - B \exp(-Cx). \qquad (11.1)$$

As shown in Fig. 11.6a, A is the maximum (asymptotic) yield attainable, $(A - B)$ is the yield without fertilizer, and C is the rate of change of y with x.

Particularly with N, a decline in yield is sometimes observed at high fertilizer rates, which has led to the fitting of quadratic or linear intersecting models to such fertilizer–response curves. However, whichever model is chosen, the precision in predicting the amount of fertilizer needed to attain maximum yield is poor because of factors other than N supply, such as climate, affecting

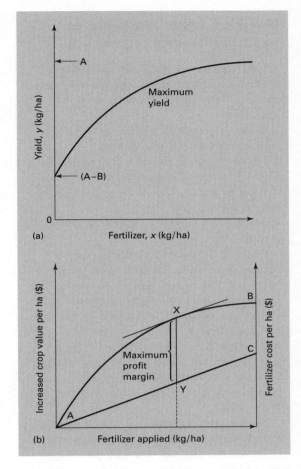

Fig. 11.6 (a) A typical fertilizer–response curve obeying Equation 11.1. (b) Maximum profit per hectare from fertilizer investment.

yield. More important is the prediction of fertilizer required to produce an optimum yield, which may be defined as the yield at which profit per ha is greatest (Figure 11.6b). For example, when the value of the extra yield due to fertilizer (curve A–B) is compared with the fertilizer cost line (A–C), the vertical X–Y indicates the fertilizer rate that gives the biggest value-cost differential per ha. If the crop's value, or the fertilizer's cost changes, X–Y slides to the right or left to a new point of maximum profit. Other approaches to estimating fertilizer requirements are outlined in Box 11.4.

Box 11.4 Other approaches to estimating fertilizer requirements.

A crop's requirement for a particular nutrient in fertilizer can be estimated using a nutrient balance approach. This requires knowledge or an estimation of:

1 External inputs of the nutrient to the soil (in residues and from the atmosphere);

2 transformations in the soil leading to net gains to or losses from the labile pool;

3 losses by leaching, erosion, in gas form and by product removal; and

4 the desired crop yield or number of grazing animals to be supported.

Point 4 quantifies the total demand for the nutrient that must be balanced against the sum of points 1–3. The shortfall determines how much of the nutrient must be supplied in fertilizer. This is the principle behind the OVERSEER™ model for N, P, K and S fertilizer recommendations for pasture land in New Zealand (Ledgard et al., 1999), and also the SUNDIAL model (Simulation of Nitrogen Dynamics in Arable Land) for N fertilizer recommendations for crops in the UK (Smith et al., 1996).

These approaches produce a 'blanket' fertilizer recommendation for an area (a field or perhaps a farm). Another approach is to map the variation in crop yield accurately at a scale of 1–2 m, as the crop is being harvested, and to use this information to adjust subsequent fertilizer applications. If yield maps are constructed over a number of years (to take account of year-on-year variation in weather, pests and disease), those areas of a field which consistently yield higher or lower than the field average can be identified. If the 'high yield' areas correspond to areas of high soil fertility, the amount of fertilizer applied to these areas may be decreased and that applied to the 'low' areas increased to achieve an overall increase in yield, while any off-site environmental effects through excess fertilizer use are minimized. This approach is a basis for site-specific fertilizer management or precision agriculture, which is now being introduced for cereal cropping and viticulture (Cook and Bramley, 1998).

11.3 Soil acidity and liming

Sources of acidity

A soil becomes acidic if Ca^{2+}, Mg^{2+}, K^+ and Na^+ ions are leached from the profile faster than they are released by mineral weathering, and H^+ and Al^{3+} ions become the predominant exchangeable cations (Sections 7.2 and 9.2). The intensity of the acidity, measured by the soil pH, is the result of an interaction between soil mineral type, climate and vegetation. The main sources of net inputs of acid are as follows:

• The dissolution of CO_2 in the soil water to form carbonic acid (H_2CO_3), which dissociates according to the reactions

$$CO_2 + H_2O \leftrightarrow H_2CO_3 \leftrightarrow H^+ + HCO_3^- \leftrightarrow 2H^+ + CO_3^{2-}. \tag{11.2}$$

Pure rain water in equilibrium with CO_2 in the atmosphere has a pH of 5.65. However, soil respiration increases the partial pressure of CO_2 in the soil air, so driving Reactions (11.2) to the right and producing more H^+;

• the accumulation and humification of soil organic matter, producing humic residues with a high density of carboxyl and phenolic groups that dissociate H^+ ions. The balancing alkalinity is not released until the organic matter decomposes, releasing HCO_3^- ions. Removal of C in plant and animal products has a similar acidifying effect;

• nitrification of NH_4^+ ions, producing H^+ ions and NO_3^-, which is susceptible to leaching (Equation 10.2). Removal of N in plant and animal products also has an acidifying effect because there is no opportunity for this organic N to be ammonified and release OH^- ions to the soil (Equation 10.1);

• inputs of H_2SO_4, HNO_3 and $(NH_4)_2SO_4$ from the atmosphere or 'acid rain' (Sections 10.2 and 10.3). The acidifying effect of H_2SO_4 is not necessarily eliminated by reaction with NH_3 to form $(NH_4)_2SO_4$, because H^+ ions are released again when NH_4^+ is oxidized in the soil. Even when the NH_4^+ ions are taken up by plant roots there can be an acidifying effect on the rhizosphere because H^+ ions are released to balance any net surplus of cations over anions absorbed, when N is absorbed

Table 11.4 Sensitivity of cultivated plants to soil acidity (high exchangeable Al^{3+}).

Pasture species	Crop species	Tolerance to Al
Cocksfoot grass	Rye Potato Oats Triticale	Highly tolerant
Ryegrass Subclover White clover	Wheat Maize	Tolerant
Phalaris grass	Canola (rape)	Sensitive
Lucerne Trefoil	Barley Sugar beet	Highly sensitive

as NH_4^+ and not NO_3^-. With atmospheric accessions of NH_4^+, volatilization of NH_3 becomes an acidifying process because there are no neutralizing OH^- ions present (Equation 10.8).
• in soils formed on marine muds, or coal-bearing sedimentary rocks, the oxidation of iron pyrites FeS_2 (formed under intense reducing conditions in the original deposit) gives rise to Acid Sulphate Soils (Box 9.5).

Plant response

Under natural conditions, plants have adapted to the prevailing soil pH through evolution. Plants tolerant of acidity are called calcifuges (actually, intolerant of lime), in contrast to those that are intolerant of acidity or 'lime-loving', which are called calcicoles. Note that sensitivity to soil acidity is primarily sensitivity to high exchangeable Al^{3+} concentration (Section 7.2). Cultivated plants also exhibit considerable diversity in their response to acidity, as the examples of crop and pasture species in Table 11.4 show.

Within the legume family, tropical species are more tolerant of acidity than most medics and clovers of temperate regions. The micronutrient Mo is often of low availability in acid soils, a factor that inhibits the ability of legumes to nodulate effectively and fix atmospheric N_2. However, this condition is easily corrected by liming (see below).

Accelerated soil acidification

Soil acidification is a natural process, but the rate can be accelerated by human activities, as indicated in 'Sources of acidity' above. For example, in large areas of southern and eastern Australia the productivity of agriculture has been raised by ley-arable farming and improved pastures based on subclover. At the same time, soil organic matter has increased, as has the removal of C and N in crop and animal products, and the leaching of NO_3^-. Thus, compared with unimproved systems, net inputs of acid to the soils have increased by approximately 1–2 kmols H^+/ha/yr.

The effect of increased acid inputs on soil pH is determined by the soil's pH buffering capacity (Section 7.2). Whereas the additional input of 1–2 kmol H^+/ha/yr is small compared with atmospheric acid inputs in the Northern Hemisphere, in southern Australia most of the podzolic soils are naturally acidic and have low pH buffering capacities in the A horizon. Thus, pH in these A horizons has declined at the rate of 1 pH unit per 30–50 years, and some 24 million ha of agricultural soils in southern Australia are now classed as strongly acidic (pH(H_2O) < 5.6 or pH($CaCl_2$) < 4.8 (Hajkowicz *et al.*, 2003). Neutralization of this extra acidity would require an annual application of 50–100 kg $CaCO_3$ per ha. However, liming is expensive, particularly for extensive pastoral enterprises, so other strategies are being employed such as the use of acid-tolerant species and sowing deep-rooted perennial grasses to minimize NO_3^- leaching. Such strategies can only slow the rate of acidification, and ultimately lime will be needed to neutralize the extra acidity (see below).

Blake *et al.* (1994) estimated the critical load (the level at which adverse environmental effects might occur) for atmospheric inputs in southern England to be *c.* 1 kmol H^+/ha/yr, which is generally exceeded by atmospheric inputs of NH_4-N alone (Section 10.2).

Liming

Problems associated with soil acidity – slow turnover of organic matter, poor nodulation and N_2 fixation by legumes, Ca and Mo deficiencies, Al and Mn toxicities – can be remedied by liming.

Ground limestone, chalk, marl and basic slag are used as liming materials, the active constituent being primarily $CaCO_3$, with some burnt lime (CaO) and hydrated lime ($Ca(OH)_2$).

Lime applied to moist soil slowly dissolves by hydrolysis to produce an alkaline pH, according to the reaction

$$CaCO_3 + 2H_2O \leftrightarrow Ca(OH)_2 + H_2CO_3. \quad (11.3)$$

However, the strong base $Ca(OH)_2$ reacts with dissolved CO_2 from the soil air to form $Ca(HCO_3)_2$ so that the net reaction is

$$CaCO_3 + CO_2 + H_2O \leftrightarrow Ca(HCO_3)_2. \quad (11.4)$$

The final pH attained can be predicted from the equation

$$pH = K - \tfrac{1}{2}\log_{10}(P_{CO_2}) - \tfrac{1}{2}\log_{10}(Ca), \quad (11.5)$$

where P_{CO_2} is the partial pressure of CO_2, (Ca) is the activity of Ca^{2+} ions in solution, and $K = 4.8$

Table 11.5 Predicted soil pH of calcareous soil for a CO_2 partial pressure ranging from that in the atmosphere to that under pasture.

Ca activity in the soil solution (M)	Soil pH		
	CO_2 partial pressure (kPa)		
	0.036	0.36	1.0
0.001	8.42	7.92	7.70

if the carbonate has the solubility of pure calcite. For soil carbonates, which are more soluble than pure calcite, $K \cong 5.2$. The predicted pH for soil containing free $CaCO_3$ that is in equilibrium with various partial pressures of CO_2 is shown in Table 11.5.

Assessing the amount of $CaCO_3$ required to neutralize soil acidity or raise the soil pH to a desired value for crop growth involves measuring a soil's lime requirement (Box 11.5).

Box 11.5 Estimating the lime requirement.

The lime requirement depends on a soil's pH buffering capacity (Section 7.2), and is expressed as the quantity of ground limestone or chalk (t/ha to a depth of 15 cm) that is required to raise the soil pH to a desired value. Methods of measuring a soil's lime requirement range from equilibrating a sample with a single buffer solution, to titration with $Ca(OH)_2$, to calculation from a model based on estimated rates of Ca loss in the field. Details of laboratory methods are given in Rayment and Higginson (1992). Laboratory analysis is also necessary to measure the neutralizing value (NV) of different liming materials. The standard in Australia is pure $CaCO_3$. The NV of liming materials ranges from c. 150 for burnt lime down to c. 70–85 for agricultural lime. The effective neutralizing value (ENV) depends on the average particle size of the material, which should be < 0.3 mm diameter. Note that gypsum ($CaSO_4.2H_2O$) has no liming value.

pH 6.5 (in H_2O) is recommended for temperate crops, but liming of pastures to pH 6 (in H_2O) is preferred because grasses are more tolerant of acidity and the possibility of inducing deficiencies of Fe, Mn, Cu and Zn is minimized. Tropical species are more acid tolerant and furthermore, liming to pH 6–6.5 may reduce P availability in soils high in exchangeable Al^{3+} (Section 7.3), and destabilize the structure of kaolinitic clay soils (Section 7.4). In such soils, liming sufficient to hydrolyse the exchangeable Al^{3+}, normally achieved at pH 5.5, is preferable. Depending on a soil's pH buffering capacity and the target pH, the lime requirement may range from 2 to 10 t/ha on average.

Finely ground $CaCO_3$ made to adhere to legume seeds – a process called lime pelleting – greatly improves nodulation of acid-sensitive species in low pH soils, especially when an effective strain of *Rhizobium* bacteria is included in the coating.

11.4 The importance of soil structure

Structure-dependent properties

Good soil management should aim to create optimum physical conditions for plant growth, as shown by:

1 Adequate aeration for roots and micro-organisms;

2 adequate available water;

3 ease of root penetration, permitting thorough exploitation of the soil for water and nutrients;

4 rapid and uniform seed germination; and

5 resistance of the soil to slaking, surface-sealing and accelerated erosion by wind and water.

Two of the most useful indices of structure are bulk density, which is inversely related to soil porosity (Section 4.5), and aggregate stability, which is assessed by the coherence of soil aggregates in water (Box 4.4). Bulk density measurements are more relevant to points 1, 2 and 3 above, whereas aggregate stability is more relevant to points 4 and 5. Organic matter, soil texture and land use modify these properties. For example, bulk density has been found to increase as organic matter content decreases, and to be highest in sandy texture classes. Aggregate stability, on the other hand, decreases with declining organic matter and is least for soils in the silty and fine sandy loam classes.

Aeration and water supply depend on a soil's pore size distribution, the key indices being the air capacity ε_a (Section 4.5) and the available water capacity (AWC) (Section 6.4). Using these two properties, classes of soil droughtiness at one extreme and susceptibility to waterlogging at the other may be separated, as illustrated for surface soils in England and Wales that regularly experience a summer soil water deficit > 100 mm (Fig. 11.7). Soils with < 10% AWC are droughty, and those with ε_a < 5% are likely to be waterlogged and anaerobic. The limiting AWC value is slightly lower, and the limiting ε_a value $\cong$ 10%, for soils of wetter regions. Irrespective of the individual values of ε_a and AWC, a storage pore space value (the sum of AWC and ε_a) < 23% is undesirable; hence the truncation of the curve separating the 'poor' and 'moderate' classes in Fig. 11.7.

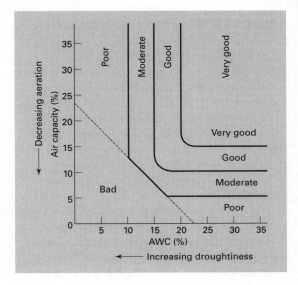

Fig. 11.7 Classification of the structural quality of topsoils (after Hall et al., 1977).

The effect of land use

Grassland vs arable

Soil physical conditions are usually best under permanent grassland (or forest) and deteriorate at a rate dependent on climate, soil texture and management when a soil is cultivated.

When land is ploughed, disruption of aggregates exposes previously inaccessible organic matter to attack by micro-organisms, and populations of structure-stabilizing fungi and earthworms decrease markedly. Removal of the protection of a permanent canopy of vegetation exposes the soil to the direct impact of rain, and the loss of surface litter reduces the detention time of water before surface runoff and consequent erosion begins. Shearing forces generated by tractor wheels and ploughs can rupture macro- and micro-aggregates. Structure is most vulnerable to damage when soil is wet (θ > FC) because the distance between clay particles in the domains and quasi-crystals is greatest then. An applied force can cause deflocculation and re-orientation of the deflocculated clay particles to form a dense layer of slow permeability to gases and water, called a

Box 11.6 Detecting soil compaction.

Compaction decreases porosity and so can be detected from an increase in soil bulk density. The bulk density might increase from the acceptable range of 1.0–1.3 Mg/m³ to > 1.5 Mg/m³. Bulk density is easily measured for surface soil by driving in a steel cylinder of known volume, extracting an intact soil core, oven-drying and weighing. The diameter of the cylinder should be at lease 10 times the thickness of the cylinder wall to avoid additional compaction of the soil core.

Compacted soil shows an increase in strength that can be measured by the soil's resistance to deformation or rupture when a shearing force is applied. Tests of soil strength can be done on individual aggregates (assessing the 'grade' of an aggregate, as described in Box 4.2), or the bulk soil strength can be measured from the resistance offered to a penetrometer. This instrument consists of a steel rod about 1 m long, with a conical tip and a proving ring and strain gauge at the top (Fig. B11.6.1). The rod is graduated to indicate the depth of insertion of the tip in the soil. Readings are most reliable when the rate of entry of the rod is constant, which can be achieved using a tractor-mounted hydraulic system. Penetration resistance is very variable spatially so a large number of readings is required. Because soil strength depends on soil water content, the water content at which measurements are made should be known. A soil strength > 2 MPa at field capacity is generally considered undesirable for root penetration. When used in a comparative rather than absolute sense,

penetrometer measurements can identify areas of compacted soil between vine rows in vineyards, or near gateways and headlands in arable fields, which may be correlated with areas where crop yield is depressed (Box 11.4).

Fig. B11.6.1 An example of a hand-held penetrometer (after Davies *et al.*, 1993).

plough pan. The detection of a plough pan or other compacted layers in a soil profile is discussed in Box 11.6.

Measurements of water-stable aggregates show that structural degradation is most rapid in the first year after grassland is ploughed (Fig. 11.8). Total porosity and the AWC of the plough layer also decline as cultivation is prolonged. These trends can be reversed, however, by introduction of a pasture phase (a ley) into the cropping cycle. The restorative effect of the ley depends on its

duration – this is illustrated by a comparison of measurements made on a continuously cultivated (fallow-wheat) soil put down to pasture for various lengths of time (Fig. 11.9). Note that the improvement in aggregate stability is largely confined to the surface 10 cm where the grass roots are concentrated. Improved aggregation following short-term leys (< 4 years) is primarily due to the polysaccharides and mucilage produced by grass roots and their associated rhizosphere micro-organisms (Fig. 4.13). The resulting

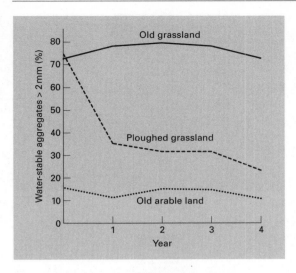

Fig. 11.8 Aggregation of the fine-earth fraction of a soil under different land uses (after Low, 1972).

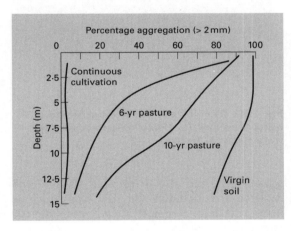

Fig. 11.9 The effect of leys of different duration on water-stable aggregation in a cultivated soil in southern Australia (after Graecen, 1958).

aggregation is sometimes called 'soft' because the aggregates do not survive wet-sieving. Longer leys, with an increase in the contribution from humified organic matter, are needed to produce the more stable 'hard' aggregation. For British soils, 2% organic matter ($\cong$ 1% C) has been suggested as the critical lower limit to maintain aggregate stability under cultivation.

Tillage methods

The traditional method of cultivation or tillage in western agriculture has been to thoroughly disturb the top 20–25 cm of soil, usually with inversion of the furrow slice by mouldboard ploughing (Fig. 11.10). Residues of the previous crop and volunteer weeds are buried, and after repeated passes with discs and harrows the soil is reduced to a fine tilth suitable for seeding. This method was exported to North America, Africa and Australasia, often with disastrous results in terms of soil structural degradation and erosion.

Traditional methods of cultivation in the tropics were very different. Following the clearing of the natural vegetation (with or without a burn), the first crop was sown with minimal soil disturbance (Section 11.1). Any plant debris left on the surface protected the soil from erosion until the crop was well-established. Intercropping using tall and

Fig. 11.10 Inverted furrow slices flowing mouldboard ploughing (courtesy of D. E. Patterson).

short crops was common so that little bare ground was exposed to the erosive tropical rain.

However, since the 1970s in developed countries, various factors have led to a reappraisal of traditional ploughing methods, notably:

• Experiments at Rothamsted in the 1930s had shown that a fine seedbed tilth did not produce higher yields than less elaborately prepared seedbeds, provided that weeds were controlled;
• human labour and fuel became more costly; and
• the adverse effects of soil erosion, not only on agricultural productivity but also in polluting the environment, gave increasing cause for concern.

New tillage methods were devised that were a blend of modern and ancient practices, combining the retention of residues, minimal disruption of the natural soil structure, and the use of selective herbicides to control weeds. These practices generally fall within the concept of conservation tillage, where at least 30% of the soil surface remains covered by residues after planting, to protect the soil from erosion (Section 11.5) and reduce evaporative losses (Section 6.5). In the USA, the two most popular conservation tillage practices are:

• *Mulch tillage*, where the soil is disturbed only before planting with a chisel or disc plough; and
• *no-tillage* (also called zero tillage or direct drilling) where the soil is left undisturbed between harvest and planting (except for fertilizer injection), and seed is sown into a narrow drill slot or behind a row chisel.

Another conservative practice is reduced or minimum tillage, which in some countries connotes tillage involving fewer and shallower passes with tines or discs than conventional ploughing. In the USA now, 'reduced till' refers to tillage that leaves 15–30% of residue cover on the soil after planting as a protection against water erosion, or < 560 kg/ha of residues from small grains during periods critical for wind erosion.

Direct drilling or no-till

This practice has been applied for many years in countries such as China, in regions where at least two crops are grown annually (e.g. winter wheat followed by maize, soybeans or rice). In recent decades, no-till has gained popularity in the USA, particularly in the southern and mid-western corn

(a)

(b)

Fig. 11.11 (a) Maize sown into wheat stubble on the North China Plain. (b) Mechanized direct-drilling into wheat stubble in Kentucky.

(maize) belt, and also in tropical Africa, South America and Australia. For row crops, such as maize, the seed is almost always sown directly into a grass sod that has been killed with herbicide, or into the residues of a previous crop. In the wheat-maize system of the North China Plain, for example, maize seed is sown by hand or by a small row planter into wheat stubble (Fig. 11.11a). But in North America, large multi-row drills sow seed directly into the stubble of winter cereals (Fig. 11.11b). Preserving plant residues on the soil surface is a major benefit of no-till agriculture because:

• The mulch formed shades the soil and reduces water loss by evaporation;
• the temperature of the surface is kept at tolerable levels; and

Table 11.6 Average soil losses under different soil management in Illinois, USA (annual precipitation c. 1300 mm). After Gard and McKibben, 1973.

	Average soil loss (t/ha/yr)	
Management system	5% slope	9% slope
Conventional tillage, maize and wheat	7.6	21.5
No-till, maize and wheat	0.9	1.3
No-till, continuous maize	0.6	0.9

• surface runoff and erosion are greatly reduced. This effect is well-documented in many countries, as illustrated by the data in Table 11.6.

There is also an energy saving of 7–18% for no-tillage compared with conventional ploughing, the higher figure occurring with direct-drilled legumes: slightly higher amounts of N fertilizer are required for many direct-drilled non-legumes, which offsets some of the energy savings on fuel.

Despite the benefits, the area of cropland under no-till in the USA had peaked at 15% by the late 1990s; similarly in Britain only c. 30% of the cereal land is considered suitable for this system. These figures illustrate that the successful application of no-till depends on the right combination of climate and soil type, as discussed in the following section.

Interactions of climate and soil under no-till

Under no-till, organic matter gradually increases in the top 5 cm and the water stability of surface aggregates is enhanced. On many soils, bulk density is increased down to the old plough depth so that total porosity decreases, mainly at the expense of macropores. Nevertheless, water infiltration and percolation do not necessarily suffer because earthworms multiply two- to threefold, especially the deep burrowing *Lumbricus terrestris*, and the predominantly vertical orientation and continuity of their holes provides a pathway for rapid water flow. Furthermore, the higher bulk density improves the soil's strength and hence its 'trafficability' when wet, which is important in cool humid climates.

Changes also occur in a soil's chemical properties. Phosphorus and K become more concentrated in the 0–5 cm layer, which is no disadvantage provided the soil stays moist (under a mulch for example); but these elements may become less available to the plant when the soil dries out. Total N follows the trend in organic C, but soil NO_3^- concentrations are often lower. Nitrate leaching is not increased because the NO_3^- is generated mainly within soil aggregates where it is better protected from leaching. But there tends to be a 'trade-off' between leaching and denitrification losses in soils of humid regions (Fig. 10.7), so denitrification generally increases within the larger aggregates of no-till soils. Lower exchangeable Ca and pH values in the 0–5 cm layer of some no-till soils may be a function of an increase in surface organic matter, so that additional lime is required. Other possible disadvantages under no-till are:

• The slow warming of wet soil in spring, which retards crop growth (particularly in northern Europe, northern USA and Canada);
• the production of volatile fatty acids around fermenting crop residues in the drill slit, which may inhibit seed germination (Section 8.3); and
• an increase in perennial weeds, especially grass weeds in cereals, which are difficult to control with herbicides alone (in some cases due to the development of herbicide resistance).

Overall, however, the advantages of no-till agriculture outweigh the disadvantages and are summarized as follows:

• Improved surface soil structure with less surface crusting and better infiltration of water;
• control of erosion by wind and water;
• conservation of soil water;
• lower surface soil temperatures (for warm and hot climates only); and
• lower energy requirements.

Figure 11.12 shows a beneficial surface mulch on a silt loam soil after several years of wheat followed by no-till maize in Kentucky, USA.

Soil conditioners

Because plant and microbial polysaccharides are known to improve soil aggregation, synthetic macromolecules, called soil conditioners, have

Fig. 11.12 Wheat stubble residues under no-till maize in Kentucky.

bitumen emulsions to achieve maximum aggregate stability. Other soil conditioners are long-chain polyanions, such as 'Krilium', which bond to cations on clay minerals (Table 4.1). A problem with all soil conditioners is that only a shallow depth of soil can be easily treated. For this reason and reasons of cost, they tend to be used in special situations, for example:

• Stabilizing sand dunes against wind erosion;
• stabilizing the bare surface of rehabilitated mine spoil, road cuttings and on landfills to protect the surface from erosion until vegetation is established; and
• protecting the fine surface tilth of seed beds, especially for high-value vegetable crops and sugar beet.

The beneficial effect of a soil conditioner on surface structure is illustrated in Fig. 11.13.

been used for the same purpose. The main types of soil conditioners are shown in Table 11.7. Application is at the rate of 0.2 g (soluble polymer) or 10 g (emulsifiable polymer) per kg of soil. Moist conditions are necessary to enable the soluble polymers to diffuse into the larger pores (> 30 μm diameter) where they are most effective. The soil should be allowed to dry after application of

Table 11.7 Summary of the main types of soil conditioners.

Soluble polymers (hydrophilic)	Emulsifiable polymers (hydrophobic)
Polyvinyl alcohol (PVA)	Bitumen
Polyacrylamide (PAM)	Polyvinylacetate (PVAc)
Polyethyleneglycol (PEG)	Polyurethane

Fig. 11.13 Stabilization of surface soil structure – the darker plots have been treated with PVA solution (courtesy of J. M. Oades).

11.5 Soil erosion

Influence of climate, human activities and scale

The term erosion describes the transport of soil constituents by natural forces, primarily water and wind. As indicated in Section 5.2, past geologic periods of erosion led to the formation of sedimentary deposits that are the parent materials of many present-day soils; but in the short term, accelerated erosion due to human activities, especially agriculture, is of considerable concern. Note that erosion and deposition are complementary processes because soil removed from steep slopes (natural or constructed) is deposited to a variable extent on footslopes and flood plains. Similarly, soil eroded from one area by wind is usually deposited in another part of the landscape (unless blown out to sea). Thus, soil erosion rates expressed per unit area are highest for individual fields, and decrease progressively when calculated on the scale of a whole farm, to a whole catchment, to a whole continent. Erosion rates for agricultural land are measured mainly at a field scale and tend to be high: for example, on cropland in the USA, average rates are:

- 20% of the area loses more than 20 t/ha/yr;
- 50% loses between 7.5 and 20 t/ha/yr; and
- 30% loses less than 7.5 t/ha/yr.

Similarly in Australia, erosion rates from agricultural land range from:

- < 0.5 t/ha/yr for good pasture;
- c. 10 t/ha/yr for summer crops; and
- 10–50 t/ha/yr on recently cleared cropland in northern Australia.

Long-term natural rates of erosion are estimated as < 1 t/ha/yr (Anon., 1995). According to a global study of catchments, these losses may be compared with stream sediment yields of < 1–4 t/ha/yr (Fig. 11.14). The solid line in this figure shows the general relationship between erosion by water and effective precipitation – defined as the precipitation required to generate a known amount of runoff under specified temperatures. Erosion increases as effective precipitation increases to a peak between 300 and 400 mm because in arid and semi-arid regions, the vegetation is sparse and the rainfall tends to show great variability, with flooding rains often following prolonged dry

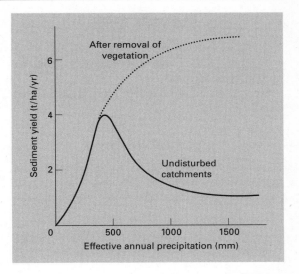

Fig. 11.14 Sediment loads in rivers in relation to annual precipitation (adjusted for differences in temperature) (after Langbein and Schumm, 1958).

periods. As the climate becomes wetter and vegetation more abundant and stable, erosion losses decrease (although there is some evidence that at effective precipitation > 1500 mm, erosion begins to increase again even under natural vegetation). Certainly, the erosive power of high rainfall is attested by the broken curve of Fig. 11.14, which applies to sub-humid and humid regions when the natural vegetation is removed. This graph also demonstrates the vulnerability of land in tropical regions to erosion by water when the natural forest is clear-felled for timber harvesting, or felled and burnt for agriculture, as discussed in Section 11.1.

Erosion is very damaging to soil fertility because mainly nutrient-rich surface soil is removed, and of that, predominantly the fine and light fractions – clay and organic matter – leaving behind the more inert sand and gravel. The fine and light fractions remain in suspension longer (in air or water) and so are carried further. Suspended stream sediment is eventually deposited in lakes, reservoirs or marine estuaries where it is lost to the land and may cause water quality problems (Section 12.3).

Erosion by water

Soil loss by erosion depends on:
• The potential of rain to erode, defined by the rainfall *erosivity*; and
• the susceptibility of soil to erosion, defined by a soil's *erodibility*.

Rainfall erosivity

To erode soil, work must be done to disrupt aggregates and move soil particles. The energy is provided by the kinetic energy (KE) of falling rain-

Fig. 11.15 (a) Soil surface just before raindrop impact. (b) Soil surface immediately after raindrop impact showing splash erosion (courtesy of D. Payne).

drops and flowing water. The principal effect of raindrops is to detach soil particles, whereas that of surface flow is to transport the detached particles. Calculations show that the KE of rain is ~ 200 times greater than the KE of runoff, and field experiments confirm that erosion from a protected surface can be < 1/100th that from a bare surface under the same rainfall (Hudson, 1995).

Raindrop impact initiates splash erosion (Fig. 11.15). On a sloping surface the downhill momentum transmitted to soil particles is greater than the uphill component, so there is net splash erosion down the slope. Slope also accelerates runoff velocity and the two processes (detachment and transport) combine to produce wash or interrill erosion. Where runoff water is concentrated, for example in cultivation furrows or tractor wheel marks, rill erosion is initiated (Fig. 11.16). Rills can be eliminated by normal cultivation, but if left alone, they may deepen to

Fig. 11.16 Rill erosion caused by surface runoff on cultivated soil (courtesy of A. J. Low).

Fig. 11.17 Gully erosion in pasture on a Sodosol (ASC) in northeast Victoria.

form gullies so large that they cannot be crossed by farm machinery. Similar processes operate on pasture land that is overgrazed, whereby wash erosion that starts in bare areas may develop into severe gully erosion (Fig. 11.17).

Raindrops vary in size up to *c.* 5 mm diameter. The median drop diameter, defined as the diameter for which half the rain in a storm falls in drops of a smaller size, increases with rainfall rate or intensity up to 25–50 mm/h above which there is little change. Raindrop KE is given by the equation

$$KE = \frac{1}{2}\{mass \times (terminal\ velocity)^2\}. \quad (11.6)$$

Terminal velocity increases with drop size (mass) so the KE of rain increases sharply up to moderate intensities (25–50 mm) and then more gradually, as illustrated in Fig. 11.18.

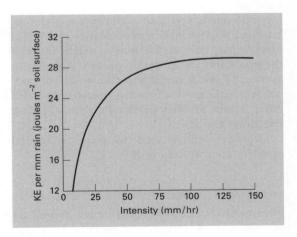

Fig. 11.18 Kinetic energy per mm of rain in relation to rainfall intensity (after Hudson, 1981).

Box 11.7 Rainfall erosivity indices.

One index of rainfall erosivity that has been widely used is the EI_{30} value, calculated as the product of the KE of a storm and its maximum 30-minute intensity. The latter is the maximum rainfall in any 30 minute period, expressed in mm/h. An alternative is the KE > 25 index which sums the KE values of a storm for only those 15-minute intervals when the rainfall intensity is > 25 mm/h, which is considered as the threshold intensity separating non-erosive from erosive rain. Both indices are empirical and should be tested by correlation against actual soil losses for individual soils under a range of rainfall intensities. Such testing has suggested that the EI_{30} value should be calculated only for storms yielding ≥ 2.5 mm of rain with a maximum 5-minute intensity > 25 mm/h. Further, the KE > 25 index seems best suited to tropical regions and a lower threshold intensity of 10 mm/h is more appropriate in temperate regions (Morgan, 1986). The unmodified EI_{30} index is used in the Universal Soil Loss Equation (see below).

Another important characteristic of rainfall is the frequency of high-intensity storms, for which there is a striking difference between tropical and temperate regions. About 95% of temperate rain, but only 60% of tropical rain falls at intensities < 25 mm/h, which are considered non-erosive. Furthermore, the maximum intensity of tropical rainfall may exceed 150 mm/h, which is twice the maximum for temperate rainfall. The greater average KE and generally greater amounts of tropical rain make erosion by water potentially a much more serious problem in tropical than in humid temperate regions.

Empirical indices used to quantify rainfall erosivity are discussed in Box 11.7.

Soil erodibility

Intrinsically, a soil's erodibility must depend on the complex interaction of factors such as clay content and type of exchangeable cations, organic matter content, structure, infiltration capacity and shear strength. However, no simple erodibility index based on soil properties measured in the laboratory or field has yet been devised. Instead, the susceptibility of a soil to erosion is defined in practical terms by the equation

$$\frac{A}{R} = K, \quad (11.7)$$

where A is the mean annual soil loss in t/ha, R is the rainfall erosivity index, and K is the soil erodibility factor. Given that the values of A and R for a particular soil and site are known, Equation 11.7 can be solved to give the value of K under standard conditions. The standard conditions are a slope length of 22.6 m and gradient of 9% for bare soil that is ploughed up and down the slope. When the actual conditions of slope, surface coverage and land management differ from the standard, the predicted soil loss can be estimated with an expanded and rearranged form of Equation 11.7. Extra terms are added to give

$$A = R \times K \times L \times S \times P \times C, \quad (11.8)$$

where the symbols L, S, P and C represent factors for slope length, slope angle, conservation practice and crop management, respectively. Equation 11.8 is called the Universal Soil Loss Equation (USLE). The values of L, S, P and C are unity when the standard conditions apply; if not, they change in a way that reflects the effect of the changed conditions in either increasing or decreasing erosion. The effect of these factors on A are briefly examined as follows.

Soil factor K. Of the intrinsic soil properties affecting K, structure is the most integrative, and also the most amenable to change through tillage practices and the use of leys or permanent pasture (Section 11.4). For example, the dramatic effect of ploughing old grassland on water-stable aggregates was shown in Fig. 11.8. Experiments in the USA on different soil types, but under standard conditions of L, S, P and C, produced a range of K values between 0.03 and 0.69. Unfortunately, such detailed information is not commonly available in other countries.

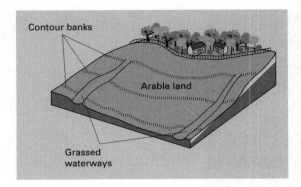

Fig. 11.19 Perspective of crop land with contour banks and grassed waterways.

Slope factor LS. Both steepness (*S*) and length (*L*) of slope affect erodibility: the former because of the greater potential energy of soil and water high on a slope, which can be converted to kinetic energy; the latter because the longer the slope, the greater the amount of surface runoff and its downhill speed. When soil conservation measures such as contour terracing or strip cropping are introduced, the distance *L* of vulnerable soil between structures is adjusted according to the slope *S*, so that a composite factor *LS* is used in USLE.

Conservation practice P. Ploughing up and down a slope is the worst soil management practice (*P* = 1). Ploughing along contours can reduce *P* by 10–50%, the greatest effect being achieved on slopes between 2 and 7%. More elaborate procedures range from contour listing – small temporary ridges thrown up by cultivation, to contour ridges or contour banks – permanent banks with graded channels above to collect water flow, to bench terraces, which convert a steep slope into a series of steps. Generally these structures are designed to:
• Reduce the length of slope from which runoff is generated;
• hold water on a slope longer so that more of it infiltrates the soil, and
• divert surplus water into safe spillways down which it may escape with minimum destructive effect, such as the grassed waterways shown in Fig. 11.19.

Much earth movement and grading may be required, so that the final design of any soil conservation measure is a compromise between the cost incurred, including the loss of cropping flexibility, and the value of the land and its produce.

Crop management C. Manipulation of this factor offers the greatest scope for reducing erosion loss because *C* can vary from 0.001 for well-kept woodland, to 0.05 for continuous forage cropping, to 1.0 for continuous bare fallow. Practices aimed at reducing *C* focus on not leaving the soil bare when the rainfall erosivity is high, and growing high-density, healthy crops so that the soil surface is protected and binding of soil by roots is strong.

Permanent pasture, leys, minimum tillage and conservation tillage all have merit for erosion control. Stubble mulching is a specific example of conservation tillage where crop residues (often a cereal crop) are chopped up and scattered over the soil surface to protect it during a long fallow period. Another example is the trash blanket used on sugar cane soils in north Queensland – the cane is harvested green and most of the leaf material chopped up and spread several cm deep over a field, through which the ratoon cane crop emerges (Fig. 11.20). Other conservative crop management practices are:
• *Strip cropping* – strips of arable land and pasture alternate down a slope, thereby reducing the length of slope most vulnerable to erosion; and
• *alley cropping* – advocated for steep slopes (> 10%) under high rainfall in the tropics, whereby shrubs or trees are grown in thin strips (*c.* 1 m wide) across the slope and cash crops are grown in the 'alleys' between (*c.* 5 m wide). If possible, the tree or shrub grown is a legume so that N_2 is fixed, and prunings from the strips and stubble from the crop (e.g. maize, pulses) are retained as mulch on the soil surface.

The uses of USLE and other more physically based models for water erosion are discussed in Box 11.8.

Erosion by wind

By analogy with water erosion, the severity of wind erosion depends on the potential of wind to erode and the susceptibility of soil to erosion.

Fig. 11.20 Trash blanket on soil after 'green cane harvesting' in north Queensland.

Although not as widespread as water erosion, the effects of wind erosion are often more dramatic in that the quantities of soil moved per hour across the edge of one hectare can be of the same order as that moved by water erosion in 1 year. Not only can wind erosion adversely affect the pro-ductivity of farmland, but dust storms resulting from such erosion cause serious air pollution problems, as occurs in and around Beijing City in China. An example of a dust storm blowing into the rural city of Horsham in northwest Victoria is shown in Fig. 11.21.

Box 11.8 Models for predicting soil erosion by water.

Because USLE is an empirical model, it works best in the regions for which it was originally developed (the USA east of the Rocky Mountains). If the maximum average soil loss that can be tolerated in a production system is known (set equal to A in Equation 11.8), various combinations of L, S, P and C can be tried such that the product of terms on the RHS of the equation does not exceed A. When used outside the USA, the appropriate values of L, S, P and C for soil conservation and cropping practices on the local soils, under local rainfall conditions, should be established.

USLE applies to agricultural land only and cannot be used to predict sediment yields and other off-site effects of erosion. With the acquisition of additional erosion data since its first development, USLE has been updated to the Revised Universal Soil Loss Equation (RUSLE). Elements of USLE have also been combined with physically-based concepts of runoff, soil detachment, deposition and transport to produce the CREAMS model – Chemicals, Runoff and Erosion from Agricultural Management Systems. This is a dynamic model with a daily time step (average annual erosion rates are obtained by summing daily values), which has given reasonably good predictions in several countries (Morgan, 1986). However, CREAMS is being displaced in the USA by WEPP (Water Erosion Prediction Project), a model designed for all land surfaces (Blackburn et al., 1994), which is entirely processed-based and designed to predict runoff, erosion and sediment concentrations in receiving channels for single rainfall events.

Fig. 11.21 A dust storm approaching Horsham in northwest Victoria (courtesy of Geoff Evans, Fire Management Officer, Department of Sustainability and Environment, Horsham).

Wind action

The soil surface exerts a frictional drag on the wind so that an atmospheric boundary layer forms, approximately 1 m thick. Any surface roughness increases the drag on the wind and creates turbulence in the air flow, which can impart an upward velocity component to loose particles or small aggregates (Fig. 11.22a). Airborne particles gain a forward velocity, and because of the increase in wind speed away from the soil surface (Fig. 11.22b), the higher the jump, the greater the particle's momentum. However, surface roughness, through its drag effect, also decreases the gradient in forward wind speed over the soil and this reduces the wind's ability to erode.

Particle or aggregate size influences surface roughness, and particle density obviously determines how much energy must be expended to lift a particle of a particular size off the ground. The interaction of these factors determines in a complex way the threshold wind speed, that is, the minimum speed required to move a particle. Once a particle has been moved, there are three possible modes of subsequent transport:

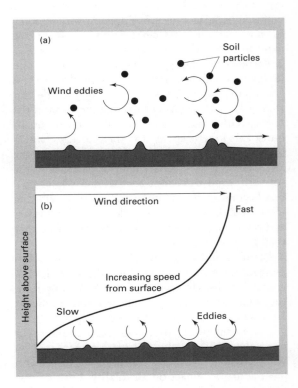

Fig. 11.22 (a) Turbulent wind flow over a rough soil surface. (b) Variation in wind speed with height above a rough soil surface (after Hudson, 1981).

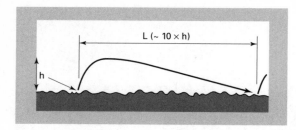

Fig. 11.23 Jump and glide path of saltating soil particles (after Hudson, 1981).

1 Particles < 0.1 mm diameter that rise more than about 30 cm into the air may stay suspended for considerable time. The smaller the particle, or the lower its density, the more likely are wind currents to be able to counteract the force of gravity and keep the particle suspended. The resulting aerial dispersion, if sufficiently concentrated (> 0.2 g/m^3), becomes a dust storm. In the past, especially during the Pleistocene glaciations, strong winds in periglacial regions dispersed vast amounts of fine particles (0.02–0.06 mm) to form the extensive loess deposits described in Section 5.2.
2 Particles and aggregates between 0.1 and 1 mm diameter do not usually rise more than 30 cm and their jump is followed by a long gentle trajectory back to earth. On striking the surface, the particle either bounces off or comes to rest, transmitting its kinetic energy to other particles, which are knocked in various directions. Once in the air, there is a net displacement of particles in the forward wind direction, so that the whole process becomes self-sustaining and is called saltation (Fig. 11.23). Saltating particles can also impart sufficient energy to smaller or lighter particles that they go into suspension.
3 Particles and aggregates between 1 and 2 mm diameter move by rolling along the surface. The energy for dislodgement may come directly from the wind or indirectly from saltating particles. This mode of transport is called creep.

Movement by saltation and creep accounts for more than 50% of the soil mass transported by wind erosion; however, transport in suspension has the major impact on air quality. Little can be done to lessen the erosive potential of wind other than by reducing its speed by means of wind breaks, and by shortening its fetch, that is, the distance of unhindered travel over erodible land.

Soil and crop factors

Moist soil supporting healthy vegetation is the most effective deterrent to wind erosion. Cultivated soil is vulnerable when the surface is dry and the structure weak, especially if most of the aggregates are < 1 mm in diameter. Often coarse and fine-textured soils are more erodible than loams, which have the right proportions of clay, silt and sand to form stable aggregates of a non-erodible size. The fen peat soils of eastern England are particularly vulnerable to wind erosion once drained and cultivated because of the lightness of the organic-rich surface horizon.

With respect to surface roughness, although slight irregularities initiate saltation as the wind speed rises, very rough or cloddy surfaces protect against erosion by reducing the wind's speed near ground level and trapping saltating and creeping particles in minor depressions. Vegetation, even if dead, has a similar effect, which is directly proportional to its coverage of the surface and height above the soil – standing crop residues are up to six times more effective than the same mass of material spread flat on the ground (Fryear, 2000). Key crop management practices (also discussed under 'Erosion by water') that are valuable in controlling wind erosion include:
• *Stubble mulching*, pioneered on the wind-swept High Plains of the USA and recommended for the light-textured cropping soils of the Mallee district in Australia;
• *trash farming*, whereby a large one-way disc plough cuts through the stubble but does not bury it; and
• *direct drilling* of a new crop through the residues of a previous crop.

Estimating soil losses due to wind erosion

Wind tunnel experiments have been used to study some of the processes involved in wind erosion, but losses need to be measured in field experiments, which is a costly and time-consuming exercise. Alternatively, erosion models, once calibrated against field data, can be used to estimate soil

Box 11.9 Empirical and process-based models of wind erosion.

Empirical models such as the Wind Erosion Equation (WEQ) use data for weather, soil erodibility (derived from soil physical and chemical properties), surface roughness, fetch and crop residues to calculate annual soil losses. Although widely used in the USA, the original WEQ had limited field verification and did not describe wind erosion in high or low rainfall regions well. Because of these shortcomings, scientists have developed a Revised Wind Erosion Equation (RWEQ), which is more physically-based than WEQ, or they have attempted to develop a fully process-based model, with a daily time step, such as the Wind Erosion Prediction System (WEPS). With recent advances in field measuring equipment, model outputs for soil loss under a range of conditions can be compared with measured losses, with varying success as described by Fryear (2000).

losses. Some of these models are briefly described in Box 11.9.

11.6 Summary

Clearing soil of natural vegetation interrupts the cycling of nutrients and may lead to a decline in fertility. The traditional method of farming in the tropics – shifting cultivation or slash-and-burn agriculture – can be sustainable if the fallow period, when the native vegetation is allowed to re-grow, is long enough for nutrients to be re-plenished from rock weathering and the atmosphere. In temperate regions, traditional rotational cropping and mixed farming – involving arable–ley pasture rotations – will generally maintain fertility provided that fertilizers or organic manures are used to replace nutrients removed in farm produce. However, because of population pressure, the cost of labour and the demand for land for non-agricultural use, these traditional methods have gradually been displaced by intensive mono-cultural systems, such as continuous cereal cropping, and livestock farming has become separated from cropping. This trend has resulted in varying degrees of soil nutrient depletion, soil structural decline, salinization, accelerated acidification and erosion in many countries.

Modern intensive farming requires substantial inputs of fertilizers and organic manures (including sewage sludge) to maintain soil fertility. Green manuring may also be advocated to prevent leaching of surplus soil NO_3^- during winter or the non-cash crop period. Soil nutrient deficiencies can be assessed directly by soil testing or indirectly by plant analysis. A rational approach to soil testing is to apply the quantity/intensity (Q/I) concept to measuring a soil's nutrient supplying power; but in practice a variety of empirical extractants are used that provide a composite estimate of both Q and I factors. Whichever approach is adopted, soil or plant tests must be calibrated against crop response to the nutrient in question by a fertilizer rate trial, carried out on a range of soils. The critical value for the nutrient is defined as the concentration (in soil or plant) below which a plant will respond to an increase in the supply of that nutrient.

In soil and plant testing, the sample chosen for analysis must be representative of the area tested. Fertilizer requirements can also be assessed using a nutrient balance method, which may involve models of nutrient inputs, losses and transformations in the soil. Instead of 'blanket' recommendations for a nutrient, fertilizer applications can be adjusted to the specific need of parts of individual fields, based on crop yield maps or spatially-referenced data for soil tests. This approach is the basis of precision agriculture.

Soil acidity and the consequent problems of low pH, Al and Mn toxicity, Ca and Mo deficiency are remedied by the application of lime. The amount needed (the lime requirement) depends on a soil's pH buffering capacity. Accelerated soil acidification that arises from high atmospheric H^+ inputs, removal of C and N in farm products, and increased NO_3^- leaching, affects many modern farming systems. Management of acidification requires a decrease in acid inputs and nutrient losses, as well as liming to correct the soil acidity.

A good soil structure provides a favourable environment for root growth and beneficial microbial activity. Soil structural quality can be measured in terms of available water capacity, aeration and water-stable aggregation. Deterioration of soil structure is revealed through inadequate soil aeration, decreased available water, impedance to root penetration because of soil compaction, and surface sealing. A grass ley following arable cropping is very effective in restoring the percentage of water-stable aggregates, especially in the surface 5 cm or so. The improvement is time-dependent, at least 4 years being required in temperate regions to achieve an improvement in 'hard' aggregation. However, the effect of the ley is relatively short-lived once the land is ploughed. A similar improvement in aggregation can be obtained with synthetic organic polymers called soil conditioners, but these are expensive.

New tillage methods designed to maintain soil structure and minimize erosion losses are a blend of modern (use of rapid-kill, non-persistent herbicides) and ancient (sowing seed with minimum soil disturbance, retaining plant residues) agricultural practices. They have gained widespread popularity compared with conventional ploughing and bare fallow, mainly on soils where high surface temperatures, high evapotranspiration, surface sealing, and water and wind erosion pose problems. These methods range from reduced or minimum tillage (fewer passes with cultivating implements) to direct drilling or no-till, where the only disturbance is injecting fertilizers and sowing seeds. Conservation tillage is a generic term denoting conservation of water and soil, not necessarily by less tillage, but by keeping at least 30% of the surface covered by crop residues.

Erosion is very damaging to soil fertility because mainly the nutrient-rich surface soil is removed. Water erosion occurs in almost any environment when the soil is bare and its structure unstable, the magnitude of the loss depending on rainfall erosivity and soil erodibility. This relationship is quantified by assigning values to variables in the Universal Soil Loss Equation (USLE)

$$A \quad = \quad R \quad \times (K, \ LS, \ P, \ C).$$
(Soil loss) (Erosivity)   (Erodibility)

Conservation measures aim at decreasing slope steepness S and length L; also modifying the soil management factor P (through contour banks and terraces) and the crop management factor C (by direct drilling, stubble mulching, strip cropping and alley farming). Although USLE has been very successful in identifying management practices that will decrease soil loss per ha to an acceptable value, it is not applicable to non-agricultural land nor can it be used to predict the off-site effects of erosion. For this reason, USLE is being superseded by process-based erosion models, such as the Water Erosion Prediction Project (WEPP) model in the USA, which apply to any form of land use.

Wind erosion is confined mainly to arid and semi-arid regions. The erosivity of wind depends on the wind speed and fetch (the distance travelled unhindered over land). The susceptibility of soil to wind erosion depends on the wetness and roughness of the surface, the diameter and density of surface particles and aggregates, and the degree of protection afforded by vegetative cover. Per kg of dry matter, standing vegetation is much more effective than material lying flat on the ground.

References

Agricultural Development & Advisory Service (ADAS) (1982) *The Use of Sewage Sludge on Agricultural Land.* Booklet 2409. MAFF Publications, Northumberland.

Anon. (1995) *Data Sheets on Natural Resource Issues.* Land & Water Resources Research & Development Corporation Occasional Paper No. 6/95, Canberra.

Blackburn W. H., Pierson F. B., Schumann G. E. & Zartman R. (1994) *Variability in Rangeland Water Erosion Processes.* Soil Science Society of America Special Publication No. 38. Soil Science Society of America Inc, Madison, WI.

Blake L., Johnstone A. E. & Goulding K. W. T. (1994) Mobilization of aluminium in soil by acid deposition and its uptake by grass cut for hay – a chemical time bomb. *Soil Use and Management* **10**, 51–55.

Brady N. C. (1996) Alternatives to slash-and-burn: a global imperative. *Agriculture, Ecosystems and Environment* **58**, 3–11.

Bruinsma J. (Ed.) (2003) *World Agriculture: towards 2015/2030.* Food and Agriculture Organization of the United Nations, Rome.

Chadwick D. R. & Chen S. (2002) Manures, in *Agriculture, Hydrology and Water Quality* (Eds P. M. Haygarth & S. C. Jarvis) Commonwealth Agricultural Bureau International, Wallingford, England, pp. 57–82.

Cook S.E. & Bramley R.G.V. (1998) Precision agriculture – opportunities, benefits and pitfalls. *Australian Journal of Experimental Agriculture* 38, 753–63.

Cooke G. W. (1982) *Fertilizing for Maximum Yield*, 3rd edn. Granada, London.

Davies D. B., Eagle D. J. & Finney J. B. (1993) *Soil Management*, 5th edn. Farming Press, Ipswich.

Epstein E. (2003) *Land Application of Sewage Sludge and Biosolids*. Lewis Publishers, Boca Raton, Florida.

FAO (1995) *World Agriculture: Towards 2010*. Food and Agriculture Organization of the United Nations, Rome.

Fryear D. W. (2000) Wind erosion, in *Handbook of Soil Science* (Ed. M. E. Sumner) CRC Press, Boca Raton, Florida, G195–G216.

Gard L. E. & McKibben G. E. (1973) No-till crop production proving a most promising conservation measure. *Outlook on Agriculture* 7, 149–54.

Graecen E. L. (1958) The soil structure profile under pastures. *Australian Journal Agricultural Research* 9, 129–37.

Hajkowicz S., Hatton T., McColl J., Meyer W. & Young M. (2003) *Futures. Exploring Future Landscapes: a Conceptual Framework for Planned Change*. Land and Water Australia, Canberra.

Hall D. G. M., Reeve M. J., Thomasson A. J. & Wrights V. F. (1977) *Water Retention, Porosity and Density of Field Soils*. Soil Survey of England and Wales, Technical Monograph No. 9. Harpenden, England.

Hudson N. W. (1981) *Soil Conservation*, 2nd edn. Batsford, London.

Hudson N. W. (1995) *Soil Conservation*, 3rd edn. Batsford, London.

Hullugalle N. R. (1994) Long term effects of land clearing methods, tillage systems and cropping systems on surface soil properties of a tropical Alfisol in S. W. Nigeria. *Soil Use and Management* 10, 25–30.

Juo A. S. R. & Mann A. (1996) Chemical dynamics of slash-and-burn agriculture. *Agriculture, Ecosystems and Environment* 58, 49–60.

Langbein W. B. & Schumm S. A. (1958) Yield of sediment in relation to mean annual precipitation. *Transactions of the American Geophysical Union* 39, 1076–84.

Ledgard S. F., Edgecombe G. A. & Roberts A. H. C. (1999) Application of the nutrient budgeting model OVERSEER™ to assess management options and Regional Council consent requirements on a Hawke's Bay dairy farm. *Proceedings of the New Zealand Grassland Association* 61, 227–31.

Low A. J. (1972) The effect of cultivation on the structure and other physical characteristics of grassland and arable soils (1945–1970). *Journal of Soil Science* 23, 363–80.

Morgan R. P. C. (1986) *Soil Erosion and Conservation*. Longman Scientific and Technical, Harlow, England.

Nye P. H. & Greenland D. J. (1960) *The Soil under Shifting Cultivation*. Commonwealth Bureau of Soils, Technical Communication No. 51, Harpenden, England.

Oldeman L. R., Hakkeling R. T. A. & Sombroek W. G. (1991) *World Map of the Status of Human-induced Soil Degradation*. An explanatory note 34. International Soil Reference Group Centre, UNEP, Wageningen.

Rayment G. E. & Higginson F. R. (1992) *Australian Laboratory Handbook of Soil and Water Chemical Methods*. Australian Soil and Land Survey Handbook, Inkata Press, Melbourne.

Smith J. U., Bradbury N. J. & Addiscott T. M. (1996) SUNDIAL: a PC-based system for simulating nitrogen dynamics in arable land. *Agronomy Journal* 88, 38–43.

Westerman R. L. (Ed.) (1990) *Soil Testing and Plant Analysis*, 3rd edn. Book Series No. 3. Soil Science Society of America Inc, Madison, WI.

Further reading

Asher C., Grundon N. & Menzies N. (2002) *How to Unravel and Solve Soil Fertility Problems*. ACIAR Monograph No. 83, Australian Centre for International Agricultural Research, Canberra.

Carter M. R. (Ed.) (1994) *Conservation Tillage in Temperate Agroecosystems*. Lewis Publishers, Boca Raton.

Havlin J. L. & Jacobsen J. S. (Eds) (1994) *Soil Testing: Prospects for Improving Nutrient Recommendations*. Soil Science Society of America Special Publication No. 40. Soil Science Society of America, Madison, WI.

Pierce F. J. & Frye W. W. (Eds) (1998) *Advances in Soil and Water Conservation*. Ann Arbor Press, Chelsea, Michigan.

Troeh F. R. & Thompson L. M. (1993) *Soils and Soil Fertility*, 5th edn. Oxford University Press, New York.

Example questions and problems

1 (a) Name three types of organic manures.
   (b) What is an essential feature of a green manure crop?
   (c) What was the 'Green Revolution'?

2 (a) Name two methods for diagnosing a nutrient deficiency in plants.
   (b) Define the critical value for an essential element.
   (c) Define the optimum yield of a crop.

3 A vegetable grower has two soils on his property – one is a Podosol where the crop roots are stunted and have brown tips; the other is a Dermosol where the plant roots are healthy. The grower knows the pH of the Dermosol, but not the Podosol, so he decides to test this soil for acidity. The Dermosol pH was measured in a 1 : 5 soil : water suspension, whereas the new measurements for the Podosol are a 1 : 5 soil : 0.01 M $CaCl_2$ suspension. The results were as follows.

Soil	Mean pH (water)	Mean pH ($CaCl_2$)
Dermosol	6.2	
Podosol		4.5

   (a) Complete the expected pH values in this table, using your knowledge of the relationship between pH (water) and pH ($CaCl_2$). (Hint – see Box 7.5).
   (b) (i) Indicate which of these soils is strongly acidic, and (ii) which ion toxicity may have damaged the roots.
   (c) For the strongly acidic soil, the grower decides to apply lime to raise the pH. His objective is to achieve pH 5.5 (in $CaCl_2$). Given a pH buffering capacity of this soil of 24 kmols H^+/pH unit/ha, calculate how many kg of lime would be required to ameliorate 1 ha of soil. (The atomic masses of Ca, C and O are 40, 12 and 16 g, respectively, and the lime has an ENV of 75%.)

4 Virgin land has been cleared for cropping. The measured soil pH (0–10 cm depth) averaged 5.5

(in 0.01 M $CaCl_2$) and the pH buffering capacity was 0.05 mol H^+/pH unit/kg soil. Acid inputs due to NO_3^- leaching and removal of crop products are expected to be 3 kmol H^+/ha/yr.
   (a) How many years would elapse under cropping before the soil pH has fallen by 1 unit?
   (b) How much lime (t/ha) with an ENV of 85% would be required per ha to prevent this fall in soil pH? (Assume 1.33×10^6 kg soil per ha to a depth of 10 cm).

5 A manager of a grazing enterprise (prime lambs) needs to make a decision about P fertilizer. There are two major soil types – a Vertosol on flat land and a Chromosol with a sandy loam A horizon on the slopes and ridges. Recent Olsen P soil test values were 25 and 5 mg P/kg soil for the Vertosol and Chromosol, respectively. The critical Olsen P value for the pasture is 15 mg P/kg soil. Consider the following.
   (a) Is either of these soils deficient in P for this enterprise? (Give a reason for your answer.)
   (b) The soil test laboratory also carried out P sorption measurements to determine the P sorption index – the amount of sorbed P required to attain the optimum soil solution P concentration for pasture growth. Where the Chromosol had been eroded (on ridges and steep slopes, phase A), the index was 30 mg P/kg soil, but where the topsoil was intact (the lower slopes, phase B), the index was 6 mg P/kg soil. How much superphosphate should be applied (kg/ha) to phase A and B to achieve an optimum P supply? (Assume 2×10^6 kg soil per ha to 15 cm depth and that superphosphate contains 9% P.)

6 (a) Would you expect to find more earthworms per square metre in a no-till soil than a ploughed soil?
   (b) What role do earthworms play in improving soil drainage?
   (c) What is the minimum percentage of crop residues that is desirable in conservation tillage?

7 (a) What is the main cause of soil detachment during erosion by water?

(b) Give the size range of soil particles that are commonly dispersed by wind.

(c) Name the process whereby wind-blown particles bounce over the soil surface.

8 Soil loss on a wheat farm in undulating country under traditional ploughing averages 12 t/ha/yr. The farmer wants to reduce this to < 4 t/ha/yr as his contribution to improved catchment management. The soil's erodibility factor K, for use in the USLE, is 0.2 under standard conditions. Installation of contour banks will decrease the combined LS term in USLE to 0.2.

(a) Would this change be sufficient for the farmer to reach his soil loss objective?

(b) If the land were returned to permanent pasture (without contour banks), the C factor in the USLE would drop to 0.01. Estimate the likely annual soil loss under this management practice.

Chapter 12

Fertilizers and Pesticides

12.1 Some definitions

The prime purpose for the use of agricultural chemicals (sometimes referred to as agrichemicals) is to increase the quantity and improve the quality of agricultural products. The broad categories of agricultural chemicals are:

• *Fertilizers*, introduced in Chapter 11, which are used to correct nutrient deficiencies and imbalances in plants and animals and so improve growth and product quality; and

• *insecticides*, *herbicides* and *fungicides* (also called pesticides or plant protection chemicals), which are used to protect crops from pests and disease or to eliminate competition from weeds. This protection extends to the crop products after harvest and during storage.

Much of the increase in world agricultural production and production per capita (Fig. 11.2) is due to expansion in the use of fertilizers, up from 16 Mt (N, P_2O_5 and K_2O*) in 1950 to 138 Mt in 2000–01 (International Fertilizer Association, 2002), and plant protection chemicals (*c*. 2.5 Mt of active ingredient in 1998). To achieve their purpose, fertilizers and other chemicals must be used according to best scientific principles so that the response of soil, plant or animal is optimal, and their impact on the wider environment is minimal. The principles of nutrient supply and

loss introduced in previous chapters are further developed here.

Fertilizers and plant protection chemicals may be applied as solutions, suspensions, emulsions or solids to soils and plants. When applied as a foliar spray, the properties of a chemical (such as its volatility), the nature of the plant surfaces and the atmospheric conditions determine the efficacy of the application and the subsequent fate of the chemical in the environment. In other cases, the type of chemical, its method of application and the reactions undergone in the soil are of prime importance. In this chapter, emphasis is placed on chemical forms and soil reactions, starting with N, P, K and S fertilizers.

12.2 Nitrogen fertilizers

Forms of N fertilizer

Soluble forms

The composition of the common water-soluble N fertilizers is given in Table 12.1. Some of these, such as NH_4NO_3 and urea, supply a single macronutrient element (N); others, such as KNO_3, $(NH_4)_2SO_4$, monoammonium phosphate (MAP) and diammonium phosphate (DAP), provide more than one macronutrient and are classed as multinutrient or compound fertilizers. Mixed fertilizers, which may be solid or liquid (fluid), are made by mixing single or multinutrient fertilizers and are usually identified by their N : P : K ratio. For example, a mixed fertilizer of NPK 5–5–8 can be made by dissolving NH_4NO_3, urea, ammonium polyphosphate and KCl in water.

---

* In the trade, fertilizer analyses have been given as a ratio of the percentages of $N : P_2O_5 : K_2O$ (although this is now changing). For simplicity in this book, analyses are given as percentages of the elements N, P, K and S unless otherwise stated.

Table 12.1 Forms of soluble N fertilizer.

Compound	Formula	N content (%)
Solids		
Sodium nitrate	$NaNO_3$	16
Potassium nitrate	KNO_3	13
Calcium nitrate	$Ca(NO_3)_2$	13–14
Ammonium sulphate	$(NH_4)_2SO_4$	21
Ammonium nitrate	NH_4NO_3	34
Urea	$(NH_2)_2CO$	46
Monoammonium phosphate (MAP)	$NH_4H_2PO_4$	11–12
Diammonium phosphate (DAP)	$(NH_4)_2HPO_4$	18–21
Liquids		
Aqua ammonia	NH_3 in water	25
Nitrogen solutions – urea ammonium nitrate	NH_4NO_3, $(NH_2)_2CO$ in water	30
Anhydrous ammonia	liquefied NH_3 gas	82

Highly soluble solid fertilizers can pose handling and storage problems because of their hygroscopicity (absorption of water vapour). This difficulty is largely overcome by granulation, which involves heating the fertilizer slurry in a rotating drum at the time of manufacture. As excess water is driven off, the small fertilizer particles coagulate to form granules that harden on air-drying. Ideally, a granulated fertilizer has uniform granules, roughly spherical, and c. 3–4 mm in diameter. The ultimate granules are the very regular, spherical prills of NH_4NO_3 or urea. The process of granulation is intended to:

• Reduce the contact area between individual particles and also the area available for water absorption; and

• prevent the fertilizer from breaking down into a fine dust that makes spreading difficult and which may create a fire hazard (e.g. during aerial application) or a health hazard.

When solid fertilizers are made by mixing the ingredients in a slurry before granulation, they are homogeneous. Other fertilizers made by mixing individual granulated fertilizers – a process of bulk blending – may not be so homogeneous because granules of different size and density may segregate in the mixture. The chemical incompatibility of some fertilizers, such as KNO_3 and urea, also precludes their bulk-blending.

The choice of the most suitable N fertilizer depends on a balance of factors: the cost per kg of N (including transport and application costs), effects on plant growth (both beneficial and detrimental), and the magnitude of N loss through leaching, volatilization and denitrification. A high N content and ease of handling as prills have made urea the most popular form of solid N fertilizer in the world. However, in the USA anhydrous NH_3 is very popular because of its high N analysis and the precision with which it can be applied. The advantage of a very high N content per unit weight afforded by anhydrous NH_3 is partly offset by higher application costs, since it must be kept under pressure and injected at least 10 cm below the soil surface. Other effects are discussed below.

Slow-release and controlled-release fertilizers

To obviate problems arising from the high solubility of many N fertilizers and their potential vulnerability to leaching, especially in the NO_3-N form, a range of slow-release fertilizers (SRFs) and controlled-release fertilizers (CRFs) has been developed (Table 12.2). SRFs include synthetic and natural materials for which the rate of

Table 12.2 Forms of slow-release (SRF) and controlled-release (CRF) N fertilizer. After Shaviv, 2001.

Fertilizer	Composition	N content (%)
SRFs		
Shoddy	Wool waste	2–15
Blood and bone meal	By-product of meat processing	5–6
Hoof and horn meal	By-product of meat processing	7–16
Ureaform	Ureaformaldehyde polymers	37–40
IBDU	Isobutylidene diurea	32
CRFs		
SCU	Sulphur-coated urea	31–38
PSCU	Polymer-coated SCU (resin or polyurethane-like coatings)	< 30

Box 12.1 Nitrification inhibitors.

Interest in nitrification inhibitors followed the development of the compound nitrapyrin (2-chloro-6 trichloromethyl pyridine), commonly called 'N-serve', which inhibits the nitrification of NH_4-N fertilizers. Although the degree of inhibition is highly variable, it is rarely greater than 25% under field conditions. Nitrapyrin must be applied at least annually (at a rate of c. 1% by weight of N applied) because it is decomposed by soil micro-organisms (see Box 8.4). In regions where the winters are cold, nitrapyrin has some effect in delaying the nitrification of NH_4-N fertilizer, applied in the autumn, until the crop begins to grow rapidly in spring.

Other nitrification inhibitors include ATC (4-amino-1,2,4-triazole), DCD (dicyandiamide), C_2H_2 (acetylene) and CS_2 (carbon disulphide). With gases such as CS_2 and C_2H_2, the problem is to keep a sufficiently high concentration (c. 1% by volume) in the soil air at the site of the fertilizer granule. A novel way of solving this problem has been to make encapsulated calcium carbide (CaC_2) pellets that can be mixed with fertilizer to produce a controlled release fertilizer. CaC_2 particles (1–2 mm diameter) are successively coated with waxes containing CaC_2 to produce a pellet of 1 part carbide and 1 part coating. On contact with water, the pellet slowly disintegrates and C_2H_2 is released according to the reaction

$$CaC_2 + 2H_2O \rightarrow C_2H_2 + Ca(OH)_2. \qquad (B12.1.1)$$

This technique has proved effective in partially inhibiting the nitrification of NH_4-N fertilizers in paddy rice and irrigated wheat. When nitrification is inhibited, there is a greater chance of NH_4^+ being immobilized by micro-organisms, or lost by volatilization of NH_3 in high pH soils. To minimize the latter process, the NH_4-N fertilizer and inhibitor should be placed at least 5 cm below the soil surface.

nutrient release is slow and not well controlled (primarily because release depends on microbial action in the soil). Examples of synthetic SRFs are urea formaldehyde (UF) and isobutylidene urea (IBDU). Note that the natural SRFs are usually by-products of animal processing and are called *organic fertilizers*, to distinguish them from the organic manures discussed in Section 11.1. CRFs are synthetic fertilizers for which nutrient release is slow and well controlled, because the soluble core is dispersed in an inert matrix or the core is coated with a hydrophobic layer. Examples are sulphur-coated urea (SCU) and polymer-coated SCU (PSCU) (Table 12.2).

Use of these fertilizers in broad-acre agriculture is very limited, primarily because of their high cost relative to soluble forms and also because plant response is usually inferior, per kg of N. They are more widely used for sports turf applications and vegetable growing and are especially useful for plant establishment on difficult sites, such as reclaimed mine workings, where a slow but assured release of N over an extended period is required. An alternative solution to the NO_3^- leaching problem associated with soluble N fertilizers is discussed in Box 12.1.

Reactions in the soil

Scorch and retarded germination

There is a possibility of retarded germination, injury to young roots and leaf scorch* because of the concentrated salt solution around dissolving fertilizer granules, particularly if they are drilled into the soil with the seed (Fig. 12.1). Generally, fertilizers containing NO_3-N are more harmful than NH_4-N fertilizers; for example $Ca(NO_3)_2$ is satisfactory at < 75 kg N/ha while $(NH_4)_2SO_4$ is safe up to 125 kg N/ha. Urea is exceptional in being injurious at rates as low as 35 kg N/ha. Urea is best placed in a band, slightly below and to one side of the seed (Fig. 12.1), where damage due to high salt concentration and losses by

---

* Chlorosis followed by leaf death, usually extending from the tip backwards.

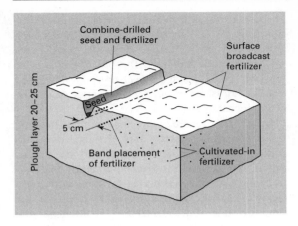

Fig. 12.1 Ways of applying fertilizer to soil.

volatilization of NH_3 are minimized. Mixed fertilizers containing N are less injurious to plants than straight N fertilizers at comparable rates.

pH effects

Ammonia solutions and anhydrous NH_3 produce an alkaline reaction in soil. Urea, whether from animal excreta or fertilizer, is rapidly hydrolysed to $(NH_4)_2CO_3$ (Equation 10.6), which in turn hydrolyses to NH_4OH, CO_2 and H_2O (Equation 10.7). The soil pH around granules of dissolving urea may therefore rise above 9 at which level NH_4OH is unstable and decomposes to NH_3 and H_2O (Equation 10.8, for which the $pK_a \cong 9.5$). As the NH_3 volatilizes, the H^+ ions released neutralize the OH^- ions from the hydroxide so that the pH stabilizes and then slowly falls as the remaining NH_4^+ ions are oxidized to NO_2^-.

A pH rise to ≥ 8 in the vicinity of ammonium or urea fertilizer granules will inhibit *Nitrobacter* organisms (Section 8.3). *Nitrosomonas* organisms are not so sensitive and function up to pH 9, so that NO_2^- accumulates in the soil. Although oxidation of NH_4^+ to NO_2^- causes the pH to fall, the high concentration of NO_2^- may continue to inhibit *Nitrobacter*.

Much of the N from ammonium fertilizers, including urea, which remains in the soil is eventually oxidized to NO_3^-. Nitrification is a potentially acidifying reaction (Equation 10.2). When NH_4^+ is produced by ammonification of soil organic N, 1 mole (net) of H^+ is produced per mole of NH_4^+

oxidized. But if NH_4-N is supplied from a fertilizer, 2 moles (net) of H^+ are produced per mole of NH_4^+ oxidized. The acceleration of acidification in agricultural soils when N inputs are increased, and significant amounts of NO_3-N are leached from the root zone, is discussed in Section 11.3. A similar effect occurs in soils heavily fertilized with NH_4-N fertilizers, when the recovery of N by the crop is usually only 50% or less, so that regular applications of lime are required to prevent an undesirable fall in soil pH.

According to the reaction

$$CaCO_3 + 2H^+ + 2NO_3^- \leftrightarrow Ca(NO_3)_2 + H_2CO_3, \tag{12.1}$$

the neutralization of 2 moles of H^+ ion requires 1 mole of $CaCO_3$. Thus, the theoretical lime equivalent of a fertilizer such as $(NH_4)_2SO_4$, where all the N is in ammonium form, is 100 kg $CaCO_3$ per 14 kg N, or 7 kg $CaCO_3$ per kg N. However, because some of the NO_3-N is taken up by the crop, in practice it is found that the lime equivalent for $(NH_4)_2SO_4$ is *c.* 5.4 : 1 and that of NH_4NO_3 is *c.* 1.8 : 1. One way of countering the acidifying effect of ammonium fertilizers is to produce mixed fertilizers of NH_4NO_3 and $CaCO_3$, such as 'Nitrochalk' (26% N) and limestone ammonium nitrate (20% N), but these are not as popular as the higher analysis, straight N fertilizers or compound N fertilizers. Alternatively, the rate of oxidation of NH_4^+ in soil can be slowed down with a nitrification inhibitor (Box 12.1).

Nitrification and consequent N losses

Nitrate produced by nitrification of NH_4-N fertilizers is susceptible to loss by leaching out of the root zone (Section 10.2) and by denitrification (Section 8.4). In addition to biological denitrification, there is the possibility of chemo-denitrification in N-fertilized soils. Chemo-denitrification appears to involve the reaction of nitrous acid (HNO_2) with soil organic matter to release gaseous N compounds. The most likely reactive organic components are phenolic compounds that form nitrosophenols with HNO_2, which is mainly undissociated under acid conditions. The nitrosophenols then decompose to N_2O and N_2. Direct leaching of soluble N fertilizer is only likely to occur when

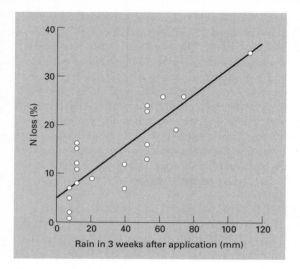

Fig. 12.2 The relationship between the loss of N fertilizer and the amount of rain in the three weeks after application (after Powlson *et al.*, 1992).

heavy rain falls soon after fertilizer application, which has been observed to occur with arable crops and N-fertilized grassland in spring in the UK. Research in southern England with winter wheat on several soil types suggested a direct relationship between the percentage of the fertilizer

N lost and rainfall in the 3 weeks after application (Fig. 12.2). Part of this loss was by leaching and part by denitrification. Overall, experiments on cereal crops indicated that the *direct* loss of fertilizer N by leaching was unlikely to exceed *c.* 6% (Addiscott *et al.*, 1991). Losses from row crops such as potatoes can be higher.

Even after fertilizer N has entered the soil biomass-organic matter pool through immobilization, the N becomes vulnerable to leaching as organic N compounds are subsequently mineralized, and especially when a 'flush' of mineralization follows the rewetting of a dry soil by rain or irrigation (Section 10.2). Leaching of soil-generated NO_3^- (and also NO_3^- from surface-applied fertilizer) is exacerbated when heavy rain falls on well-structured soils because of the preferential flow of water and NO_3^- down macropores into the drainage water (Fig. 10.8b). Thus, wherever agricultural productivity has been increased over several decades through the use of N fertilizers, there has been an indirect effect of rising NO_3^- concentrations in surface waters and underground aquifers (Fig. 12.3). In surface waters, high NO_3^- concentrations can accelerate eutrophication (Section 12.3), and in both surface and ground waters high concentrations may be detrimental to human health (Box 12.2).

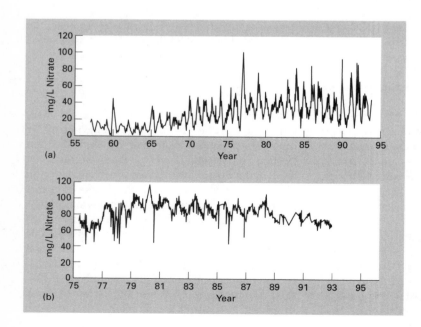

Fig. 12.3 (a) Variations with time in the NO_3^- concentration of the Great Ouse at Bedford, UK (after Croll, 1994). (b) Variations with time in the NO_3^- concentration in groundwater in limestone at Waneham Bridge, UK (after Croll, 1994).

Box 12.2 Nitrate and human and animal health.

The condition of acute infant well water methaemoglobinaemia (or 'blue-baby disease') has been linked to the use of high NO_3^- well waters to make bottled-milk feed for babies (mainly under 3 months old). This condition is induced by bacterial reduction of NO_3^- to NO_2^- in the bottled milk or the infant's stomach, and the absorption of excess NO_2^- in the blood where it interferes with the O_2-carrying capacity of haemoglobin. Most cases of methaemoglobinaemia arose from bacterially contaminated well water of NO_3^- concentration > 22.6 mg NO_3-N/L. However, with the improvement in well water quality in recent years, the disease has virtually disappeared from developed countries (Laegreid et al., 1999).

High NO_3^- concentrations in vegetables used for baby food, and bacterial reduction of NO_3^- to NO_2^- in the food, can cause methaemoglobinaemia. The disease can also occur in ruminant animals fed on herbage high in NO_3^- because of strong reducing conditions in the rumen. It has been argued that NO_3^- (reduced to NO_2^-) combines with amines in food to produce carcinogenic nitrosoamines. However, NO_2^- in the stomach of anyone other than a very young baby is rapidly reduced to NO, which is beneficial in killing pathogenic bacteria. Nor is there any sound epidemiological evidence of a positive correlation between NO_3^- intake and stomach cancer in humans.

Against this background of possible health risks, the World Health Organization (WHO) and national governments have set conservative limits for NO_3^- concentrations in drinking water. These limits are 11.3 mg NO_3-N/L (50 mg NO_3/L) for WHO and in the EU, and 10 mg NO_3-N/L in the USA and Australia. The response to these limits in the EU has been broadly twofold:
• Changes in land management. Nitrate Sensitive Areas have been identified where surface waters or groundwater used for potable supplies have ≥ 50 mg NO_3/L. In the UK, farmers in such areas (mainly over the Chalk, limestone and sandstone aquifers that provide 40% of the potable water) receive financial incentives to change their management to minimize the NO_3^- available for leaching in autumn and winter;
• treatment of water for domestic use. Here the preferred option is to blend high and low NO_3^- concentration water.

By the mid-1990s, the rising trend in NO_3^- concentrations in rivers and groundwater in the UK had levelled out through changes in land management in the previous decade (Fig. 12.3), but the concentrations in many sources are still above the EU limit and may remain so for some time.

Efficiency of utilization

Rate and method of application

N recovery. The recovery of fertilizer N in a crop or animal product depends on the climate, the kind of crop, and on the rate, method and timing of fertilizer application. Potentially the best recovery is achieved when temperature and other factors allow the crop to be grown during the wetter part of the year, as can be done in much of the tropics. For a given climate, N recoveries are usually lowest for plantation crops and short-season arable crops, and highest for permanent grass. Bananas in the Ivory Coast, for example, recovered only 14% of the 430 kg N/ha applied, but grass recovered 53% at rates up to 700 kg N/ha. Maize grown at different locations in the USA, Central America and Brazil recovered about 56% of fertilizer N at application rates up to 150 kg N/ha. In Britain, where leaching losses in the wet winters are appreciable, recovery of fertilizer N by crops ranges between 50 and 80% and recovery by grass fertilized at rates up to 400 kg N/ha/yr ranges between 50 and 65%. However, when grass is grazed by domestic animals, recovery of N in animal products is only 5–15% of the input – the other 85–95% is returned to the soil in dung and urine from which, under intensive grazing management, much is lost by leaching, denitrification and volatilization.

Timing of N application. Timing is crucial for best utilization by a crop. Spring applications in Britain give better results than those in autumn, especially in areas with annual precipitation > 675 mm (and excess winter rainfall* > 125 mm). Depending on the soil type, little NO_3^- remains in the soil by spring when the excess winter rainfall exceeds 150–250 mm. During the grass-growing season, N is best applied as split dressings, say 4–5 times during the season, or after each cut for hay or silage. In this way, the supply of N is better matched to the growth cycle of the grass and losses by leaching are minimized. Similarly, in the southern maize belt of the USA, a delay in N application for 3–4 weeks after germination avoids leaching losses from late spring rains, and provides N at the stage of growth when the crop's demand is highest.

Residual effects

A residual effect occurs when sufficient of a fertilizer element applied to one crop remains available in the soil for the growth of a succeeding crop. For N fertilizer, residual effects depend on the rate of N applied, the crop, the amount of rain falling between crops, and the soil type. Provided that the applied fertilizer N is reasonably well matched to plant demand, the proportion of actual fertilizer N remaining in mineral form in the soil by harvest time is very small (Powlson, 1997). However, plant growth (except under irrigation) is affected by the uncertainty of weather, and farmers tend to apply extra fertilizer as 'insurance' to minimize the risk of diminished yields. Also, a significant proportion of the N fertilizer is converted into labile organic N from which it is readily mineralized (Box 10.1). Thus, field and lysimeter studies in many countries have reported NO_3^- concentrations in drainage from cropped soils and grazed pasture in the range 5–50 mg N/L (and up to 100 mg N/L for intensively managed pasture). It is therefore possible to construct a schedule of probable N leaching

---

* Defined as the excess of precipitation over evaporation when the soil is at field capacity during winter. An estimate of excess winter rainfall is a substitute for measured soil drainage.

Table 12.3 Estimated N losses by leaching in relation to excess winter rainfall (or drainage).

Mean NO_3^- concentration in drainage water (mg N/L)	Quantity of N (kg/ha) leached by an excess winter rainfall (mm) of				
	100	200	300	400	500
5	5	10	15	20	25
10	10	20	30	40	50
15	15	30	45	60	75
20	20	40	60	80	100
30	30	60	90	120	150
50	50	100	150	200	250

Table 12.4 Residual effects of N-fertilized crops and legumes. After Cooke, 1969.

(a) Wheat yield (t grain per ha) following potatoes given		
0	and	185 kg N/ha
2.84		3.91

(b) Wheat yield (t grain per ha) following N-fertilized ryegrass	Clover ley
4.83	6.12

losses as shown in Table 12.3. As excess winter rainfall (or drainage) increases, one moves across the table from left to right and slightly upwards, because the NO_3^- concentration usually falls as the cumulative drainage volume exceeds 100 mm. Thus, for an excess winter rainfall of 250–300 mm and allowing for crop uptake, there will be little residual N from a fertilizer application of 150 kg N/ha to a preceding crop.

Residual effects have been observed when deep-rooting cereals, such as winter wheat, follow crops such as potatoes to which heavy N dressings have been applied (Table 12.4a). In other cases, such as cereal following cereal, part of any residual effect is undoubtedly due to the larger residue of N-rich plant material in the soil, which decomposes to release mineral N. When the preceding crop is a legume, the stimulatory effect of the residues is generally greater than that of even heavily fertilized grass (Table 12.4b).

Table 12.5 Forms of phosphate fertilizer.

Fertilizer	Composition	P content (%)
Ortho-P		
Phosphoric acid	H_3PO_4	23
Normal or single superphosphate	$Ca(H_2PO_4)_2$; $CaSO_4$	8–10
Concentrated or triple superphosphate	$Ca(H_2PO_4)_2$	19–21
Monoammonium phosphate	$NH_4H_2PO_4$	21–26
Diammonium phosphate	$(NH_4)_2HPO_4$	20–23
Monopotassium phosphate	KH_2PO_4	23
Poly-P		
Superphosphoric acid	$H_4P_2O_7$ and higher MW polymers	> 33
Ammonium polyphosphate	$(NH_4)_4P_2O_7$ and higher MW polymers	23
Insoluble phosphates		
Phosphate rocks	$Ca_{10}(PO_4)_6(F,OH)_2$ with variable SiO_2, $CaCO_3$ and sesquioxide impurities	6–18
Basic slag	Basic Ca,Mg phosphates, Fe_2O_3 and $CaSiO_3$	3–10
Organic phosphates	Bone meal, blood and bone, guano	5–13

12.3 Phosphate fertilizers

Forms of phosphate fertilizer

Phosphate fertilizers may be subdivided into:

- Water-soluble orthophosphates;
- condensed orthophosphates or polyphosphates; and
- insoluble mineral and organic phosphates (Table 12.5).

The natural rock phosphates, consisting of minerals of the apatite type with calcite, silica and other impurities, are the raw material from which acidulated and partially acidulated P fertilizers are made.

Ortho-P fertilizers

The active constituents of ortho-P fertilizers are monocalcium phosphate (MCP), formula $Ca(H_2PO_4)_2.H_2O$, and phosphoric acid. The former compound is the basis of the superphosphates formed by the dissolution of rock phosphate in H_2SO_4 or H_3PO_4, for example

$$Ca_{10}(PO_4)_6F_2 + 7H_2SO_4 + 3H_2O \rightarrow$$
$$3Ca(H_2PO_4)_2.H_2O + 7CaSO_4 + 2HF. \quad (12.2)$$
$$\text{(single superphosphate, SSP)}$$

If rock phosphate is reacted with excess H_2SO_4, wet-process H_3PO_4 is produced, which is then used to dissolve fresh rock phosphate and give higher grades of superphosphate, such as triple superphosphate

$$Ca_{10}(PO_4)_6F_2 + 14H_3PO_4 + 10H_2O \rightarrow$$
$$10Ca(H_2PO_4)_2.H_2O + 2HF. \quad (12.3)$$
$$\text{(triple superphosphate, TSP)}$$

If insufficient H_3PO_4 is used to convert all the rock phosphate to MCP, a partially acidulated phosphate rock (PAPR) fertilizer is produced. Single superphosphate has the advantage of containing 11% SO_4-S, which provides additional value on sulphur-deficient soils.

Roasting rock phosphate ore in an electric furnace with quartz and coke reduces the phosphate minerals to elemental P, which when burnt in O_2 forms the oxide P_2O_5. This oxide combines with water to form phosphoric acid according to the reaction

$$P_2O_5 + 3H_2O \rightarrow 2H_3PO_4. \quad (12.4)$$

Phosphoric acid normally contains 52–54% P_2O_5 (23–24% P). Partial neutralization of the acid with NH_3 gives the multinutrient fertilizers MAP and

DAP, discussed earlier (Section 12.2). Alternatively, ordinary superphosphate can be mixed with aqua or anhydrous NH_3 to form ammoniated superphosphate (*c.* 4% N and 7% P); and TSP can be ammoniated to give a product containing 8% N and 14% P. Unlike MAP and DAP, the ammoniated superphosphates contain some S. About half the P present is water-soluble, but all of it is soluble in neutral ammonium citrate (Box 12.3).

The relationship between these ortho-P fertilizers is summarized in Fig. 12.4.

Box 12.3 Phosphate rocks (PRs) as direct-application fertilizers.

There are many sources of PR ores in the world, but the main reserves lie in North America, North and West Africa and the Middle East. The agronomic effectiveness of these materials when directly applied depends on the chemical and physical nature of the PR, soil properties, type of crop or pasture, and climatic conditions. A critical step is the dissolution of PR in the soil according to the reaction

$$Ca_{10}(PO_4)_6F_2 + 12H^+ \leftrightarrow 6H_2PO_4^- + 10Ca^{2+} + 2F^-.$$
$$(B12.3.1)$$

This reaction is shown as being reversible because the rate of dissolution depends not only on the concentration of H^+ ions in the soil solution, but also on the concentration of the main products Ca^{2+} and $H_2PO_4^-$. The interaction of soil, plant and climatic variables in determining the rate of PR dissolution is summarized in Fig. B12.3.1.

PRs are arbitrarily divided into 'reactive' (soft) and 'unreactive' (hard) rocks on the basis of their solubility in chemical extractants, e.g. 2% citric acid (New Zealand), or M ammonium citrate at pH 7

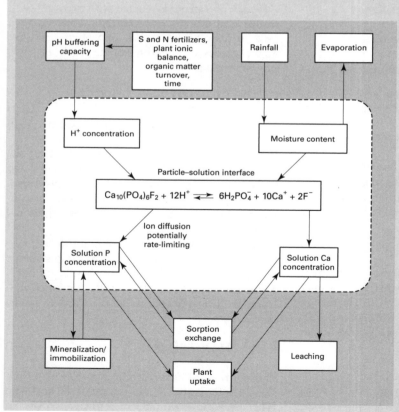

Fig. B12.3.1 Diagram showing the rate-determining factors (boxes in clear area) for PR dissolution in soil, and the variables (boxes in shaded area) that determine the magnitude and degree of interaction of the rate-limiting factors (after Bolan *et al.*, 1990).

Box 12.3 *continued*

(Australia and the USA), or 2% formic acid (European Union). Hard rocks such as Florida (USA) and Nauru (Pacific Islands) are used for the manufacture of soluble P fertilizers, but more reactive phosphate rocks (RPRs) such as North Carolina (USA), Gafsa (Tunisia) and Sechura (Peru) are used for partial acidulation (producing PAPRs) and for direct application. Given suitable crop, soil and climatic conditions, PAPRs and most RPRs can be comparable in agronomic effectiveness to soluble P fertilizers, and the RPRs in particular have the advantage of costing less per kg of P supplied. A disadvantage is that for maximum effectiveness, RPRs should be finely ground, which creates problems in their application. Agronomic

effectiveness is assessed by comparing plant performance when supplied with PR against that when a soluble P fertilizer (e.g. TSP) is used at the same rate of P per ha. According to Karama and Willett (1998) and others, the appropriate conditions for the use of RPRs and beneficiated PRs are:
• Acid soils (pH < 5.5 in water);
• adequate soil moisture (> 800 mm rain, reasonably well distributed through the year); and
• perennial pastures on soils not severely deficient in P, and plantation crops such as trees, sugar cane and cassava.

PRs can be applied at greater rates than soluble P fertilizers but less frequently, provided conditions are suitable.

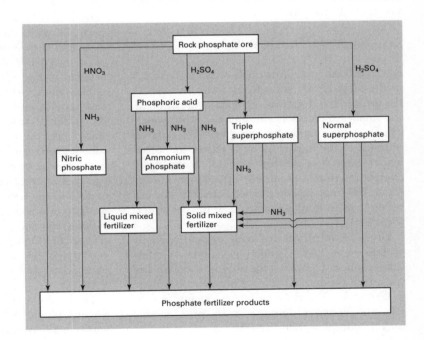

Fig. 12.4 Flow diagram of P fertilizer manufacture (after Slack, 1967).

Poly-P fertilizers

Ordinary H_3PO_4 can be upgraded to superphosphoric acid (72% P_2O_5) by decreasing the ratio of water to P_2O_5. This acid consists of a mixture of pyrophosphoric acid ($H_4P_2O_7$) and higher molecular weight polymers, for example

$$3P_2O_5 + 5H_2O \rightarrow 2H_5P_3O_{10}. \qquad (12.5)$$
(tripolyphosphoric acid)

Ammoniation of superphosphoric acid produces a very soluble fertilizer – ammonium polyphosphate (12% N and 23% P). This fertilizer is used mainly in liquid form, a form increasingly

popular for all fertilizers in the USA and Europe. Compared to solids, liquid fertilizers offer the advantages of:
• Precise application, which is important for P and the micronutrients;
• more uniform application, especially for N;
• being excellent carriers of pesticides and micro-nutrients; and
• are well adapted to irrigation systems. The use of soluble fertilizers in irrigation water is referred to as fertigation.

Insoluble phosphates

Phosphate rock (PR) ores are found in certain igneous and sedimentary rocks. The minerals present include apatites, crandallites, millisites, silica and calcite. The most important mineral apatite ranges in composition from fluorapatite $Ca_{10}(PO_4)_6F_2$ to francolite

$$Ca_{10-\alpha}(Na,Mg)_\alpha(PO_4)_{6-\beta}(CO_3)F_\gamma,$$

where α and γ are functions of β, and $\beta = 1.5$ in the most highly substituted form. In general, the greater the degree of carbonate substitution, the smaller the crystal size and the more water-soluble is the apatite. Beneficiation – the removal of much of the impurities by flotation and washing in water – is the first step in raising the P content of crushed ore to 13–18%. Finely ground PR ($< 150\ \mu m$), whether beneficiated or not, can be used as a fertilizer under certain conditions (Box 12.3).

Other PR fertilizers are made by heating finely ground ore to 1200°C and above, a process called calcination, or by heating the ore with soda ash and silica (the product is called Rhenania phosphate in Germany). The roasting ignites organic impurities and converts fluorapatite to more soluble β-tricalcium phosphate $Ca_3(PO_4)_2$.

Sometimes ground PR is mixed with basic slag, a by-product of steel manufacture from phosphatic iron ores. Basic slag must also be finely ground and is best used on acid soils where its Neutralizing Value (equivalent to approximately two-thirds its weight of ground limestone) and its micronutrient impurities are beneficial, especially for legumes.

Reactions in the soil

Dissolution of water-soluble fertilizers

A granule of MCP, the active constituent of SSP and TSP, placed in dry soil takes up water by vapour diffusion until a saturated salt solution forms that flows outwards from the granule. In moist soil, liquid water is drawn into the dissolving granule by osmosis and the nutrient ions move out by diffusion. In dry soil, the saturated solution is very acid (pH ~ 1.5) and concentrated in P (c. 4 M) and Ca (c. 1.4 M) : less so as the soil becomes moister. This acid solution reacts with the soil minerals, dissolving large amounts of Ca, Al, Fe and Mn, some of which are subsequently precipitated as new compounds – the soil fertilizer reaction products – as the pH slowly rises and solubility products are exceeded. An MCP granule 5 mm diameter dissolves incongruently in 24–36 hours to leave a residue of dicalcium phosphate dihydrate (DCPD) and anhydrous dicalcium phosphate (DCP) containing about 20% of the original P. The sequence of events is summarized in Fig. 12.5.

The rapidity of this reaction means that a plant feeds not so much on the fertilizer itself, but on the fertilizer reaction products that are metastable and revert slowly to more stable (but less soluble) products. Depending on whether the soil is calcareous or rich in sesquioxides, and on the presence of other salts such as NH_4Cl, $(NH_4)_2SO_4$ or KCl in the fertilizer granule, a range of intermediate products is formed including:
• Potassium and ammonium taranakites ($H_6K_3Al_5(PO_4)_8.18H_2O$ and $H_6(NH_4)_3Al_5(PO_4)_8.18H_2O$);
• complex calcium-aluminium and calcium-iron phosphates; and
• DCPD.

The complex Al and Fe compounds hydrolyse to amorphous $AlPO_4$ and $FePO_4$, respectively, which may in very acid soils slowly revert to the stable end forms of variscite and strengite, respectively. In acid soils, DCPD dissolves congruently, but at pH > 6.5 it hydrolyses to form a less soluble residue, probably octacalcium phosphate (OCP), which in more basic soils may revert to tricalcium phosphate (TCP) or hydroxyapatite (HAp). The availability of P in the Al and Fe reaction products relative to MCP is illustrated in

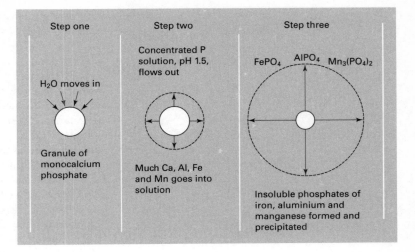

Fig. 12.5 Successive stages in the dissolution of a MCP granule in soil (after Tisdale and Nelson, 1975).

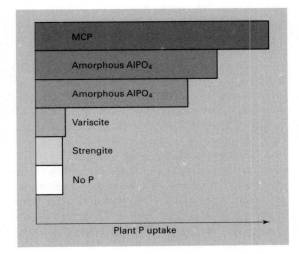

Fig. 12.6 Relative availability of different P sources to a crop (after Huffman, 1962).

Fig. 12.6. Note also the effect of soil pH on the dissolution of the main Ca, Al and Fe fertilizer reaction products, as shown in Fig. 10.11.

Powdered vs granular forms

Insoluble fertilizers, such as the basic calcium phosphates and PRs, are most effective when used in a finely divided form. With these materials, absorption of water does not cause the problem during storage that it does with soluble fertilizers, but because fine powders are difficult to apply and create a dust problem, a form of weak aggregation or 'mini-granulation' is sometimes employed. By contrast, granulation of soluble P fertilizers has the advantage, in addition to ease of handling, of confining the reaction between soil and fertilizer to small volumes around each granule. The rate of P fixation in the soil is therefore slowed (Section 10.3).

Residual effects

Typically, only 10–20% of fertilizer P is absorbed by plants during the first year: the remainder is nearly all retained as fertilizer reaction products, which become less soluble with time. A rough guideline is that two-thirds of the water and citrate-soluble P remains after one crop, one-third after two crops, one-sixth after three crops and none after four crops.

Larsen (1971) suggested that a more precise way of assessing the residual effect of a P fertilizer was to measure the rate at which the labile pool of soil phosphate (the L value) decreased after the addition of fertilizer, and to calculate the half-life for the 'decay' of the fertilizer's value to the crop. Half-lives in neutral and alkaline soils vary from 1 to 6 years. Larsen also pointed out that the less soluble P fertilizers, in contrast to soluble ones, require time to reach peak effectiveness, as well as having a longer half-life (Fig. 12.7).

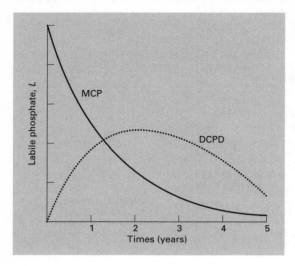

Fig. 12.7 Changes in soil labile phosphate after the addition of P fertilizer (after Larsen, 1971).

Eutrophication

Cause and effects

Eutrophication is a natural process whereby surface waters are gradually enriched with organic matter and inorganic nutrients carried in runoff from a surrounding catchment. One end point of such natural eutrophication is the formation of fen peats, in both temperate and tropical climates, where base-rich waters and sediments collect in drainage areas. Nevertheless, eutrophication can become a problem when it is accelerated by increased loads of sediment and dissolved nutrients in streams draining urban and rural areas. The resultant higher concentrations of P, N and organic matter promote the growth of aquatic plants, from blue-green algae (Cyanobacteria) to rooted higher plants. The most serious effects occur when there is a sudden and unpredictable surge in the growth of blue-green algae to create an 'algal bloom' (Box 12.4).

Sources of nutrient input

Dissolved and particulate nutrient loads arise from point and diffuse sources. Examples of point sources are effluents from factories, sewage works and intensive livestock units, such as cattle feedlots, piggeries and dairies. Diffuse sources comprise the drainage from the remaining agricultural and non-agricultural land – grassland, arable and forest – which by its very nature is difficult to monitor accurately. Overall, the total load of a nutrient entering a water body will determine whether it becomes eutrophic or not, although the criteria for levels of water quality are expressed in terms of nutrient concentrations, as indicated in Table B12.4.1.

In most countries, the bulk of the N-enrichment of surface waters comes from agricultural land. In the UK, for example, drainage from one hectare may carry between 5 and 250 kg N, depending on the climate, land use and N inputs (Table 12.3). In contrast to the mobile NO_3^- ion, ortho-P ions are strongly adsorbed and the reaction products formed by P fertilizers in soil are relatively insoluble, so that leaching (and surface runoff) from agricultural land usually contributes < 1 kg P/ha/yr. P loads to surface waters are usually much higher from intensive animal industries and urban areas, the latter through sewage effluent. In the UK, for example, 65% of the P load to surface waters is from non-agricultural sources. However, there is accumulating evidence in several countries that, with the increase in soil P contents through regular inputs of fertilizer and manure P that exceed off-take in crop and animal products, P losses from agricultural land have increased. Much of this loss is via surface runoff (Table 12.6), but losses of 1–3 kg P/ha/yr by leaching of soluble organic and colloidal P have also been recorded (Leinweber *et al.*, 2002).

Table 12.6 P loss (kg/ha/yr) from agricultural land in surface runoff. After White and Sharpley, 1996.

Land use and form of P input	Phosphorus loss	
	Dissolved P	Total P
Cereals or grass, P fertilizer	0.02–2.8	0.2–5.5
Maize or lucerne, dairy manure	0.1–4.8	0.1–7.4
Grass, pig manure	0.1–4.8	0.1–4.8
Grass, poultry manure	0.0–4.3	0.1–12.4

Box 12.4 Algal blooms and water quality.

Blue-green algae such as *Anabaena*, *Microcystis* and *Nodularia* are natural components of all freshwater ecosystems and can also occur in saline waters. However, these organisms grow excessively (bloom) in nutrient-enriched waters, said to have a high trophic status or to be eutrophic. Waters with low concentrations of nutrients are of low trophic status and are called oligotrophic. Algal blooms can have several adverse effects, for example:

• Production and release of toxins that can poison fish, waterfowl, animals and humans;

• production of foul tastes and odours that make water less palatable to drink;

• formation of surface scums that are visually displeasing and may emit offensive odours on decomposition; they may also create problems for water treatment works because of the fouling of filter beds; and

• as the bloom subsides, dead vegetation sinks to the bottom and decomposes. Large quantities of dissolved O_2 are consumed, creating anoxic conditions that can result in the death of fish and other aquatic fauna.

Algal blooms are predisposed by the following conditions:

1 High concentrations of dissolved nutrients, especially P. Threshold P concentrations considered critical for algal blooms in several countries are given in Table B12.4.1. As water becomes more eutrophic, N becomes the limiting nutrient.

Eutrophication of most marine and coastal waters is limited by N;

2 high light and high temperature conditions in summer;

3 low water turbidity, allowing greater penetration of sunlight;

4 low flow rates, so that nutrient concentrations remain high.

Most of the P entering surface waters is carried on suspended sediment, although in some heavily fertilized pasture systems a significant proportion of P in runoff is dissolved, reactive P. Much of the sediment settles to the bottom and acts as a long-term reserve for the overlying water into which P is released by reduction of Fe-P compounds under anaerobic conditions (Section 8.4), or by desorption when the sediment is re-suspended as a result of mechanical disturbance.

Table B12.4.1 Critical P concentrations (mg/L) for the incidence of algal blooms.

Country	Dissolved reactive P	Total P (dissolved and particulate)
USA – selected states	0.05	0.1
Australia (Victoria)	–	0.05
European Union (guideline)	–	0.3

12.4 Other fertilizers including micronutrient fertilizers

Potassium fertilizers

Potassium occurs naturally as KCl (muriate of potash) in salt deposits (mines and salt lakes), which also include some NaCl, K_2SO_4 and $MgSO_4$. The KCl is separated by flotation or by recrystallization, which depends on the different effect of temperature change on the solubility of these salts. Nevertheless, retention of a low Na and Mg content in K fertilizers is an advantage

for the nutrition of some crops and also grazing animals (see below).

KCl is the most widely used K fertilizer, accounting for *c.* 95% of world use. It is especially favoured in liquid mixed fertilizers because of its high solubility. Other K fertilizers, representing a wide range of K content and water solubility, are listed in Table 12.7. Potassium sulphate is generally made by reacting SO_3 gas with water and KCl, that is

$$2KCl + SO_3 + H_2O \rightarrow K_2SO_4 + 2HCl. \qquad (12.6)$$

Table 12.7 Forms of potassium fertilizer.

Fertilizer	Composition	K content (%)
Potassium chloride	KCl	32–51
Potassium sulphate	K_2SO_4	41
Potassium nitrate	KNO_3	38
Kainit	$KCl.MgSO_4.NaCl.nH_2O$	12

It is less hygroscopic than KCl and therefore easier to handle. Despite the higher cost of K_2SO_4 per kg K, it is the preferred source of K for crops that are sensitive to high concentrations of Cl, such as tobacco, potatoes and many vegetables grown in glasshouses. Potassium nitrate is also favoured for tobacco because it contains no Cl and N is present as NO_3^-, the preferred form of N for this crop. Kainit is an insoluble slow-release K fertilizer.

K balance in cropping

Because the K immediately available to plants is held as an exchangeable cation and is not readily leached, it is feasible to attempt to balance K removed by crops and in animal products with K applied in fertilizers and released from non-exchangeable sources in the soil.

K release

Slow release of non-exchangeable K depends on the content of potash feldspars and micaceous clay minerals. Soils with such reserves, even though low in exchangeable K, may supply between 20 and 80 kg/ha/yr for many years, especially to grasses that are efficient in taking up K at low concentrations; but sandy soils lacking micaceous clays are soon exhausted of K by continuous cropping. On many soils where K is adequate initially, regular N and P fertilizing may so stimulate growth that K-deficiency symptoms appear, more often on pasture land where fodder is conserved as hay or silage.

K uptake by crops

Recovery of K fertilizer during one season varies from 25 to 80%, being highest for grassland,

which may remove up to 450 kg K/ha/yr when highly productive. Clovers and root crops require higher soluble K concentrations in soil than grasses, so that the ratio of N : K fertilizer applied to such crops should be 2 : 3, compared to 1.5 : 1 for temperate grasses and 2 : 1 for tropical grasses. If too much K is used on pasture, the natural content of Mg in the fertilizer may be insufficient to prevent a K-Mg imbalance developing in the herbage, which may induce grass tetany or hypomagnesaemia in grazing animals.

Sodium can partially substitute for K in many plants and is essential for animals and humans, so the Na content of K fertilizers is beneficial. Because K specifically increases the yield and sugar content of sugar beet, the standard K fertilizer recommended for this crop in Germany, for example, contains 6% MgO and 8% Na_2O.

Sulphur fertilizers

Sulphur for fertilizers is produced by:
• Roasting iron pyrites FeS_2, or as a by-product of processing non-ferrous metal sulphides;
• recovery from natural gas and gases released during oil refining that contain H_2S; or
• mining elemental S.
The first two processes produce 'by-product' S, most of which is used to make H_2SO_4 for the fertilizer industry. For use on its own, S may be made into prills from molten S and clay (up to 90% S). In the soil, S is slowly oxidized to sulphate by *Thiobacillus* bacteria according to the reaction

$$S + 3/2O_2 + H_2O \rightarrow 2H^+ + SO_4^{2-}. \qquad (12.7)$$

Another source of S for agriculture is gypsum ($CaSO_4.2H_2O$) which is mined from saline deposits, or produced during flue-gas desulphurization (Box 12.5), or as a by-product (phosphogypsum) in the manufacture of triple and concentrated superphosphates. Ammonium sulphate, containing 24% S, is the most soluble sulphate salt in fertilizers and is used in liquid mixed fertilizers.

Crop requirements

Crop requirements for S are 20–30 kg S/ha and exceptionally up to 50 kg S/ha for some Brassicas

Box 12.5 Sulphur emissions and flue-gas desulphurization.

The emission of S to the atmosphere from the combustion of fossil fuels and the effect of this S on the magnitude of acid inputs to the soil are discussed in Sections 10.3 and 11.3. The decrease in S emissions achieved in recent years is partly due to a switch to lower S content fuels, but mainly to emission control by 'scrubbing' the flue gases from power stations through beds of crushed limestone, which produces gypsum. Gypsum produced in this way is used mainly as raw material for plaster board manufacture, but can also be used in agriculture if competitive in price with other forms of S.

Table 12.8 Application rates for micronutrients in fertilizer. After Jones, 1982.

Element	Fertilizer content (%)	Safe maximum for the element (kg/ha)
Mo	0.005	0.2
B	0.025	1.0
Mn	0.15	10
Cu	0.15	10
Zn	0.25	10

These elements can be added to the soil as sparingly soluble salts or oxides, or as chelates (Section 10.5). When used in the last mentioned form, or applied as foliar sprays of soluble salts, such as $ZnSO_4.7H_2O$, they are intended for immediate uptake by the plant. In their less soluble forms they maintain a very low concentration in the soil solution for several years, thus providing a prolonged residual effect. An increasingly popular controlled-release micronutrient fertilizer is the frit, made by fusing the element in glass that can be crushed and mixed with NPK fertilizers. Table 12.8 gives the normal concentrations of these elements in mixed fertilizers and the maximum rate of application considered safe for most crops.

(e.g. oilseed rape). With the decrease in S emissions over western Europe in recent years (Box 10.3), S deficiency in crops is becoming more widespread and is bound to increase. This is also true of grassland that is heavily fertilized with N, because the N : S ratio in herbage should be kept near 10 : 1 for the optimum nutrition of ruminant animals. Similarly, in areas of the tropics and subtropics where the soils are highly leached and atmospheric inputs of S small (e.g. northern Australia), the extent of S-deficient soils is widespread. For soils under pasture in high rainfall areas, SO_4-S is readily leached so that supplying part of the fertilizer S as finely divided elemental S (a slow-release form) has merit for decreasing the rate of SO_4^{2-} leaching. In the New Zealand, for example, SSP fortified with elemental S (up to 50%) is used on soils that are naturally very deficient in S and subject to high rates of leaching.

Micronutrient fertilizers

The micronutrients Fe, Mn, Zn, Cu, B and Mo may need to be applied individually, or in mixed fertilizers, to correct a soil deficiency. Cobalt may also be required to correct a vitamin B_{12} deficiency in ruminants and for effective N_2 fixation by legumes. Selenium and I supplements to animal diets may be needed in deficient areas.

12.5 Plant protection chemicals in soil

Environmental impact

Plant protection chemicals or pesticides include the many natural and synthetic chemical compounds that have biocidal or biostatic effects. They can be more specifically designated insecticides, miticides, nematicides, fungicides, herbicides, rodenticides or molluscicides according to the group of organisms they are intended to control. Of these, the herbicides, insecticides and fungicides account for the bulk of agricultural usage, in that order.

The ideal pesticide is one that controls only the target organism and persists long enough to achieve this purpose before degrading into harmless products. It must also be of low toxicity to mammals. In practice, however, the ideal is not always achieved and chemicals have been used in agriculture (and in public health programmes,

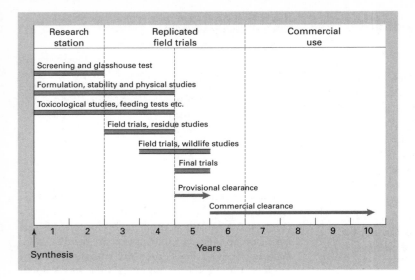

Fig. 12.8 Pesticide testing procedures in the UK (after Glasser, 1976).

such as for the eradication of malaria-carrying mosquitoes) that are 'broad spectrum' in their activity; that is, they kill harmless and beneficial organisms as well as the target organisms. Furthermore, there are pesticides, such as the organochlorine insecticides, that are very stable and have high lipid solubility. They therefore accumulate in the fatty tissues of animals, particularly predators high in natural food chains. Extensive use of these and other persistent chemicals has led to their residues and breakdown products becoming widely disseminated, and there have been many instances of their detrimental effect on beneficial insects and plants, domestic animals and wildlife.

As a result of public concern over the environmental effects of persistent and sometimes toxic chemicals, pesticide manufacturers voluntarily or in response to government legislation now undertake extensive research into the development and testing of new products before their release as pesticides, as shown in Fig. 12.8. These procedures are designed to evaluate not only the efficacy of a chemical for the control of specific pests, but also its environmental impact, that is:
• The effect of the chemical on non-target organisms – beneficial insects, indigenous flora and fauna, especially mammals; and
• its persistence in a defined compartment (soil, water, organisms or air) of the environment.
Measures of pesticide persistence are discussed in Box 12.6.

The soil has a key role in determining the fate of a pesticide in the environment. Nematicides, many insecticides and fungicides are deliberately applied to soil to control soil-inhabiting pests. There are also soil-applied herbicides (e.g. pre-emergent herbicides) and an increasing number of systemic insecticides and fungicides intended for absorption by roots and transport to plant shoots. An even wider range of compounds reaches the soil unintentionally because of rain wash from leaves and the incorporation of pesticide-treated plant residues. Much attention has therefore been paid to improving the efficiency of pesticide application to target organisms, which can be achieved through:
• Improvement in spray technology, particularly to deliver sprays of optimum drop size and maximum chemical concentration per unit volume;
• minimizing spray drift; and
• micro-encapsulation to delay the release of an active constituent (controlled-release technology).

Pesticide persistence in soil

The main factors governing pesticide stability in soil and the loss of pesticide to surface waters, groundwater and air may be summarized as follows:
1 Volatilization;
2 adsorption by soil minerals and organic matter;

Box 12.6 Measurement of pesticide persistence in soil.

A simple approach to measuring the persistence of a chemical that is subject primarily to biotic decomposition is to assume first-order decay kinetics, as for organic C compounds in soil (Section 3.5). The half-life $t_{1/2}$, defined as the time for 50% of the compound to disappear, is given by the equation

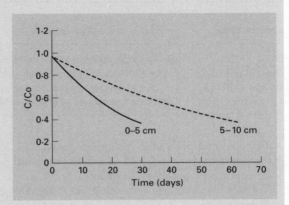

$$t_{1/2} = \frac{0.693}{k},\qquad\text{(B12.6.1)}$$

where k is the first-order decay coefficient, which can be measured experimentally in an incubation experiment. An example of first-order decay of the insecticide phosmet in soil is shown in Fig. B12.6.1. Pesticides can be ranked according to their half-lives, as discussed under 'Pesticide persistence in soil'.

Another approach to measuring persistence is to set a critical value below which the concentration of a pesticide in soil, water or air must fall before it classed as having 'disappeared'. For example, when the environmental impact of a pesticide is being tested in the EU, very low limits may be set for an acceptable concentration remaining in soil or in drainage water coming from the soil. Alternatively, in the USA the maximum persistence time was the

Fig. B12.6.1 First-order decomposition of the pesticide phosmet in soil. C/C_o is the ratio of phosmet remaining at time t relative to that present initially. Note the decay is faster in the 0–5 cm layer where the organic matter content is higher (after Suter et al., 2002).

time taken for 90% or more of a pesticide to disappear from the site of application. The maximum persistence time can be assessed from field measurements or calculated from the type of decay curves shown in Fig. B12.6.1. Examples of the maximum persistence times for groups of common pesticides are given in Fig. B12.6.2.

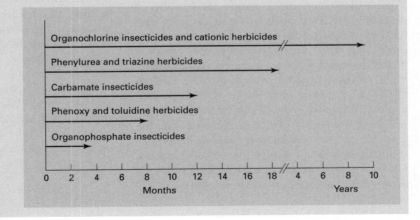

Fig. B12.6.2 Maximum persistence of pesticides in soil under a mild climate (after Stewart et al., 1975).

Table 12.9 Description and key properties of four important organochlorine insecticides. After Guenzi, 1974 and others.

Common name	Chemical name	Solubility in water (mg/L)	Vapour pressure at 30°C (mm Hg)
DDT	1,1-bis(4-chlorophenyl)-2,2,2-trichloroethane	0.001–0.04	7×10^{-7}
Dieldrin (HEOD)	1,2,3,4,10,10-hexachlorocyclopentadiene	0.1–0.25	1×10^{-5}
Lindane (γ-BHC)	γ-1,2,3,4,5,6-hexachlorocyclohexane	7.3–10.0	13×10^{-5}
Toxaphene	Chlorinated camphene containing c. 68% chlorine	0.4	No data

3 chemical and biological transformations;

4 transport in the air, liquid or solid phase; and

5 absorption by plants and animals.

The relative importance of each process depends very much on:

• The pesticide's properties – its water solubility, volatility, affinity for organic or mineral surfaces;

• its formulation – usually the pesticide is dissolved in a suitable oil or organic solvent (with an emulsifier, and diluted with water as required), or supplied as a wettable powder (pesticide is adsorbed on finely divided clay particles with a wetting agent), or as granules (pesticide is mixed with inert fillers and pelletized);

• its mode of application – whether applied to crop surfaces or cultivated into the soil; and

• the environmental conditions (temperature and rainfall), soil type and cropping system.

Volatilization

Volatilization can be a major mechanism of pesticide loss from plant and soil surfaces, and for very persistent chemicals, such as the organochlorines, the means whereby they become widely dispersed in the environment. For example, for a relatively volatile compound like lindane (Table 12.9), more than 90% of the pesticide exposed on moist surfaces can be lost to the air within 48 hours of application. Incorporating the pesticide into the soil decreases vapour losses substantially because of adsorption on soil particles, but the factors controlling volatilization of a soil-incorporated pesticide are complex (Box 12.7).

Adsorption and desorption

Pesticide molecules in the soil solution reach equilibrium with the solid phase by a variety of adsorption–desorption mechanisms, which are described for simple inorganic and organic ions in Chapter 7. Broadly, whether the equilibrium favours the solution or solid phase depends on the chemical properties of the pesticide and the type of adsorbent (see Box 2.5): in soil, the adsorbent comprises a mixture of organic and inorganic surfaces that are predominantly negatively charged. For this purpose, pesticides may be grouped as:

• Non-polar and hydrophobic – water solubility < 0.001 M;

• sufficiently polar that the solubility > 0.001 M, but not a cation or weak acid or base;

• weak bases, i.e. $RH_2^+ \leftrightarrow RH$ (pesticide) + H^+;

• weak acids, i.e. RH (pesticide) $\leftrightarrow R^- + H^+$; and

• cations.

For the non-polar, hydrophobic compounds of low water solubility (e.g. organochlorines), soil organic matter is the most important adsorbent. For pesticides that are weak bases or weak acids, the strength of adsorption depends on the soil pH because this influences the charge on both the adsorbent and pesticide molecule. Pesticides that are cations are the most strongly adsorbed. Some examples follow.

Weak bases. The triazine and pyridione herbicides are very weak bases with pK_a values ranging from 1.7 for atrazine to 4.3 for prometon. The molecules can take up H^+ ions (protonate) at pHs up to 2 units above the pK_a. The triazine simazine, for examples,

Box 12.7 Volatilization of soil-incorporated pesticide.

In soil, a pesticide is distributed through the soil solids-water-air continuum. Partitioning between the soil solution and soil air is governed by Henry's Law, that is

$$VD = hC, \qquad (B12.7.1)$$

where VD is the vapour density (mass of vapour per unit volume), C is the solution concentration and h is Henry's law coefficient. The solution concentration is governed by an adsorption-desorption equilibrium with the solid phase (see below). In moist soil, as the pesticide volatilizes from the surface a diffusion gradient develops, and the upwards diffusive flux may limit the rate of volatilization for a compound with a high h value ($> 2.65 \times 10^{-5}$). However, if water is evaporating from the surface, pesticide is also delivered to the surface by mass flow, described as the 'wick effect'. For relatively soluble chemicals such as lindane, mass flow will support a volatilization rate up to five times faster than by diffusion alone. Volatilization of pesticides with a h value considerably $< 2.65 \times 10^{-5}$ is controlled not by soil movement, but by boundary layer conditions above the soil surface.

Irrespective of whether movement through the soil is by mass flow or diffusion, once the surface has become very dry (less than the equivalent of a monolayer of water molecules on the solid surfaces), a pesticide is much more strongly adsorbed and its VD (and hence volatilization)

is decreased accordingly. The escaping tendency of a pesticide from the micro-environment of moist or dry soil is called its fugacity, a concept illustrated in Fig. B12.7.1. For a moist soil, the fugacity increases as the pesticide concentration increases up to a maximum set by the saturated vapour pressure of the compound at that temperature. The graphs also show that the fugacity in dry soil is very low and changes little with an increase in pesticide concentration.

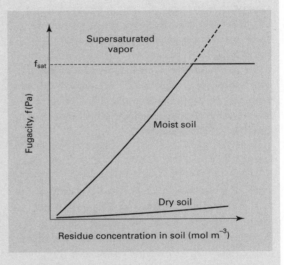

Fig. B12.7.1 The relationship between fugacity and pesticide residue concentration in moist and dry soil (after Taylor and Spencer, 1990).

protonates in acid media at one of its amino groups. This enhances its adsorption by clays.

Weak acids. Other herbicides, such as the phenoxyalkanoic acids and substituted phenols, are very weak acids. They exist as undissociated molecules or anions at normal soil pHs and hence are not appreciably adsorbed, except by non-specific adsorption or on positively charged sesquioxides. The more common representatives of this group are:

2,4-D   2,4-dichlorophenoxyacetic acid

2,4,5-T   2,4,5-trichlorophenoxyacetic acid
MCPA   (4-chloro-o-tolyl) phenoxyacetic acid

The herbicide glyphosate is an interesting example of a chemical that at normal soil pH probably exists as the zwitterion

$^{-}OOCCH_2NH_2^{+}CH_2PO_3H^{-}$.

Because the charged groups are all pH-dependent, the net charge on the molecule changes with pH, but it is strongly adsorbed over a range of pH.

Box 12.8 Sorption isotherms for pesticides.

The distribution of a pesticide between solid and liquid phases in soil can be described by a sorption isotherm. If the isotherm is linear, or approximately so, the distribution coefficient K_d is given by the equation

$$Q = K_d C, \qquad (B12.8.1)$$

where Q is the amount sorbed per unit mass of soil and C is the pesticide concentration in solution at equilibrium. K_d usually has the units of L/kg soil. Note that the larger the K_d value, the lower is the slope of the fugacity line shown in Fig. B12.7.1.

For non-ionized, non-polar molecules of low water solubility, organic matter provides the most important sorbing surface in the soil. The sorption characteristic of such chemicals is therefore better described by the equation

$$Q = K_{oc} C, \qquad (B12.8.2)$$

where the distribution coefficient K_{oc} is defined as K_d/soil organic C content (g C/kg soil). If not measured directly in soil, the relative values for K_{oc} for different pesticides can be estimated with from their octanol–water partition coefficients (K_{ow}), where

$$K_{ow} = \frac{[\text{pesticide concentration}]_{octanol}}{[\text{pesticide concentration}]_{water}}, \qquad (B12.8.3)$$

and octanol is an eight C alcohol. K_{oc} values are best used to characterize pesticide sorption when the ratio of clay content (%) to organic C (%) is < 40.

A sorption isotherm that is not linear may often be described by the non-linear Freundlich equation

$$Q = kC^n, \qquad (B12.8.4)$$

where k and n are coefficients. Note that when $n = 1$, Equation B12.8.4 is the same as Equation B12.8.1.

Cations. The bipyridyl herbicides diquat and paraquat are divalent cations that have the chemical formulas, respectively,

Because of their positive charge and size, these molecules are very strongly adsorbed. Adsorption renders paraquat inert and also induces a shift in the wavelength of maximum light absorption into the range of sunlight so that the molecule becomes unstable and is photolysed.

A quantitative approach to pesticide adsorption in soil is discussed in Box 12.8. When the mechanism of surface retention is uncertain, the term 'sorption' is frequently used to describe the process.

The combined effects of vapour pressure, water solubility and sorption affinity on the distribution of several pesticides between solids, water and air are exemplified in Table 12.10, assuming that 50% of the soil's pore space is filled with water. These data demonstrate that even for a volatile fumigant such as ethylene dibromide, the major part of the chemical is retained by the solid phase, which has important consequences for the activity and biological transformations of a pesticide, as discussed below.

Chemical and biological transformations

With the increase in pesticide potency, the quantity needed for effective pest control has decreased dramatically. For example, at the beginning of

Table 12.10 Distribution of pesticides in a soil of 50% porosity, half-filled with water. After Hartley and Graham-Bryce, 1980.

Compound	K_d (L/kg)	Air	Liquid	Solid
Ethylene dibromide	0.5	0.7	28.4	70.9
Dimethoate	0.3	1.6×10^{-7}	40	60
Simazine	1.9	1.3×10^{-7}	9.5	90.5
Monuron	2.2	2.0×10^{-7}	8.3	91.7
DDT	5.0×10^4	1.2×10^{-6}	4.0×10^{-4}	100

the 20th century inorganic pesticides, such as sodium chlorate, lead arsenate and flowers of S, were applied at rates of 10–500 kg/ha. After World War II, with the advent of organochlorines, organophosphates and selective herbicides, rates fell to 0.5–5 kg/ha. At present the synthetic pyrethroid insecticides are effective at rates of 10 g/ha or less. Thus, the concentration of pesticide in soil is very small when compared with the other reactive constituents, such as clay, organic matter etc., so that the decay of the chemical usually follows first-order kinetics (Box 12.6).

Transformations of pesticides can occur abiotically or biotically. Note that these transformations may result in a relatively harmless chemical being converted into a compound more toxic to target and non-target organisms, as well as to the breakdown of the chemical to less harmful products. The former process is called activation; the second detoxication (Alexander, 1994). However, the versatility of micro-organisms is such that eventually any pesticide will be metabolized or decomposed in some environmental niche; but the rates of decomposition can vary enormously. In this section, we are primarily concerned with transformations that lead to detoxication by decomposition:

Abiotic processes – these processes include:
• Reaction with water or hydrolysis (e.g. the organophosphate and organophosphorothioate esters, carbamates, amides and anilides – mainly by alkaline hydrolysis);
• reduction under anoxic conditions, especially in sediments (but difficult to distinguish from biotic reduction); and
• photolysis through exposure to sunlight (UV and visible radiation), most commonly in water. Some examples are paraquat (in the adsorbed state), trifluralin (a dintroaniline herbicide) and the pyrethroids. The natural pyrethroids and the early synthetic compounds decompose within a few hours of exposure to sunlight. On the other hand, later synthetic compounds, such as deltamethrin, are not only photostable but are also potent at very low concentrations. The stable synthetic pyrethroids are now being used against many pests previously controlled by dichlorodiphenyltrichloroethane (DDT) because of their potency and low environmental impact.

Biotic processes – the major process of pesticide decomposition in soil is through microbial activity. This is true of all groups of pesticides with the notable exception of DDT and the cyclodienes, such as dieldrin and lindane, which are particularly recalcitrant. The processes involved are:
• biodegradation, in which the pesticide is a substrate for microbial growth;
• co-metabolism, in which the pesticide is changed by micro-organisms but does not serve as a direct energy source for their growth;
• polymerization, in which pesticide molecules are linked with other molecules either of pesticide or naturally occurring organic molecules;
• accumulation, in which the pesticide is incorporated into micro-organisms; and
• secondary effects in which the pesticide is changed because of changes in pH, redox potential or reactive surfaces brought about as a result of microbial activity.

These processes are discussed further by Bollag and Liu (1990). Sorption usually makes a chemical physiologically inactive, although if sorption is reversible, the chemical's removal from solution by decomposition will cause more to be desorbed and become active. The long persistence of the bipyridyls (paraquat) is due to their irreversible sorption by clays when used at normal rates. But sorption of a chemical, especially on organic surfaces, may place it in closer proximity to colonies of bacteria and extracellular enzymes and hasten its decomposition. Therefore, no sound generalization on the effect of sorption on the rate of biotic decomposition of pesticides in soil can be made.

Enrichment and bioremediation – biodegradation of the phenoxyalkanoic acid herbicides illustrates a phenomenon also observed with the phenylureas, organophosphates and carbamate insecticides when they are added to a soil not previously exposed to the compound in question. An initial lag phase during which the herbicide concentration remains constant is followed by a period of rapid disappearance (curve A, Fig. 12.9). Soil micro-organisms, principally bacteria that are capable of biodegrading the foreign molecule, multiply during the lag phase. These 'adapted' organisms then enjoy a competitive advantage for substrate over non-adapted organisms, resulting in their proliferation and rapid breakdown of

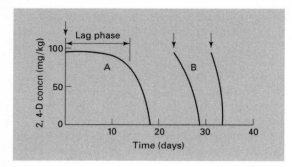

Fig. 12.9 Time course of microbial decomposition of 2,4-D is soil. Curve A – initial enrichment, curves B – addition of herbicide to enriched soil; arrows indicate time of addition (after Audus, 1960).

the herbicide. The soil is said to be enriched with adapted organisms, a condition persisting for several months in the absence of fresh herbicide. Herbicide reapplied to an enriched soil is detoxified without any lag phase (curve B, Fig. 12.9). The enhanced biodegradation of pesticides due to enrichment effects is common and can lead to an apparent decrease in the effectiveness of a pesticide in controlling its target organisms.

The phenomenon of enrichment can be used to advantage to accelerate the biodegradation of pesticides (and other undesirable organic chemicals) that are serious contaminants in soil, waste disposal sites and groundwater. This process is an example of bioremediation where the source of contamination is treated *in situ* by stimulating the growth of organisms that can decompose the contaminants.

Absorption and transport of pesticides

Uptake of pesticides by plant roots depends on their water solubility, whereas absorption through the cuticle of leaves depends more on their lipid solubility. Transport in the liquid phase mainly occurs by mass flow, and therefore depends on the chemical's water solubility and the rate of water flow. Herbicides, such as simazine, picloram, the phenylureas (monuron, fenuron) and phenoxyalkanoics, can be transported in surface runoff or leached. Transport by surface runoff is potentially more serious than leaching in its off-site effects because compounds that pass through the soil are more likely to be biodegraded. Strongly adsorbed pesticides (DDT, paraquat) are transported when soil particles are removed during erosion. As a result, residues of these persistent chemicals build up in stream and lake sediments. However, there is evidence that DDT is decomposed more rapidly to the less toxic DDD under anaerobic conditions, in the presence of plant residues, so that its disappearance from sediments may be accelerated.

A diagrammatic summary of modes of pesticide input, reactions in soil and pathways of loss to the environment is shown in Fig. 12.10.

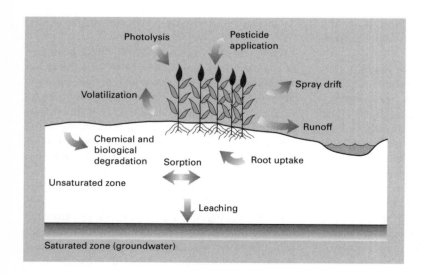

Fig. 12.10 Pesticide inputs, reaction in soil and pathways of loss (after Severn and Ballard, 1990).

12.6 Summary

Substances recognized as fertilizers and plant protection chemicals (pesticides) have been used for centuries. But in the last 60 years or so there has been a phenomenal increase in the quantity and variety of agrichemicals used to meet the need for greater production and improved quality of the crops harvested.

The main nutrients supplied are N, P, K and S either singly in straight fertilizers or as combinations in compound or multinutrient fertilizers. Most of the simple compounds of N – NH_4NO_3, KNO_3, $Ca(NO_3)_2$, $(NH_4)_2SO_4$ and $(NH_2)_2CO$ – are soluble solids, but N is also used as solutions of NH_4NO_3, urea, or NH_3 dissolved in water and as liquefied NH_3 gas. The N of these compounds is immediately available to plants, but is vulnerable to leaching and denitrification in the NO_3^- form. Provided the amount and timing of N-fertilizer input are matched to the crop's requirements, little of the fertilizer remains in a mineral form in the soil by harvest time. But repeated high N inputs as fertilizer or in manures will build up a pool of labile soil organic N that can mineralize rapidly. Thus, leaching losses of NO_3^- from intensively farmed land can range up to 250 kg N/ha/yr, depending on the rate of fertilizer (and/or manure) applied, the land use and the excess winter rainfall. Loss by leaching and denitrification can be decreased by a nitrification inhibitor (e.g. N-serve) applied with a soluble N fertilizer, or by the use of slow-release or controlled-release N fertilizers.

Naturally occurring phosphate rocks (PRs) are insoluble in water and for most purposes must be treated with strong acids (H_2SO_4 or H_3PO_4) to convert most of the P to a soluble form. An exception is the PRs consisting of francolite minerals (the reactive phosphate rocks, RPRs), which may be used for direct application to perennial pastures and plantation crops on soils of pH < 5.5 (in water) in humid climates. The RPRs are an example of slow-release P fertilizers. Monocalcium phosphate $Ca(H_2PO_4)_2 \cdot H_2O$ (MCP) is the active constituent of single and triple superphosphates (SSP and TSP, respectively), whereas higher analysis P fertilizers (the polyphosphates) are made from superphosphoric acid, formed by decreasing the amount of water in which P_2O_5 is dissolved. Soluble P fertilizers react quickly with the soil to form less soluble fertilizer reaction products on which plants feed. These fertilizer reaction products can have a long residual effect. P lost from agricultural systems can have a major effect on the eutrophication of surface waters and the outbreak of undesirable blue-green algal blooms.

The main K compounds KCl and K_2SO_4 are very soluble, but provided the soil has an appreciable CEC, K is retained as the exchangeable cation K^+ and leaching losses are not large. Prior to the 1980s, S inputs to the atmosphere from the combustion of fossil fuels in industrialized countries were large and subsequent deposition on the land was sufficient to satisfy most crop requirements. With subsequent emission controls, however, deposition from the atmosphere have declined dramatically and S deficiency has become more widespread in these countries, necessitating the widespread use of S fertilizer. Atmospheric S inputs have never been large in undeveloped regions of the tropics and subtropics, so that where SO_4^{2-} leaching occurs S deficiency is common. Sulphur is applied as elemental S or gypsum $CaSO_4 \cdot 2H_2O$, or in $(NH_4)_2SO_4$ and SSP. The micronutrients Fe, Mn, Zn, Cu, B, Mo and Co are applied at low concentrations in mixed fertilizers, or individually as chelates or soluble salts.

Ideally, a pesticide should be lethal for the target organism and remain active only long enough to control that pest. Pesticides may be lost from soil and plant surfaces by volatilization, decomposed by photolysis in sunlight, or inactivated in soil by sorption on clays or organic matter, or by chemical and biological transformations. The strength of sorption depends on the properties of the pesticide molecule and the nature of the adsorbent. Transformations can be abiotic or biotic. Biotic processes may cause chemicals to be converted to more active (and toxic) compounds – a process of activation; or cause the chemical to be rendered less active and toxic – a process of detoxication. The most important detoxication process in soil is decomposition when a pesticide becomes the substrate for microbial growth. Repeated use of a pesticide can result in soil becoming enriched with micro-organisms that rapidly decompose the pesticide. Although such an effect may decrease the pesticide's potency in controlling pests, it is valuable in the process of bioremediation for decontaminating sites *in situ*.

The relative persistence of pesticides can be compared by means of their half-life, which is the time taken for the pesticide's concentration to fall to half its original value at the site of application. In assessing environmental impacts, persistence is also expressed as the time taken for 90% of the pesticide to disappear. Of the more stable compounds, the organochlorine insecticides have become widely disseminated in the environment due to their volatilization from the soil and accumulation in animal fat tissues. More soluble compounds, such as the phenoxyalkanoic, phenylurea and some triazine herbicides, may be leached, but fortunately most of them are decomposed rapidly by soil micro-organisms.

References

Addiscott T. M., Whitmore A. P. & Powlson D. S. (1991) *Farming, Fertilizers and the Nitrate Problem.* CAB International, Wallingford, UK.

Alexander M. (1994) *Biodegradation and Bioremediation.* Academic Press, San Diego.

Audus L. J. (1960) Microbiological breakdown of herbicides in soils, in *Herbicides and the Soil* (Eds. E. K. Woodford & G. R. Sagar). Blackwell Scientific Publications. Oxford, pp. 1–18.

Bolan N. S., White R. E. & Hedley M. J. (1990) A review of the use of phosphate rocks as fertilisers for direct application in Australia and New Zealand. *Australian Journal of Experimental Agriculture* 30, 297–313.

Bollag J. M. & Liu S. Y. (1990) Biological transformation processes of pesticides, in *Pesticides in the Soil Environment: Processes, Impacts and Modeling* (Ed. H. H. Cheng), Soil Science Society of America Book Series No. 2. Soil Science Society of America Inc, Madison, Wisconsin, pp. 169–211.

Croll B. T. (1994) Nitrate – best agricultural practice for water – the UK experience. *Proceedings of the Fertiliser Society* No. 359. The Fertiliser Society, London.

Cooke G. W. (1969) Prediction of nitrogen requirements of arable crops in mainly arable cropping systems, in *Nitrogen and Soil Organic Matter.* MAFF Technical Bulletin No. 15. Her Majesty's Stationery Office, London, pp. 40–60.

Glasser R. F. (1976) Pesticides: the legal environment, in *Pesticides and Human Welfare* (Eds D. L. Gunn & J. G. R. Stevens). Oxford University Press, Oxford.

Guenzi W. D. (Ed.) (1974) *Pesticides in Soil and Water.* Soil Science Society of America Inc, Madison, Wisconsin.

Hartley G. S. & Graham-Bryce I. J. (1980) *Physical Principles of Pesticide Behaviour.* Academic Press, London.

Huffman E. O. (1962) Reactions of phosphate in soils: recent research by TVA. *Proceedings of the Fertiliser Society.* The Fertiliser Society, London, pp. 5–35.

International Fertilizer Association (2002) *Fertilizer Indicators,* 2nd edn. International Fertilizer Industry Association, Paris.

Jones U. S. (1982) *Fertilizers and Soil Fertility,* 2nd edn. Prentice Hall, Virginia.

Karama A. S. & Willett I. (1998) Agronomic effectiveness of phosphate rock materials in acid upland soils in Asia, in *Nutrient Management for Sustainable Crop Production in Asia* (Eds A. E. Johnston & J. K. Syers). CAB International, Wallingford, UK, pp. 376–7.

Laegreid M., Bockman O. C. & Kaarstad O. (1999) *Agriculture, Fertilizers and the Environment.* CAB International, Wallingford, UK.

Larsen S. (1971) Residual phosphate in soils, in *Residual Value of Applied Nutrients.* MAFF Technical Bulletin No. 20. Her Majesty's Stationery Office, London, pp. 34–40.

Leinweber P., Turner B. L. & Meissner R. (2002) Phosphorus, in *Agriculture, Hydrology and Water Quality* (Eds P. M. Haygarth & S. C. Jarvis). CAB International, Wallingford, UK, pp. 29–55.

Powlson D. S. (1997) *Integrating agricultural nutrient management with environmental objectives – current state and future prospects.* The Fertiliser Society, York, UK.

Powlson D. S., Hart P. B. S., Poulton P. R., Johnstone A. E. & Jenkinson D. S. (1992) The influence of soil type, crop management and weather on the recovery of ^{15}N-labelled fertilizer applied to winter wheat in spring. *Journal of Agricultural Science, Cambridge* 118, 83–100.

Severn D. J. & Ballard G. (1990) Risk/benefit and regulations, in *Pesticides in the Soil Environment: Processes, Impacts and Modeling* (Ed. H. H. Cheng). Soil Science Society of America Book Series No. 2. Soil Science Society of America Inc, Madison, WI, pp. 467–91.

Shaviv A. (2001) Advances in controlled-release fertilizers. *Advances in Agronomy* 71, 1–49.

Slack A. V. (1967) *Chemistry and Technology of Fertilizers.* Wiley Interscience, New York.

Stewart B. A., Woolhiser D. A., Wischmeier W. H., Caro J. H. & Frere M. H. (1970) *Control of water pollution from cropland,* Volume 1. Agricultural Research Service and Environmental Protection Agency, Washington DC.

Suter H. C., White R. E., Heng L. K. & Douglas L. A. (2002) Sorption and degradation of phosmet in two contrasting Australian soils. *Journal of Environmental Quality* **31**, 1630–35.

Taylor A. W. & Spencer W. F. (1990) Volatilization and vapour transport processes, in *Pesticides in the Soil Environment: Processes, Impacts and Modeling* (Ed. H. H. Cheng). Soil Science Society of America Book Series No. 2. Soil Science Society of America Inc, Madison, WI, pp. 213–69.

Tisdale S. L. & Nelson W. L. (1975) *Soil Fertility and Fertilizers*, 3rd edn. Macmillan, London.

White R. E. & Sharpley A. N. (1996) The fate of non-metal contaminants in the soil environment, in *Contaminants and the Soil Environment in the Australasia-Pacific Region* (Eds R. Naidu, R. S. Kookana, D. P. Oliver, S. Rogers & M. J. McLaughlin). Kluwer, Dordrecht, pp. 29–67.

Further reading

Bacon P. E. (Ed.) (1995) *Nitrogen Fertilization in the Environment*. Marcel Dekker, New York.

Havlin J. L., Beaton J. D., Tisdale S. L. & Nelson W. L. (1999) *Soil Fertility and Fertilizers*, 6th edn. Prentice-Hall, New Jersey.

Linn D. M., Carski T. H., Brusseau M. L. and Chang F. H. (Eds) (1993) *Sorption and degradation of pesticides and organic chemicals in soil*. Special Publication No. 32. American Society of Agronomy and Soil Science Society of America, Madison, WI.

Ministry of Agriculture, Fisheries and Food (1994) *Fertiliser Recommendations for Agricultural and Horticultural Crops*. Reference Book 209. Her Majesty's Stationery Office, London.

Example questions and problems

1 A compound fertilizer has a nutrient content of $10–12–16$ ($N : P_2O_5 : K_2O$). For an application rate of 75 kg/ha, calculate how much N, P and K (kg/ha) are supplied. (The atomic masses of N, P, O and K are 14, 31, 16 and 39 g, respectively).

2 (a) Name two strategies that can be used to slow the rate of dissolution of N fertilizers in soil.

   (b) What treatment can be applied with an NH_4^- N based fertilizer to slow its rate of nitrification in soil?

   (c) Name two processes by which fertilizer NO_3^- is lost from soil (other than by plant and microbial uptake).

3 (a) Give the NO_3-N concentration of drinking water considered safe for very young children in (i) the USA, and (ii) the EU.

   (b) Give the critical total P concentration considered to predispose to algal blooms in water bodies in Australia.

4 (a) Name two commonly used inorganic fertilizers that contain both soluble N and P.

   (b) What natural form of phosphate is used for the manufacture of superphosphate?

   (c) Name an organic form of P commonly used as a fertilizer.

5 (a) Which is the more soluble compound – KCl or K_2SO_4?

   (b) Name a slow-release K fertilizer.

   (c) In which form is S fertilizer taken up by plants?

   (d) Other than fertilizer, what is a major input of S to soil in industrialized parts of the world?

6 As a horticultural consultant, you have been asked to provide a fertilizer strategy for a large rose garden. Soil testing has shown the soil pH is 5.5 (in water), exchangeable K is adequate, but N and P are deficient. You decide on a strategy that will supply the roses with 40 and 4 g of N and P, respectively, per m^2 per year.

   (a) If the N is provided as NH_4NO_3 (34% N), what would be the annual use of that fertilizer for a garden of 500 m^2?

   (b) If NH_4NO_3 is used, and given that a rose is a perennial plant, what form of P fertilizer could be used?

   (c) Could the desired amounts of N and P be supplied by either MAP or DAP alone?

7 (a) Name the main physicochemical processes by which a pesticide can be inactivated at a soil surface.

   (b) What is the main process by which lindane is lost from a soil surface?

   (c) What is the basis for bioremediation of pesticide-contaminated soil?

8 (a) Insecticides A and B are applied to a soil at the rate of 5 g/ha: their K_d values are

1000 and 40 L/kg soil, respectively. Which of these two insecticides will maintain a higher concentration in the soil solution?

(b) Define the half-life of a pesticide in soil.

(c) A spray containing 10 g/L of atrazine herbicide is applied at the rate of 5 L/ha prior to sowing a crop. The half-life of atrazine in this soil is 40 days. Calculate how much herbicide (in g/ha) will remain in the soil after 60 days. (Hint – use the equation $C(t) = C(0) \exp(-kt)$, where $C(0)$ is the amount of herbicide at time zero.)

Chapter 13

Problem Soils

13.1 A broad perspective

Nutrient deficiencies, soil acidity, structural instability or a soil's susceptibility to erosion can affect the growth of natural vegetation and agricultural crops. Other conditions that can severely limit the productivity of agricultural land are a high concentration of soluble salts (salinity) and excess soil water (waterlogging).

As discussed in Sections 9.2 and 9.5, an excess of precipitation over evaporation for several months each year, impermeable subsurface layers and high groundwater tables, separately or combined, induce soil waterlogging and the attendant problems of:
• Inadequate aeration for root growth and soil microbial activity;
• poor trafficability of the soil for machinery and animals, with the danger of surface structural collapse and 'poaching'*;
• invasion by flood-tolerant weeds (e.g. sedges and rushes) and an increase in animal parasites and diseases favoured by wet conditions; and
• slow warming of the soil in spring.
The causal climatic factors are beyond control, but an improvement in soil drainage (Section 13.4) does much to mitigate these ill effects.

On the other hand, in areas of high evaporation, serious problems occur when the watertable rises to within 2 m of the soil surface and salinization occurs due to the upward movement and evaporation of saline groundwater (Sections 6.5 and 9.6). Such a situation may arise when the steady-state equilibrium of a soil's hydrology is disturbed by natural events, such as earth movements (faulting) or long-term climatic change (usually over a time span of centuries), or by man's intervention through changing the land use or supplying irrigation (usually a time span of decades). Problems associated with hydrological disturbances, and their solutions, fall into two main categories:
• Soils not initially saline, with deep watertables, which become saline under irrigation or due to a change in land use; and
• soils already salinized and of limited cropping potential, but requiring careful management to avoid chemical and physical deterioration under irrigation.

13.2 Water management for salinity control

Permeation of saline groundwater

Soil salinization is a natural process, as demonstrated by the existence of salt pans and salt lakes in the inland areas of large land masses, where regional drainage collects naturally in low-lying parts of the landscape and salts accumulate following the evaporation of water. For example, in the southwest of Western Australia, many of the valley bottoms were naturally saline and supported a scrubby heath vegetation tolerant of high salinity. Groundwater levels were kept low under the valley slopes and hill tops by deep-rooted indigenous Eucalyptus trees (Fig. 13.1a). But following forest clearance and the planting of short-rooted pasture and crop species, the average annual rate of deep percolation through the soil has increased from

* Rutting and damage to a wet soil surface by the hooves of animals; sometimes called 'pugging'.

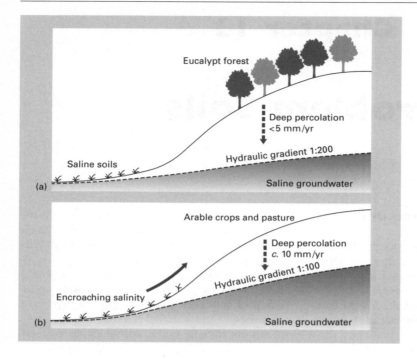

Fig. 13.1 (a) Natural landscape in southwest Western Australia in steady-state hydrologic equilibrium. (b) Same landscape with raised groundwater table after forest clearance (after Holmes, 1971).

< 5 to > 10 mm, depending on the rainfall and underlying regolith. With the increase in hydraulic gradient in the groundwater, more water has flowed to the valley bottoms causing a rise in the regional watertable and gradually the area of saline soils has crept upslope (Fig. 13.1b). This sequence of events has also occurred in extensive areas of southeastern Australia north and west of the Great Dividing Range in Victoria and New South Wales, respectively, over the past 150 years. These are examples of dryland salinization (Box 13.1).

Land affected by dryland salinity can be recognized by the type of vegetation and the appearance of bare patches, sometimes with salt crystals evident on the soil surface (Fig. 13.2).

Fig. 13.2 An area of dryland salinity in southwest Western Australia. Note the bare ground and salt-tolerant vegetation in the foreground (see also Plate 13.2).

Box 13.1 Dryland salinity in Australia.

Although dryland salinity occurs naturally in Australia, clearing land for agriculture has caused existing areas to expand and salinity to appear in areas previously unaffected. Induced dryland salinity (sometimes called secondary salinization) occurs as:

• *Salt scalds* in arid and semi-arid regions, where surface vegetative cover is lost, exposing saline subsoils that are relatively impermeable; or

• *saline seepages* – typically caused by the sequence of events illustrated in Fig. 13.1.

Hill tops and upper slopes comprise the recharge areas for groundwater. Groundwater may be naturally saline, or it becomes saline as the watertable rises and salts stored in the regolith are dissolved and mobilized. Where groundwater emerges at low points in the landscape (discharge areas), dryland salting occurs. Streams draining the land also become more saline.

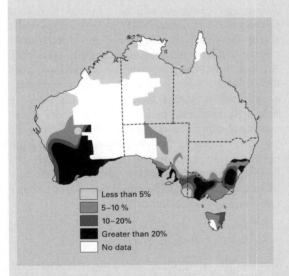

Fig. B13.1.1 Proportion of Australian farms with a salinity problem.

Legend:
- Less than 5%
- 5–10 %
- 10–20%
- Greater than 20%
- No data

Dryland salinity can be prevented (or at least minimized) by controlling recharge, through the retention of native trees or the establishment of plantation forests in recharge areas. Establishing 'break-of-slope' tree plantations can also be used to intercept subsurface lateral flow of water down slope wherever the soil is relatively shallow. However, where agriculture is already established, control and amelioration of dryland salinity can be attempted as follows:

• Replacement of short-rooted annual pastures and crops with deep-rooted perennials such as the grasses *Phalaris aquatica* and *Pennisetum clandestinum*, or lucerne (*Medicago sativa*). The choice of species depends on the climate and soil type, and whether grazing can be managed in a way that maintains the perennials;

• groundwater pumping (where aquifers are reasonably permeable), or surface drainage works to minimize accessions to groundwater. Groundwater pumping is expensive and only feasible in areas of high-value horticultural crops;

• the planting of salt-tolerant perennials such as tall wheat grass (*Lophopyrum elongatum*) and hill wallaby grass (*Danthonia eriantha*) can be combined with surface drainage to reclaim land that is salinized. Where the salt concentration is very high (approaching that of sea water), salt-loving plants such as saltbush (*Atriplex* spp) or salt grass (*Distichlis spicata*) can be grown, in which case the land is not being 'reclaimed' but is put to productive use in a saline state.

In 2001, the area affected by, or at risk from dryland salinity in Australia was estimated as 5.7 million ha, and predicted to increase to 17 million ha by 2050 (National Land and Water Audit, 2001). An indication of the distribution of dryland salinity on Australian farms is given in Fig. B13.1.1.

Soils under irrigation

Regulating the water supply

Some soils, notably those of the Nile Valley in north Africa, have been irrigated for centuries without adverse effects. During the annual Nile flood, water was led on to the land and impounded in basins of 400 to 16,000 ha for up to 40 days, during which time the soil water storage was replenished. Following that, any remaining water was released and crops were sown. The soils were

Fig. 13.3 An asphalt-lined
irrigation channel in Victoria.

naturally well-drained and the annual flooding
leached out any accumulated salts. More recently,
in order that irrigated crops could be grown
throughout the year, dams and barrages have been
built to regulate the river's flow. Productivity has
increased, but not without cost, because with the
application of 1.5–2 m of water annually, most
of which is lost by evapotranspiration, there has
been an inevitable build-up of salts in the soils.

Abundant and cheap supplies of water encourage farmers to apply more water than a crop can
use, as insurance against the risk of any water
stress that might reduce yield. The excess water
drains to the groundwater, which slowly rises.
Another problem is seepage loss from canals and
ditches that conduct water to the fields, which
is especially serious in Pakistan where extensive
waterlogging of the permeable soils of the Indus
Valley has occurred alongside unlined irrigation
canals. A similar problem occurs in Australia,
where water for agriculture and domestic use flows
northwards to the semi-arid Wimmera-Mallee District through thousands of kilometres of unlined
irrigation channels. Losses by evaporation and
seepage exceed 90% of the water leaving sources
in the Grampian Mountains to the south. Seepage
is prevented by lining canals with concrete – the
best, but most expensive material. Asphalt or plastic sheeting are cheaper options, but of limited
durability (Fig. 13.3).

Irrigation scheduling

Water application should be adjusted according
to the soil water deficit (SWD) that develops under
a crop (Section 6.4). The loss of soil water can be
measured directly using a neutron probe (Box 6.1).
Alternatively, the loss can be calculated from
measured soil water potentials (using tensiometers
or gypsum resistance blocks, as in discussed in
Box 6.4) and a knowledge of the soil's water
retention curve (Section 6.4). More appropriate
for large areas is the prediction of SWD from
evapotranspiration losses calculated from the
Penman–Monteith equation (Box 6.9). The prediction of SWDs using a simplified water balance
calculation is discussed in Box 13.2.

Irrigation methods

The method of applying water also influences the
likelihood of soil salinization. Control of infiltration and deep percolation is essential to prevent
watertables rising above the critical depth of 2 m
from the soil surface. Nearly level (< 4% slope)
or uniformly graded land can be irrigated by flood
or furrow methods as in the following examples:

• *Border check* or *border strip* flooding, in
which water is distributed along bays separated
by levees or mounds running parallel to the direction of flow (Fig. 13.4). Ideally, the bays should

Box 13.2 Prediction of SWD and irrigation scheduling.

As discussed in Section 6.5, evapotranspiration from crop land proceeds at the potential rate (E_{to}), provided that ground cover is complete and the crop is not limited for water. Under other conditions, the actual rate (E_{ta}) falls below the potential rate. The ratio E_{ta}/E_{to} for changes in canopy development for different crops is expressed as a crop coefficient C_c. The value of E_{ta} (mm/day) is given by

$$E_{ta} = E_{to} \times C_c. \qquad (B13.2.1)$$

If the soil was brought to field capacity at its last watering, the predicted SWD that develops is given by

$$SWD = \{E_{ta} \text{ (mm/day)} \times \text{days since last watering}\} - \text{rainfall (mm).} \qquad (B13.2.2)$$

Given that available water capacities (AWC) normally range from 80 mm/m for sandy soils to 220 mm/m for silty clays and clay loams, an E_{ta} of 6 mm/day in summer would remove about half the available water in 1 m of soil in 7–18 days, depending on the soil's texture. When this point is reached, re-irrigation is recommended to avoid a check to crop growth. If the crop roots grow deeper than 1 m, plant available water (PAW) will be greater (as shown in Table 6.3).

be constructed after the land has been levelled by laser-guided earth moving equipment. Sufficient depth of water (> 50 mm) must be applied at each irrigation to ensure that water reaches the bottom end of each bay. To avoid over-watering at the input end, bays should be shorter in permeable soils (< 90 m) than in less permeable clays (up to 270 m). On the lowest gradients (< 1%), construction of levees across the slope at vertical intervals of approximately 0.05 m to form rectangular basins gives more uniform water application, and therefore better control of infiltration. However, this method of *basin flooding* is more labour-intensive and usually practical only on highly productive orchards and paddy rice crops.

• *Furrow irrigation* is suited to row crops and can be used on land too steep or uneven to be flooded, provided that the furrows follow the contours. But there is the problem of salt redistribution in the soil as water moves by capillarity from the wet furrows to evaporate from the ridges where the plants are growing. The salt concentration at the apex of the ridge can be 5–10 times greater than in the body of the soil, creating an unfavourable environment for seed germination and seedling growth. This problem can be avoided by sowing seeds in double rows on broad sloping ridges (Fig. 13.5a), or in single rows on the longer slope of asymmetric ridges (Fig. 13.5b).

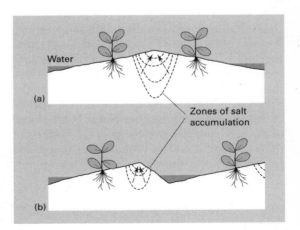

Fig. 13.5 Ridge and furrow designs to avoid salinity damage to plants. (a) Paired crop rows on broadly sloping ridges. (b) Single crop rows on asymmetric ridges.

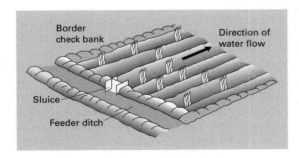

Fig. 13.4 Border check flood irrigation.

Sprinklers, sprays or trickle emitters are suitable for irregular terrain and more uniform infiltration of water than is achievable by flood or furrow methods, as in the following examples:

• *Spray* or *sprinkler* irrigation allows greater flexibility in the type of terrain and depth of water applied than either flood or furrow methods, but it requires more costly equipment. The water is delivered through upright nozzles fitted to pipes that are fixed or movable. Of the latter, self-propelled water guns and 'centre-pivot' spray systems permit a uniform distribution of water and hence improved efficiency of application, which is important for keeping the leaching fraction under irrigation to the minimum necessary to control soil salinity (see below). One drawback of spray irrigation is the evaporative loss, especially on windy days, which reduces the efficiency of application and may cause leaf scorch on sensitive crops.

• *Trickle* or *drip* irrigation is an adaptation of spray irrigation that minimizes evaporative loss and is especially suited to high-value orchard and row crops on soils of low AWC. Normally, water is delivered through flexible tubing with flow-regulating 'drippers' that are placed above ground close to the plants; but in vineyards and some vegetable crops, the drip lines may be placed below ground (subsurface drip irrigation). Salts in the soil are leached to the periphery of the wetted zone, the shape of which depends on the soil's permeability and the rate of water application (Fig. 13.6). The irrigation rate can be adjusted to create a volume of low-salinity soil large enough for most of the roots to function without ill effect. Because the soil water potential in the root zone remains high, water of greater salinity (> 3 dS/m) that could not be used for spray or sprinkler irrigation is often acceptable for drip irrigation. However, salt accumulation towards the fringe of the wet zone necessitates periodic flushing of the whole soil if salinity problems for subsequent crops are to be avoided.

Quality of irrigation water

Kinds of salts and their concentration

The major ions in surface and underground waters used for irrigation are Ca^{2+}, Mg^{2+}, Na^+,

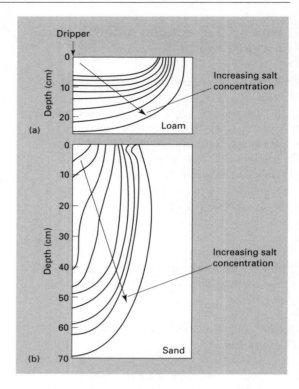

Fig. 13.6 The distribution of salt concentration during drip irrigation in soils of different texture (a) loam – input 4 L/h; (b) sand – input 4 L/h (after Bresler, 1975).

Cl^-, SO_4^{2-} and HCO_3^- with low concentrations of K^+ and NO_3^-. The total concentration of dissolved salts (TDS), sometimes referred to as the 'salinity hazard', is most conveniently measured by the specific conductance, which is the electrical conductivity (EC) of the water, independent of the sample size. Analysis of many surface and well waters has revealed the approximate linear relationship

$$TDS \ (mg/L) \cong 640 \ EC \ (dS/m). \qquad (13.1)$$

Note that EC normally has the units of dS/m (SI units), which are numerically equivalent to the older units of mmho/cm. In Australia EC is commonly quoted in 'EC units', which are μS/cm (dS/m × 1000). Equation 13.1 holds for waters of EC up to 10 dS/m. EC increases with temperature so the value should be corrected to a standard temperature of 25°C.

Experience in the western USA (Richards, 1954) suggested that EC class limits of < 0.25, 0.25–0.75, 0.75–2.25 and > 2.25 dS/m at 25°C could be used to define waters of low, medium, high and very high salinity. The EC of most irrigation waters lies between 0.15 and 1.5 dS/m, which is roughly equivalent to 1.5–15 mmol anion (−) or cation (+) charge per litre* (moles of charge and equivalents are discussed in Box 2.3). In the soil, this water is concentrated 2–20 times due to evapotranspiration, at which level many crops experience a reduction in yield through osmotic stress. The osmotic potential ψ_s of the soil solution, which is a measure of the osmotic stress a plant might experience, can be calculated from the approximate formula

$$\psi_s \text{ (kPa)} = -36 \text{ EC (dS/m)}. \qquad (13.2)$$

Measurement of salinity in the soil solution and bulk soil is discussed in Box 13.3.

Sodium hazard

Apart from total salinity, the relative proportion of Na to Ca and Mg has an important effect on the quality of irrigation water. The Gapon Equation 7.10, introduced in Section 7.2, shows that the proportion of exchangeable Na^+ to exchangeable Ca^{2+} and Mg^{2+} ions is determined by the Gapon coefficient $K_{Na\text{-}Ca,Mg}$ and the molar concentration ratio $(Na^+) : (Ca^{2+} + Mg^{2+})^{1/2}$ in the soil solution. In practice, when concentrations are expressed in mmols charge/L we obtain the sodium adsorption ratio (SAR), written as

$$SAR = \frac{[Na^+]}{\left[\dfrac{Ca^{2+} + Mg^{2+}}{2}\right]^{1/2}}. \qquad (13.3)$$

In this form the SAR is approximately equal to the soil's exchangeable Na percentage (ESP) and is an index of the sodium hazard or sodicity of the water. When the cations are expressed in cmol charge/kg, the ESP is calculated as

---

* The units of mmol charge (+) or (−) per litre are equivalent to the older units of milliequivalents per litre (meq/L).

$$ESP = \frac{100 \, Na^+}{\sum \left(Ca^{2+}, Mg^{2+}, K^+, Na^+, Al^{3+}\right)}. \qquad (13.4)$$

In the USA and several other countries, an ESP $\geq 15\%$ (or SAR $\cong 15$ mmol charge$^{1/2}$/L$^{1/2}$) has generally been accepted as the criterion of a sodic soil, which is a generic term for soils that exhibit poor physical properties due to the influence of Na^+ ions (Section 13.3). In Australia, the critical ESP for a sodic soil is $\geq 6\%$.

Prediction of a soil's ESP is crucial for irrigation management because swelling pressures are increased and clay deflocculation more likely as the ESP rises, especially if the total salt concentration of the soil solution is low. To predict the likely soil ESP under irrigation, knowledge of the SAR of the irrigation water is essential. As irrigation water is concentrated in the soil by evapotranspiration, assuming insoluble salts are not precipitated, the concentrations of all ions in solution increase in the same proportion, but the SAR increases disproportionately. For example, a twofold increase in the ion concentrations causes the SAR to increase by $2/\sqrt{2} = 1.4$ (Equation 13.3).

Box 13.3 Measurement of soil salinity.

Traditionally, soil salinity has been measured by the EC of the soil solution. Sufficient deionized water is mixed with an air-dry soil sample to make a glistening paste and a saturation extract is then obtained by filtering under suction or centrifuging. The EC of the saturation extract (EC_e) is measured with a conductivity meter. The EC_e value is 2–3 times less than the true EC of the soil solution before dilution with water. In the USA and many other countries, the distinction between saline and non-saline soils is drawn at an EC_e of 4 dS/m (Section 9.6). Very sensitive crops can be adversely affected at an $EC_e < 2$ dS/m.

In Australia, EC is measured in a 1 : 5 soil : water suspension. After mixing air-dry soil and deionized water, the suspension is allowed to settle before the electrodes of the conductivity meter are inserted into the supernatant liquid. $EC_{1:5}$ is less

Box 13.3 *continued*

Fig. B13.3.1 An example of a portable EM38 meter.

than EC_e for the same soil because of the dilution effect (between 5 and 13 times less for a soil texture ranging from heavy clay to loamy sand). However, any solid salts present (e.g. gypsum) will dissolve and increase the EC value, so this method is of little value in such soils. An approximate value for the soil's TDS (%) can be obtained by multiplying $EC_{1:5}$ (dS/m at 25°C) by 0.34.

Several field methods have been developed to measure the bulk soil EC (solid and solution phases combined) using a soil's response to pulses of electromagnetic radiation. The technique of time domain reflectometry (TDR) (see Box 6.1) for measuring soil water content *in situ* can be adapted in saline soils (at constant water content) to measure bulk soil EC. Attenuation in the amplitude of an electromagnetic pulse transmitted down waveguides in the soil is related to the EC of the surrounding soil. A similar principle is used in portable EM meters (e.g. EM31 and EM38) (Fig. B13.3.1). A transmitter in the meter generates 'loops' of current in the soil, the size of which is directly proportional to the EC. Secondary electromagnetic fields set up by the current flows are intercepted by a receiver in the instrument and the resultant signal transformed into a voltage that is proportional to the bulk soil EC. Depending on whether an EM38 is placed horizontally or vertically on the soil, the effective depth of measurement is approximately 1 or 2 m, respectively. These depth-weighted EC values indicate the relative salinity of soil profiles, provided texture and water content are reasonably constant. For estimates of absolute salinity content, the instrument should be calibrated for a given soil type using depth-averaged EC_e measurements.

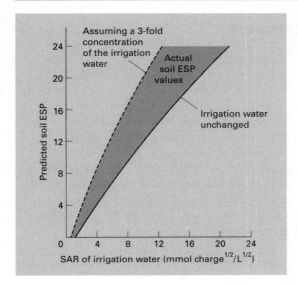

Fig. 13.7 Relationship between surface soil ESP and SAR of irrigation water (after Richards, 1954).

Exchange then occurs between Na^+ ions in solution and Ca^{2+} and Mg^{2+} ions on the colloid surfaces so that a new equilibrium is attained at a higher ESP value. Because the quantity of exchangeable cations in a given soil volume is usually much greater than the quantity of cations in solution, the changes in ESP are small when compared with the change in SAR of the irrigation water as it evaporates. Nevertheless, continued use of water of high SAR eventually leads to a greater ESP. This difference in the cation buffering capacities of the solution and solid phases explains why, for a given SAR, waters of high salinity are considered to be of greater sodium hazard than waters of low salinity.

In Fig. 13.7, the solid line represents the relationship between predicted ESP and the SAR of a number of irrigation waters in western USA. The dashed line shows the predicted relation if the irrigation water were to be concentrated threefold in the soil. The actual ESP for a number of surface soils fell between the two lines, indicating reasonable agreement between prediction and measurement. However, ESP values at depth in the soil profile were less well predicted, probably because the simple ESP–SAR relation does not take account of (1) the precipitation of insoluble carbonates

(Box 13.4), and (2) differences in ion-pair formation and ionic strength effects on the activities on the Na^+, Ca^{2+} and Mg^{2+} ions as the water becomes more concentrated through evapotranspiration.

Salt balance

Even good quality irrigation water contains some salts. Because crops take up little of this salt (about one-tenth) but transpire nearly all the water, salts inevitably accumulate in the soil under sustained irrigation. Good irrigation practice seeks to minimize this salt build-up.

A complete salt balance for an irrigated soil in a given time interval can be written as

$$S_p + S_{iw} + S_r + S_d + S_f = S_{dw} + S_c + S_{ppt}. \quad (13.5)$$

In this equation, the cumulative contribution of S_p (atmospheric inputs), S_r (residual soil salts), S_d (salt released by weathering) and S_f (fertilizer salts) is usually small when compared with S_{iw} (salts in the irrigation water) and tends to be balanced by S_c (crop removal of salt). Usually, the effect of S_{ppt} (precipitation of carbonates and sulphates in the soil) has been discounted and management practices have concentrated on the relationship between S_{iw} and S_{dw} (salt removed in drainage water). This approach leads to the simplified equation

$$EC_{iw} \times d_{iw} = EC_{dw} \times d_{dw}, \quad (13.6)$$

where the subscripts iw and dw refer to irrigation and drainage water, respectively, d is the volume of water applied per unit area (in mm), and EC measures the salt concentration. Rearranging Equation 13.6 gives

$$\frac{EC_{iw}}{EC_{dw}} = \frac{d_{dw}}{d_{iw}} = \text{leaching requirement (LR)}. \quad (13.7)$$

The LR is therefore defined as the fractional depth of water, of given EC_{iw}, which must pass through the soil to maintain EC in the upper two-thirds of the root zone below a specified value. This critical value of EC, expressed as EC_e, is set by a crop's tolerance to salinity, which can be interpolated from diagrams of the kind shown in

Box 13.4 Carbonate precipitation in irrigated soils.

Precipitation of $CaCO_3$ in soil can be predicted from the Langelier saturation index (LSI), defined as

$$LSI = pH_a - pH_c, \qquad (B13.4.1)$$

where pH_a (= 8.4) is the pH of water just saturated with respect to $CaCO_3$ at atmospheric pressure of CO_2 (Table 11.5), and pH_c is given by the equation

$$pH_c = (pK_2 - pK_{SP} + p\gamma_{HCO3} + p\gamma_{Ca}) + p[Ca] + p[HCO_3]. \qquad (B13.4.2)$$

The terms inside the round brackets are the negative logarithms (base 10) of the second dissociation constant of H_2CO_3, the solubility product of $CaCO_3$, and the activity coefficients of HCO_3^- and Ca^{2+} ions, respectively. Because the sum of these terms is relatively constant, pH_c is influenced mainly by the concentration Ca^{2+} and HCO_3^- ions in solution: as these increase (and $CaCO_3$ is more likely to precipitate), p[Ca] and p[HCO_3] in Equation B13.4.2 will decrease and therefore pH_c decreases. Thus, as pH_c falls below 8.4, $CaCO_3$ is likely to precipitate. This is illustrated in Fig. B13.4.1, which also shows that the smaller the fraction of applied water draining through a soil (the leaching fraction, LF), the greater is the precipitation of $CaCO_3$.

Empirical equations have been used to predict the effect of LF and LSI on the SAR of irrigation water, and hence the expected ESP, such as

$$ESP \cong SAR_u = SAR(1 + LSI), \qquad (B13.4.3)$$

for the upper (u) root zone, and

$$ESP \cong SAR_l = kSAR_u \qquad (B13.4.4)$$

for the lower (l) root zone. The coefficient k depends on the LF and rate of soil mineral weathering (Rhoades, 1968). Alternatively, dynamic models such as UNSATCHEM (Suarez, 2001) can be used to simulate the effect of variables such as soil CO_2 partial pressure, $CaCO_3$ solubility and LF on the soil solution SAR, and hence ESP, at different depths and times.

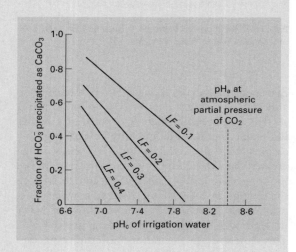

Fig. B13.4.1 Precipitation of insoluble carbonates in relation to pH_c of the irrigation water and LF (after Bower et al., 1968).

Fig. 13.8 (note that a different set of tolerance values applies if $EC_{1.5}$ is used, as shown in Sumner et al. (1998)). LR values usually lie between 0.05 and 0.3.

Equation 13.7 can be rewritten by replacing the term d_{dw} by $(d_{iw} - d_{cw})$, where d_{cw} is the total depth of irrigation water required to satisfy the crop's consumptive use (evapotranspiration), so that

$$EC_{iw} \times d_{iw} = EC_{dw} \times (d_{iw} - d_{cw}), \qquad (13.8)$$

from which the depth of irrigation water to be applied is calculated as

$$d_{iw} = \frac{d_{cw}}{1 - LR}. \qquad (13.9)$$

However, good salinity management must consider not only the soil salt balance, but also the off-site effects of salt (Box 13.5).

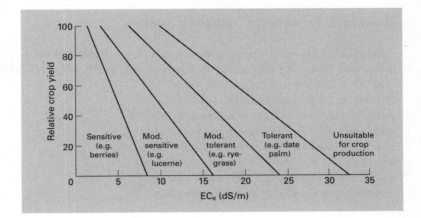

Fig. 13.8 Salt tolerance classes for crops based on the EC_e of the saturation extract (after Maas and Hoffman, 1977).

Box 13.5 Total salinity management.

Unnecessary leaching of salts from soil into the irrigation 'return flow' creates salinity problems for water-users downstream. The Murray-Darling Basin of Australia provides an unfortunate example of steadily increasing salinity proceeding downstream in the Murray River as successive irrigation areas discharge their return flows into the river and its tributaries. There is also a significant contribution to stream salinity from dryland salinity areas, as discussed in Box 13.1. Proper irrigation management must therefore focus on the dual objectives of:
• Minimizing saline return flows; and
• avoiding an accumulation of soluble salts in the soil and in groundwater immediately below the soil.

In this context, the distinction between LR and LF, the actual fractional depth of the applied water

that passes through the root zone, is important. Ideally LF should equal LR, but because of non-uniformity in the application and infiltration of water on a field scale, the average LF is generally > LR. Thus, crucial aspects of salinity management under irrigation are to:
• Keep the LR as small as possible ($\cong 0.05$) by using water of low EC_{iw}; and
• optimize the uniformity of water application and infiltration into the soil so that LF is as small as possible ($\cong$ LR).

For irrigation waters rich in Na, a high pH_c which promotes the precipitation of $CaCO_3$ (Fig. B13.4.1) may be a disadvantage; but for non-sodic waters, maximizing the precipitation of insoluble salts in the profile can be an advantage for off-site salinity control.

Specific ion effects

In addition to the osmotic effect of high salt concentrations on plant growth, specific toxicity symptoms arise when individual elements exceed certain concentrations. For example, fruit trees are likely to show marginal leaf burn and necrotic spots when Na and Cl concentrations in the dry matter exceed 0.2 and 0.5%, respectively. Lithium can prove toxic at concentrations > 0.1 mg/L in the soil solution and irrigation water with > 0.3 mg B/L is suspect for use on sensitive crops like citrus.

13.3 Management and reclamation of salt-affected soils

Permeability changes on leaching

As discussed in Section 13.2, the evaporation of water from the soil-plant system leads to an increase in the soil solution SAR and eventually an increase in ESP. The effect is exacerbated when high-salt irrigation water is used. A saline soil ($EC_e \geq 4$ dS/m) is called saline-sodic when the ESP is ≥ 15. A more detailed description of saline-sodic and sodic soils is given in Box 13.6.

Box 13.6 Saline-sodic and sodic soils.

Australia and Africa have large areas of soils that are naturally sodic. The definition of sodicity is based on a soil's ESP and ranges from ≥ 6 in Australia to ≥ 15 in the USA. The lower ESP threshold for Australian soils is attributed to the lower electrolyte concentration of tap waters used to assess permeability changes in soils of different ESP, compared with those used in the USA (Sumner et al., 1998). Rengasamy and Olsson (1991) have proposed a classification of Australian sodic soils based on:

• Sodium adsorption ratio (SAR);
• a threshold electrolyte concentration (TEC) (measured as $EC_{1:5}$); and
• soil pH (1 : 5 soil : water suspension).

SAR is a surrogate for ESP and when measured in a 1 : 5 soil/water extract is approximately *half* the corresponding soil ESP. The TEC is defined as the minimum concentration of salts required to maintain the soil in a permeable condition. The TEC is not a fixed value because the balance between a stable (and permeable) structure and an unstable structure depends on the SAR, as seen in Fig. 13.9. Nevertheless, based on the Rengasamy and Olsson

Table B13.6.1 A simple classification of Australian sodic soils. After Rengasamy and Olsson, 1991.

Saline-sodic ($SAR_{1:5} > 3$, $EC_{1:5} > TEC$)		Sodic ($SAR_{1:5} > 3$, $EC_{1:5} < TEC$)
Alkaline sodic (pH > 8.0)	Neutral sodic (pH 6.0–8.0)	Acidic sodic (pH < 6.0)

principles, a simple classification of sodic soils is as given in Table B13.6.1.

Transformation from a saline-sodic soil to an acidic sodic soil can occur naturally, as shown in Fig. 9.16. In addition, change from the saline-sodic to sodic category can occur under irrigation, as discussed below. Sodic soils occupy some 60% of the area of Victoria, Australia, with their distribution reflecting the influence of parent materials and effective rainfall. Few sodic soils occur in the wetter, mountainous country to the east and the alkaline-sodic soils occur mainly in the drier northwest (Fig. B13.6.1).

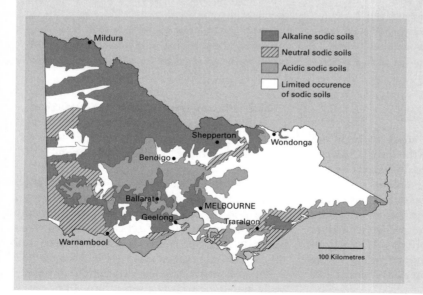

Fig. B13.6.1 Major sodic soil classes in Victoria.

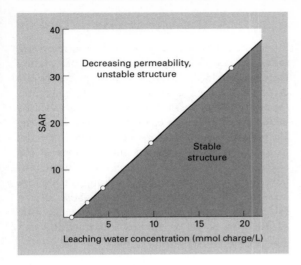

Fig. 13.9 Interactive effect of ESP and concentration of leaching water on soil permeability (after Quirk, 1994).

Saline-sodic soils need very careful management because, although the structure remains stable as long as the soil solution EC > TEC, when soluble salts are removed by leaching, adverse changes in structure occur. Fig. 13.9 illustrates the interactive effect of the SAR and salt concentration of percolating water on a saline-sodic soil's permeability. For example, for an SAR of 17 (ESP ≅ 17), the threshold concentration below which the hydraulic conductivity falls by 15% is 10 mmol charge/L. The decrease in hydraulic conductivity is caused by swelling of the clay domains and quasi-crystals (Section 7.4), which decreases pore diameters and hence slows the rate of water flow. At lower concentrations of the percolating water, swelling may be great enough to cause deflocculation of clay particles, which may then be eluviated and block the water-conducting pores. The concentration at which deflocculated clay appears in the percolating water is called the 'turbidity concentration' and is about 25% of the threshold concentration (Quirk, 2001). Although swelling of flocculated clay is reversible (by increasing the salt concentration), deflocculation *and* clay translocation are not, and they can cause a permanent deterioration in soil structure (Section 9.6).

The critical SAR for structural stability, derived from the relationship in Fig. 13.9, is given by the equation

$$SAR = 1.79C - 0.6, \tag{13.10}$$

where C is the salt concentration of the percolating water (mmol charge/L). Note that Equation 13.10 does not include any solid-phase properties. However, it has been found that, for the same soil profile, the surface soil (of high organic matter) requires a higher salt concentration to remain stable at a given SAR than the subsoil (of low organic matter). Further, a large iron oxide content can lower the salt concentration required for stability at a given SAR.

Reclamation of saline-sodic soils

Reclamation of sodic soils requires leaching with water of SAR low enough to initiate Ca^{2+} exchange for Na^+, but of sufficiently high total salt concentration to preserve the soil's permeability. As the ESP is gradually decreased, the salinity of the leaching water can also be decreased until reclamation is complete. The right conditions can be created by adding soluble calcium salts, usually gypsum $CaSO_4.2H_2O$, which has a solubility of 30 mmol charge/L in distilled water. Gypsum can be spread on the soil or dissolved in the irrigation water, its solubility increasing in more concentrated solutions due to the effect of ionic strength and ion-pair formation in reducing the activity of the Ca^{2+} and SO_4^{2-} ions. The gypsum required (GR) to lower the ESP of 1 ha of soil in a specific depth can be calculated from the approximate equation (Oster and Jayawardane, 1998):

$$GR \cong 0.0086d_s\rho_bCEC(ESP_i - ESP_f). \tag{13.11}$$

For example, suppose the soil's bulk density (ρ_b) is 1.3 Mg/m³, the CEC is 200 mmol charge/kg soil and the soil depth (d_s) to be ameliorated is 0.15 m, the GR (t/ha) for a 10 unit change from the initial ESP (ESP_i) to final ESP (ESP_f) is 3.35 t/ha. If the gypsum is dissolved in the irrigation water, less is required. Calcium carbonate, if present in the soil, is normally too insoluble to lower the SAR of leaching water significantly; but in soils containing *Thiobacillus* bacteria the addition of elemental S enhances the dissolution of $CaCO_3$ due to the acidity generated when S is oxidized to H_2SO_4. Similarly, a green manure crop that is ploughed

Table 13.1 Reclamation of a saline-sodic soil with water of stepped-down salt concentration and SAR. After Reeve and Bower, 1960.

Dilution ratio (seawater : river water)	Water properties		Soil properties	
	TDS* (mmol charge/L)	SAR† (mmol charge$^{1/2}$/L$^{1/2}$)	Hydraulic conductivity (m/day)	ESP†† (%)
Salton Sea	562	57	–	–
1 : 3	149	27	0.13	28
1 : 15	45.4	12	0.12	14
1 : 63	19.6	5	0.11	6
0 : 1	11.0	2	0.12	5
Colorado River	11.0	2	0.005	5

* Total dissolved salts.
† Sodium adsorption ratio.
†† Exchangeable sodium percentage.

in may raise the CO_2 concentration in the soil air sufficiently to increase the dissolution of $CaCO_3$.

Metal sulphides in marine sediments also oxidize when first exposed to air (Box 9.5), a reaction that has benefited the reclamation of Dutch polders. The H^+ and SO_4^{2-} ions produced react with indigenous $CaCO_3$ to produce gypsum sufficient to saturate the soil solution for several years as the reclaimed land is leached by rainwater.

Blending of low and high salinity waters

The theory of saline-soil reclamation was put to an interesting test with a saline-sodic soil from the Coachella Valley, USA. The soil, which had an ESP of 39 and an EC_e of 4.4 dS/m, was leached with Salton seawater that was progressively diluted with good quality Colorado river water. Each water mixture, of which the range in composition is given in Table 13.1, was leached through the soil until the soil was in equilibrium with that mixture. With four stepwise reductions in the leaching water concentration, the soil permeability did not fall below 0.12 m/day and ESP was lowered to 5 in only 12 days. However, when the soil was leached with Colorado river water alone, permeability fell to 0.005 m/day and reclamation took 120 days.

Method of leaching

Intermittent leaching is more efficient in removing salts than continuous ponding. Water percolating through ponded soil tends to flow preferentially down the larger channels between aggregates (Section 6.3), bypassing much of the salt located in the small intra-aggregate pores. Breaks in the leaching process therefore allow time for salts to diffuse into the depleted outer regions of the aggregates, whence they are removed during the next leaching period. Thus, although 1000 mm of water may be required to leach 80% of the salt from 1 m of soil under ponding, as little as 300 mm of water applied intermittently can achieve the same result.

Wastewater management

There is a growing requirement for effluent water from sewage treatment works and processing plants, such as canneries, wineries, dairy factories and paper mills, to be discharged on to land before entering rivers and other water bodies. This policy is being adopted to save irrigation water where fresh supplies are scarce, and also to decrease nutrient loadings to freshwater, which predispose to eutrophication (Box 12.4). However, a problem with these wastewaters is the relatively high salt concentration (800–1800 mg/L) and high SAR values (3–11 mmol charge$^{1/2}$/L$^{1/2}$). The combination of high salt concentration and SAR can lead to the disposal soil becoming sodic, with attendant permeability problems that are compounded by organic colloids in the wastewater, which can form a 'scum' on the soil surface and block pores. Ideally, soils chosen for wastewater disposal should have a low ESP and structures stabilized by kaolinitic clay, in association with Fe and Al oxides at pH < 6. The tendency for ESP values to increase can be controlled through the use of gypsum.

Removal of surplus water

It is a corollary of good irrigation management that surplus water, arising out of a leaching requirement or through seepage from a shallow

groundwater table, should be disposed of efficiently. In permeable soils, the natural hydraulic properties may be adequate to cope with the excess water: where this is not the case, soil drainage must be improved. This can be achieved by:
• *Surface drains* designed to remove as quickly as possible irrigation water and any surplus rainfall that has not infiltrated the soil; or
• *underdrainage* – pipes installed underground through which water is rapidly led away (see below).
Groundwater levels can be controlled by pumps, but such technology is outside the scope of this book.

13.4 Soil drainage

The purpose of drainage

Irrigation is practised predominantly in dry climates on soils that should be reasonably permeable. Because large quantities of water are applied (up to 1500 mm/yr in the case of paddy rice), disposal of salts and control of the regional groundwater table are prime objectives in the management of surplus water. Achieving these objectives necessitates a drainage approach different from that for soils in humid temperate climates, where the natural permeability may be inadequate to prevent soil from becoming waterlogged for unduly long periods (> 1 day), when precipitation exceeds evaporation. Irrigation schemes are also very capital-intensive, producing high cash returns per hectare, so that more expensive drainage systems (and groundwater pumps) may be justified on irrigated, but not non-irrigated, land. Drainage for these two broad categories of soils is therefore discussed separately.

Irrigated soils and underdrainage

The variables that interact to determine the watertable height at steady-state equilibrium in a soil with underground (pipe) drainage are:
• The saturated hydraulic conductivity K_s (Section 6.3);
• the depth of drains d and their height H above any impermeable layer in the soil; that is, a layer with a K_s value ≤ 0.1 of the layer above;
• the distance L between drains; and
• the mean water flux J_w through the soil to the watertable.

The direction of flow to the drains through the unsaturated zone is predominantly vertical. Below the watertable, flow becomes two-dimensional (both vertical and horizontal) and then radial in the zone around the drain, as illustrated in Fig. 13.10. The relative extent of these different flow zones depends on the magnitude of L, H and h, the last being the overall difference in hydraulic head between the watertable and drain level. When the flux J_w to the watertable is equal to the drain discharge rate, the system is in steady-state and Hooghoudt's equation can be used to estimate the required value of L for given values of d, h, H and pipe diameter (Box 13.7). For the more complex situation of non-steady flow of water to drains, the reader is referred to the mathematical analysis by Youngs (1999).

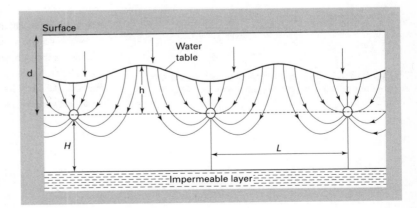

Fig. 13.10 Idealized flow lines of water moving to pipe drains after rain or irrigation.

Box 13.7 Calculation of optimal drain spacing and depth.

Hooghoudt's modelling approach was to reduce the complex flow pattern around pipe drains to the equivalent of horizontal flow to vertical trenches, at the same spacing as the pipes, but filled with water to a lesser height H_e above an impermeable layer (Fig. B13.7.1). Because $H_e < H$, there is less cross-sectional area available for horizontal flow, so the head drop is greater than in the real case. The difference in head loss is just equal to the head loss in radial flow in Fig. 13.10, so that the total head loss in the simulated and real cases is the same. The average thickness through which water flows in the equivalent case is $H_e + h/2$, from Fig. B13.7.1, and the equation giving the discharge rate J_w is

$$J_w = 8K_s h(H_e + h/2)/L^2. \qquad (B13.7.1)$$

Rearranging Equation B13.7.1 gives

$$L = \sqrt{8K_s h(H_e + h/2)/J_w}. \qquad (B13.7.2)$$

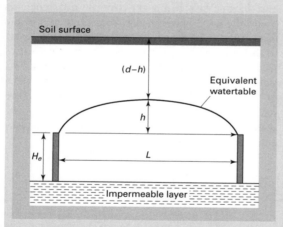

Soil surface

$(d-h)$

Equivalent watertable

h

H_e

L

Impermeable layer

Fig. B13.7.1 Diagram for Hooghoudt's 'equivalent flow' model to drains (after Smedema and Rycroft, 1983).

The effective height H_e above an impermeable layer is a function of the true height H, L, and p, the wetted perimeter of the drain. Equation B13.7.2 must therefore be solved by successive approximations using trial values of L and H_e until the calculated value of L is very close to the trial value chosen. The procedure is as follows:

1 Choose the design criteria for J_w and the mid-drain watertable height $(d - h)$. To prevent salt accumulation in an irrigated soil, J_w should be ≥ 0.001 m/day. To prevent the capillary rise of groundwater bringing salts to the surface (Section 6.5), $(d - h)$ should be ≥ 2 m;
2 choose a drain depth d consistent with (1);
3 establish the values of K_s and p;
4 knowing H, the true depth to an impermeable layer, find from tables values of H_e for trial values of L. Alternatively to determine L, appropriate values for the variables can be entered into the widely-used DRAINMOD model, based on Hooghoudt's equation (Skaggs, 1999).

Note that the important drainage variable K_s is likely to change by one to two orders of magnitude in a field. Because of this variability and the fact that subsoil K_s is difficult to measure by well permeameter or pumping methods, K_s is often estimated using surrogate variables (Ahuja et al., 1999). For example, K_s has been found to be related to the effective porosity ε_e through the equation

$$K_s = B\varepsilon_e^n, \qquad (B13.7.3)$$

where ε_e is the difference between the total porosity and the volumetric water content (θ_{33}) at -33 kPa matric potential, and B and n are empirical coefficients. Total porosity can be estimated from the bulk density (Section 4.5) and θ_{33} from the soil water retention curve (Section 6.4).

Drainage of clay soils and duplex soils with clay subsoils presents a different problem in that the impermeable layer usually lies at or just below drain depth ($H \to 0$), so that deep flow to the drains is inhibited. In this case, Equation B13.7.2 simplifies to

$$L = 2h\sqrt{K_s/J_w}. \tag{13.12}$$

Often, the calculation of theoretical drain spacings is vitiated by the imprecision with which K_s can be measured or estimated, so that predictions based on Hooghoudt's equation or any other model of flow to drains must be tempered by experience and practical judgement. Nevertheless, Equation 13.12 demonstrates that drain spacing can be increased if K_s increases or J_w decreases, and also if h increases. To increase h, either the watertable at the mid-drain point must be allowed to rise or, more usually, drain depth must be increased. There are, however, practical limits to the depth at which drains can be placed, particularly when the subsoil is poorly permeable.

Soils of humid temperate regions

In reasonably permeable soils (including fen peats), drainage is sometimes achieved by a network of surface ditches, as illustrated in Fig. 13.11. This technique has also been used to remove surplus water in flood irrigation schemes, such as in much of the irrigation area of the Murray-Darling Basin of Australia. Design of such systems is essentially an engineering exercise to ensure adequate control of the regional watertable and flow rates in the ditches or channels. However, the drainage of wet, impermeable soils of humid regions generally requires a system of pipe (or tile) drains, with

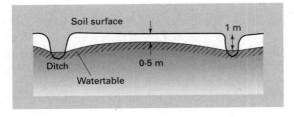

Fig. 13.11 Drainage ditches in a reasonably permeable soil with a high watertable.

or without supplementary mole drains. A widely accepted design criterion for draining soils in these regions is to lower the watertable to 0.5 m below the surface at the mid-point between drains within 24 h after rainfall. This condition creates a matric potential of −5 kPa at the surface, corresponding to the FC of a naturally drained soil in temperate regions. There are two major constraints on the effectiveness of such drainage:

- The pore volume drained at −5 kPa matric potential, because this determines the maximum air-filled porosity of the drained soil (the drainable porosity). Its value depends on the soil water retention curve (Section 6.4), but if it is < 10%, the improvement in aeration that drainage can achieve will be limited;
- the hydraulic conductivity K_s of the saturated soil, which determines the maximum rate of drainage when the hydraulic head gradient is one. The conductivity of clay soils in Britain may be 1–10 m/day in the dry state, if deep fissures are prominent, but only 0.01–0.001 m/day when the soils are saturated and fully swollen. Given that the drainage system should be capable of removing 0.01 m of surplus water per day, one can calculate from Equation 13.12 that for an impermeable layer close to the drain depth, the theoretical drain spacing for a soil of K_s between 0.001 and 0.1 m/day is as small as 0.3–3 m. Such close spacing of tile drains is rarely economically feasible.

Underdrainage in practice

Drain depth

In order that the watertable does not come within 50 cm of the soil surface, tile drains need to be placed more deeply, the actual depth depending on the position of any impermeable layer and the adequacy of the outfall into the collection ditch. Drains are laid with a gentle gradient at a depth between 70 and 120 cm. The deeper the drains, the drier the soil in the plough layer once hydraulic head equilibrium is established; but this condition is not always attained in clay soils in wet winters due to their small K_s values.

Tile drains are either short lengths of porous clay pipes, usually 7.5 cm in diameter, laid end-to-end, or continuous plastic pipe of the same

1

1

1

1

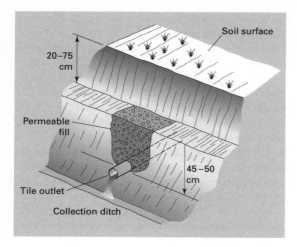

Fig. 13.12 Diagram of a tile drain imbedded in permeable fill.

diameter which is corrugated for strength and perforated to allow water entry. Because water enters the pipe from all sides, the drain's performance is much improved if it is imbedded in permeable filling material (Fig. 13.12). Permeable fill, usually consisting of gravel or crushed stone of size range 0.5–5 cm, is of benefit for three reasons:

1 It filters out very small soil particles that may otherwise block the drain;

2 it reduces the head loss caused by the restricted number of entry points for water into the pipe; and

3 it serves as a connection between the tile drains and any mole drains drawn at shallower depths.

If the hydraulic conductivity of the permeable fill is at least 10 times the K_s of the surrounding soil, the head entry loss of the drain should be close to zero. Because gravel permeable fill is relatively expensive, well-structured soil from the A horizon of the trench dug for the tile drains can be used as 'permeable fill'.

Drain spacing

Clearly, laying tile drains at the spacing necessary to drain effectively soils with K_s values as low as 0.01–0.001 m/day is very expensive. The practical alternative is to choose an economic spacing between 20 and 80 m, commonly 20–40 m, and

to increase water flow to the drains by secondary treatments, of which the most important are:

• *Moling* – this provides cheap drains at the spacing required for efficient drainage in impermeable soils. The depth of the mole drains determines the steady-state watertable position and the spacing governs its rate of rise and fall during and after rainfall. Closer spacings delay the rise and hasten the fall of the watertable;

• *subsoiling* – this has the aim of improving the soil's hydraulic conductivity to the point where a drain spacing of 20–40 m provides efficient drainage.

As Equation 13.12 shows, a 10-fold increase in drain spacing requires a 100-fold increase in K_s for the same values of h and J_w. This increase in K_s is not easily achieved because the soil at depth is often too moist for effective subsoiling. Nevertheless, subsoiling can be effective where a plough pan markedly reduces the permeability of the soil above the drains.

Methods of moling and subsoiling

Moles are drawn by the implement illustrated in Fig. 13.13. The 'bullet' of 7.5 cm diameter is pulled through the soil at a depth of 40–50 cm and a spacing of 2–3 m. Behind the bullet, the slightly larger 'expander' consolidates the walls of the channel and seals the slit left by the vertical blade. Passage of the bullet and blade through the soil should fracture the structure to some extent and improve flow to the mole drain. Mole drains should lie at right angles to the tiles, provided their gradient can be kept between 2 and 5%, and pass through the top of the permeable fill. Whereas tile drains should last for 50 years or more, moles should be redrawn every 5–7 years for best results.

In subsoiling, a wedge-shaped share is pulled at right angles to the tile drains at a depth of 40–50 cm. Ideally, at this depth and a spacing of c. 1.3 m, the structure of all the upper soil profile should be disturbed by fracturing and fissuring (Fig. 13.14). The effect of subsoiling is usually more transient than moling. The conditions for effective moling and subsoiling are discussed in Box 13.8. Subsoiling is also effective in breaking up shallow impermeable layers such as a plough pan (Section 11.4).

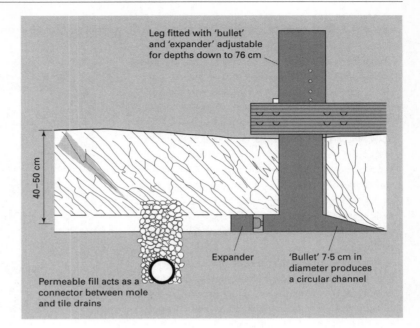

Leg fitted with 'bullet' and 'expander' adjustable for depths down to 76 cm

40–50 cm

Expander

'Bullet' 7·5 cm in diameter produces a circular channel

Permeable fill acts as a connector between mole and tile drains

Fig. 13.13 Diagram of a mole drain plough in operation (after MAFF, 1980).

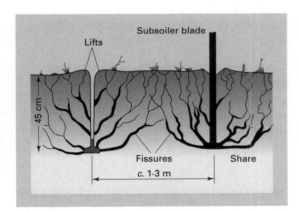

Subsoiler blade

Lifts

45 cm

Fissures

Share

c. 1·3 m

Fig. 13.14 Soil fissuring induced by effective subsoiling (after MAFF, 1981).

13.5 Summary

Excess soil water due to a consistently high P : E ratio, an impermeable subsoil or an elevated watertable imposes severe limitations on land use.

In arid areas, groundwater is often saline and irrigation, if not properly managed, tends to raise the watertable. Capillary rise of salts and salinization then occur, with adverse effects for plant growth. Because most of the water added to soil is lost by evapotranspiration, but only about one-tenth of the salts is taken up by the crop, irrigation itself increases soil salinity. The salinity hazard of irrigation water is determined by its total dissolved salts (TDS). TDS is conveniently measured by the electrical conductivity (EC) of the water, i.e.

$$\text{TDS (mg/L)} \cong 640 \text{ EC (dS/m)}.$$

The concentrating effect of evaporation on the soil solution also causes the sodium adsorption ratio (SAR), defined by

$$\text{SAR} = \frac{[\text{Na}^+]}{\left[\dfrac{\text{Ca}^{2+} + \text{Mg}^{2+}}{2}\right]^{1/2}},$$

to increase. The equilibrium between exchangeable and solution cations then changes so that adsorbed Ca^{2+} and Mg^{2+} ions are replaced by Na^+ and the exchangeable sodium percentage (ESP) of the soil increases. Thus, SAR can be used to assess the sodicity of irrigation waters. The rate at which the actual ESP increases with SAR depends on the total salinity of the irrigation water.

Box 13.8 Conditions for effective moling and subsoiling.

The success of moling and subsoiling depends largely on the soil's consistence (Box 4.2) at the time of the operation. Consistence describes the change in a soil's physical condition with change in water content, and reflects both the cohesive forces between particles (ped shear strength) and the resistance of the soil mass to deformation (bulk shear strength). Bulk shear strength is measured with a penetrometer (Box 11.6). The changes in ped or aggregate shear strength and bulk shear strength with change in water content θ for a clay soil are shown in Fig. B13.8.1. Ideal conditions for moling occur when:

• The soil at mole depth is near the lower plastic limit (maximum bulk shear strength and therefore stability of the mole channel when drawn), while the soil above is drier (in a friable state) and so tends to shatter as the mole is drawn;

or for subsoiling when

• the soil to 40–50 cm depth is friable, near to the shrinkage limit, when the bulk shear strength is less and the soil should shatter as the subsoiler blade passes through.

These conditions are generally satisfied if the SWD in the whole soil to 1 m depth is at least 50 mm (for moling) and 100 mm (for subsoiling). The effect of subsoiling is usually more ephemeral

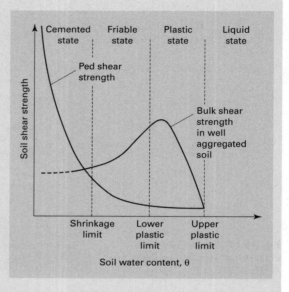

Fig. B13.8.1 Variation in soil shear strength and consistence with water content (after MAFF, 1975).

than that of moling. Mole channels tend to have poor stability in silty soils, and moling is quite unsuited to soils with sodic subsoils, unless gypsum is injected into the mole-channel walls to stabilize the clay.

In the USA, a soil is classed as saline if the EC of its saturation extract (EC_e) ≥ 4 dS/m. If $EC_e \geq 4$ dS/m and ESP ≥ 15, a soil is classed as saline-sodic. In Australia, the classification of saline-sodic and sodic soils is based on the SAR, a threshold electrolyte concentration TEC (measured as EC in a 1 : 5 soil : water extract), and the soil pH. In this case, saline-sodic soils have an SAR > 3 and EC > TEC, and sodic soils have an SAR > 3 and EC < TEC. The SAR (measured in 1 : 5 soil : water ratio) in mmol charge$^{1/2}$/L$^{1/2}$ is approximately half the predicted ESP.

Saline-sodic soils require careful management to avoid:

• Swelling and loss of permeability; and possibly
• clay deflocculation and translocation,

when leached with good quality water. The first process is reversible; the second may lead to an irreversible deterioration of soil structure (solonization → sodic soils). Sodic soils are widespread in Australia and generally require gypsum to supply Ca^{2+} in solution and displace some of the exchangeable Na^+. In reclaiming saline or saline-sodic soils, leaching water of high EC and low SAR should be used and progressively diluted as salts and Na^+ ions are removed.

Drainage is often necessary to remove excess salts from irrigated soils. The ratio of drainage water to applied irrigation water that will maintain a salt balance in the soil is called the leaching requirement, LR. The ratio of *actual* drainage to irrigation water is the leaching fraction, LF.

Ideally, LF = LR, but usually LF > LR because of unevenness in the application of water, particularly by flood or furrow methods. Spray irrigation is more efficient but the equipment is expensive. Drip irrigation is the most efficient and allows water of a higher salinity than normal to be used. All irrigation should be scheduled according to water consumption by the crop (allowing also for any leaching requirement).

Too much drainage of salty water from irrigation schemes creates problems for water-users downstream. Seepage of saline water to streams and rivers can also arise from dryland salinization, induced by large-scale changes in land use (e.g. deep-rooted native forests replaced by short-rooted annual crops and pastures), which disturb the regional hydrology. Total salinity management requires the monitoring and control of salt build-up in soils, subsoils and drainage waters.

In cool humid climates, problems of wetness, inadequate aeration, poor trafficability, invasion by flood-tolerant weeds, pests and disease, and low soil temperatures in spring can be mitigated by the installation of pipe or tile drains to create a matric potential at least ≤ -5 kPa at the soil surface at the mid-drain point, 24 hours after rain. The efficacy of this treatment will be limited if the drainable porosity at -5 kPa is < 10%, or the saturated hydraulic conductivity K_s is < 0.01 m/day. The drain spacing necessary to achieve the necessary drainage rate can be calculated from Hooghoudt's equation, provided that certain soil characteristics (especially K_s) are known.

When K_s is so small that the pipe drains need to be placed at uneconomically close spacings (1–2 m), the practical solution is to use wider spacings (20–40 m) with secondary treatments such as moling and subsoiling to improve flow to the pipes. The consistence of the soil at moling or subsoiling is critical for the success of these operations. Mole drains are drawn above the pipes at intervals of 2–3 m, preferably when the soil water deficit (SWD) is at least 50 mm in the top metre (and the soil at mole depth is near the lower plastic limit). Embedding the pipes in 20–25 cm of permeable fill promotes peripheral flow into the pipe and provides a good connection with the moles above. Subsoiling is cheaper than moling, but its effect does not last as long (5–7 years) unless it is performed under optimum

conditions – a SWD at least 100 mm, and friable soil to 40–50 cm depth. Subsoiling is also effective in breaking up shallow impermeable layers such as a plough pan.

References

Ahuja L. R., Rawls W. J., Nielsen D. R. & Williams R. D. (1999) Determining soil hydraulic properties and their field variability from simpler measurements, in *Agricultural Drainage* (Eds R. W. Skaggs & J. van Schilfgaarde). Agronomy Monograph No. 38. Soil Science Society of America, Madison WI, pp. 1207–33.

Bower C. A., Ogata G. & Tucker J. M. (1968) Sodium hazard of irrigation waters as influenced by leaching fraction and by precipitation or solution of calcium carbonate. *Soil Science* **106**, 24–34.

Bresler E. (1975) Two-dimensional transport of solutes during non-steady infiltration from a trickle source. *Soil Science Society of America Proceedings* **39**, 604–13.

Holmes J. W. (1971) Salinity and the hydrologic cycle, in *Salinity and Water Use* (Eds T. Talsma & J. R. Philip). Macmillan, London.

Maas E. V. & Hoffman G. J. (1977) Crop salt tolerance – current assessment. *Journal of Irrigation and Drainage Division. Proceedings of the American Society of Civic Engineers* **103**, 115–34.

Ministry of Agriculture, Fisheries and Food (1975) *A Design Philosophy for Heavy Soils*. Field Drainage Experimental Unit, Technical Bulletin 75/5, Cambridge.

Ministry of Agriculture, Fisheries and Food (1980) *Mole Drainage*. Field Drainage Leaflet No. 11. MAFF Publications, Middlesex.

Ministry of Agriculture, Fisheries and Food (1981) *Subsoiling as an Aid to Drainage*. Field Drainage Leaflet No. 10. MAFF Publications, Northumberland.

National Land and Water Audit (2001) *Australian Dryland Salinity Assessment 2000: Extent, Impacts, Processes, Monitoring and Management Options*. Land and Water Australia, Canberra.

Oster J. D. & Jayawardane N. S. (1998) Agricultural management of sodic soils, in *Sodic Soils* (Eds M. E. Sumner & R. Naidu). Oxford University Press, New York, pp. 125–47.

Quirk J. P. (1994) Interparticle forces: a basis for the interpretation of soil physical behavior. *Advances in Agronomy* **53**, 121–83.

Quirk J. P. (2001) The significance of the threshold and turbidity concentrations in relation to sodicity and microstructure. *Australian Journal of Soil Research* **39**, 1185–217.

Reeve R. C. & Bower C. A. (1960) Use of high-salt waters as a flocculant and source of divalent cations for reclaiming sodic soils. *Soil Science* **90**, 139–44.

Rengasamy P. & Olsson K. A. (1991) Sodicity and soil structure. *Australian Journal of Soil Research* **29**, 935–52.

Rhoades J. D. (1968) Mineral-weathering correction for estimating the sodium hazard of irrigation waters. *Soil Science Society of America Proceedings* **32**, 648–52.

Richards L. A. (Ed.) (1954) *Diagnosis and Improvement of Saline and Alkali Soils*. United States Department of Agriculture, Handbook No. 60.

Skaggs R. W. (1999) Drainage simulation models, in *Agricultural Drainage* (Eds R. W Skaggs & J. van Schilfegarde). Agronomy Monograph No. 38. Soil Science Society of America, Madison WI, pp. 469–500.

Smedema L. K. & Rycroft D. W. (1983) *Land Drainage*. Batsford Academic, London.

Suarez D. L. (2001) Sodic soil restoration: modelling and field study. *Australian Journal of Soil Research* **39**, 1225–46.

Sumner M. E., Rengasamy P. & Niadu R. (1998) Sodic soils: a reappraisal, in *Sodic Soils* (Eds M. E. Sumner & R. Naidu). Oxford University Press, New York, pp. 3–17.

Youngs E. G. (1999) Non-steady flow to drains, in *Agricultural Drainage* (Eds R. W Skaggs & J. van Schilfegarde). Agronomy Monograph No. 38. Soil Science Society of America, Madison WI, pp. 265–96.

Further reading

Rhodes J. D. & Miyamoto S. (1990) Testing soils for salinity and sodicity, in *Soil Testing and Plant Analysis*, 3rd edn (Ed. R. L. Westerman). Soil Science Society of America Book Series No. 3, Madison WI, pp. 299–336.

Ritzema H. P. (Ed.) (1994) *Drainage Principles and Applications*. International Institute for Land Reclamation and Improvement Publication No. 16, Wageningen.

Smart P. & Herbertson J. G. (Eds) (1992) *Drainage Design*. Blackie & Son, Glasgow.

Sumner M. E. (1993) Sodic soils: new perspectives. *Australian Journal of Soil Research* **31**, 683–750.

Thomasson A. J. (Ed.) (1975) *Soils and Field Drainage*. Soil Survey of England and Wales, Technical Monograph No. 7. Harpenden, UK.

Example questions and problems

1 (a) What is EC_e?
  (b) (i) What is $EC_{1.5}$ and (ii) how does it differ from EC_e?
  (c) Give the definition of a saline soil.

2 The following is the summary of a soil test report.

  EC_e 1.5 dS/m
  Exchangeable cations
  Calcium   5.00 cmol charge (+)/kg
  Magnesium   3.60 cmol charge (+)/kg
  Sodium   1.81 cmol charge (+)/kg
  Potassium   1.00 cmol charge (+)/kg
  Aluminium   0.68 cmol charge (+)/kg

  Use these results to determine whether the soil is likely to require gypsum, and give reasons for your conclusion.

3 A producer is growing potatoes under irrigation. He knows that the soil's volumetric water content at field capacity is 0.45 m³/m³ and that the soil should not dry to less than 0.38 m³/m³

while the potatoes are growing actively. Calculate the following.

  (a) How much water (in mm) should be applied at each irrigation, given the effective rooting depth of the crop is 50 cm?
  (b) How often should irrigation be applied, given that the average evaporation rate during the growing season is 5 mm/day?
  (c) The bulk density of the top 50 cm of soil is 1.2 Mg/m³. (i) Calculate the soil's air-capacity, and (ii) state whether the aeration is adequate at the field capacity (assume a particle density of 2.65 Mg/m³).

4 Irrigated lucerne is grown for hay on a clay loam soil. The EC of the irrigation water is 0.5 dS/m. The maximum acceptable EC of water in the root zone of lucerne is 5 dS/m.

  (a) Calculate the leaching requirement (LR) for maintaining a salt balance in the soil.
  (b) If crop water use (d_{cw}) averages 6 mm per day over a growth period of 60 days,

calculate the irrigation water to be applied (d_{iw}), consistent with maintaining a salt balance (note: $d_{iw} = d_{cw}/(1 - LR)$).

(c) If the area under lucerne is 10 ha, calculate the volume of water required to grow the crop (in megalitres).

(d) The lucerne area is drained with tile drains. Drain discharge was measured as 80 mm over 60 days. Calculate the actual leaching fraction (LF).

(e) By comparing LF with LR, comment on the efficiency of irrigation application and crop water use.

5 You are asked to advise on the management of a saline soil with a clay loam subsoil that is being irrigated for grape production. The water has the following cation composition – 690 mg Na/L, 200 mg Ca/L and 72 mg Mg/L (as chloride and sulphate salts).

(a) Calculate the sodium adsorption ratio (SAR) of the water, given that the atomic masses of Na, Ca and Mg are 23, 40 and 24 g, respectively.

(b) The measured EC_e of the soil is 5 dS/m. Comment on whether you think the permeability of this soil will remain satisfactory under this irrigation regime.

(c) If the soil were leached by rain during a series of wet winters, so that the EC_e decreased to 2 dS/m, what would you expect to happen to the soil's permeability?

6 Young fruit trees under drip irrigation need 50 L water per tree weekly. The total period of irrigation is 16 weeks in summer each year.

Because of a water shortage, the orchard manager is considering the use of recycled water from a nearby sewage works. This water has an average salt concentration of 1400 mg/L.

(a) Calculate the annual salt load per tree from this water (assume no irrigation water percolates below the rooting depth).

(b) If the trees are planted at a 3×3 m spacing, what is the total amount of salt (kg) applied per ha per year?

(c) If the mean concentration of salts in the harvested fruit is 3%, and the expected yield is 30 t/ha, how much salt will be removed in the fruit each year? (Assume prunings are returned to the soil).

(d) Calculate the expected annual salt accumulation in the soil.

(e) Given the soil's field capacity is 0.35 m^3/m^3 and the average rooting depth is 0.6 m, estimate how much excess winter rainfall is needed to leach the accumulated salts below rooting depth (note – the trees are dormant in winter).

7 A cooperative irrigation company in New South Wales provides water to shareholders through 3300 km of open unlined channels. In the irrigation period from September to May, the company supplied 1230 gigalitres (GL) of water. The management has decided to audit water losses by seepage and evaporation from the channels.

(a) If the average width of water in a channel is 1.5 m, estimate the evaporation loss (in GL) from the channel system, given the following evaporation rates E_o from a Class A pan.

Month	Sept	Oct	Nov	Dec	Jan	Feb	Mar	Apr	May
E_o (mm/month)	45	63	90	160	205	165	150	130	92

(b) If the average wetted perimeter of a channel is 2.4 m, and the average saturated hydraulic conductivity of the soils is 120 mm/day, estimate the seepage loss from the whole channel system during the period 1 September to 31 May the following year.

(c) Express the seepage loss as a percentage of the water supplied from September to May.

(d) Which pathway of loss is the more serious?

Chapter 14

Soil Information Systems

14.1 Communication about soil

Soil variability

The point is made in Chapters 1 and 5 that the distinctive character of a soil is shaped by a complex interaction of many physical, chemical and biotic forces. The result is a remarkable range of soil types in the landscape, as discussed in Box 1.1. The existence of soil variability has two notable consequences:

1 Soil scientists have been stimulated to create order out of disorder by collecting, collating and codifying soil information – the process of classification. The main purpose has been to facilitate the communication of existing knowledge and encourage the acquisition of new knowledge about soils and their behaviour;

2 information about soils tends to be site-specific. Many soil users – farmers, foresters, orchardists, landscape gardeners – acquire this site-specific information primarily from their own practical experience of 'working the land'. Other users – engineers, hydrologists, ecologists, planners – do so in a more abstract way. They require generic information either about the properties and behaviour of individual soil types (broadly defined), or the effect that a particular soil property (e.g. infiltration rate, pH, clay content and type of clay) has on the behaviour of a system they are studying (e.g. runoff in catchments, distribution of plant species, choice of a waste disposal site and so on).

Hopefully, the activities referred to in (1) should benefit soil scientists individually and collectively, and satisfy the needs of the non-specialist users

identified in (2). But this outcome has not been achieved consistently and effectively for two main reasons:

a Soil scientists have communicated mainly with each other, or with scientists in closely related disciplines. Their language has not been readily understood by non-soil scientists and even less by lay people;

b in many cases, the end-users do not know enough about the influence of soil in their systems to ask the right questions of soil scientists.

With respect to (a), classification has been seen as a prerequisite for effective communication between soil scientists. But the traditional approach to classification (see below) has difficulties for the reasons outlined in Box 1.2 – a soil has no fixed inheritance and is continuously variable in space and time. As a result, many classifications have evolved, in some cases several in one country. In this chapter, the principles of soil classification are outlined and some of the main products discussed.

However, in the age of information technology, many of the constraints that diverted soil scientists into traditional classification methods no longer apply. Soil information can be collected in digital form, or converted from analogue to digital form, and stored in huge spatially referenced databases (Geographic Information Systems, GIS). These databases, with their layers of information about soil properties and the soil 'environment', can be used to create digital soil maps that are either generic in nature or provide specific soil information in response to a user request. No longer is it necessary to generalize soil information through soil classes that may be unfamiliar to a non-specialist user. The primary data can be accurately referenced in space using a Global Positioning System (GPS), stored, easily retrieved and rapidly disseminated. With respect to point (b), therefore, soil information can be made widely available in forms intelligible to both specialist and non-specialist users. Some of the principles of modern soil information systems are outlined in this chapter.

14.2 Traditional classification

Faced with variability in natural populations, the traditional approach to classification usually involves three steps:
1 Identification of the full range of variability in the population;
2 creation of groups or classes within the population according to the similarity between individuals; and
3 prediction of the likely behaviour of an unidentified individual from knowledge of the characteristics of the class as a whole.
Step (2) – the creation of classes – is called classification. Soil classification is more difficult and contentious than the classification of other natural populations because a soil lacks the hereditary characteristics by which individuals within one generation are distinguished, and which are transmitted from one generation to the next (Box 1.2). The soil population presents a continuum of variation so that arbitrary judgements are inevitable in the creation of classes.

Definition of a soil entity

Soil variability can be crudely partitioned according to the distance over which discernible changes in properties occur. Differences associated with soil faunal and microbial activity and the distribution of pores, for example, achieve maximum expression within 1 m (short-range variation). Since as much as half of the total variation in any property can occur as short-range variation, it is impossible to define rigorously a fundamental soil unit or entity. Attempts to do so have resulted in the introduction of esoteric terms such as the 'pedon' and 'pedounit' (Box 5.1); but whatever the theoretical merits of these terms, the field

Fig. 14.1 A selection of augers and core samplers used in soil survey.

worker describes a soil from a profile in a pit (usually 1×0.5 m in area, dug to the parent material or to 1.5 m, whichever is the shallower), or from a freshly exposed road cutting or quarry face. Because the excavation of pits is laborious and expensive, the assessment of soil variability is usually based on an examination of soil samples collected by means of augers or coring devices, inserted vertically into the soil (Fig. 14.1). Details of the observations made in soil survey are given by Soil Survey Division Staff (1993), Hodgson (1978) and McDonald *et al.* (1998).

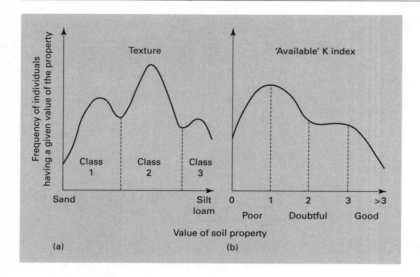

Fig. 14.2 (a) and (b) Subdivision of the variation in a soil property to define classes for a special-purpose classification.

Kinds of classification

Another problem confronting the soil classifier is the choice of properties by which to separate classes. Many properties can be evaluated qualitatively in the field – soil colour, structure, type of humus and degree of gleying, for example. Other properties, such as pore-size distribution and the organic C content, clay or calcium carbonate, must be measured in the laboratory. Practical constraints prevent all the soil properties being measured and of those measured, not all are used. Some are selected as the definitive properties for class separation, the choice of which depends on the aim of the classification as discussed below.

Special-purpose classification

This kind of classification is made with a specific aim in mind and is based on one or a few soil properties. The texture classifications of Fig. 2.3 and the classification of soils for droughtiness and poor aeration (Fig. 11.7) are two examples. There are many other examples, especially relating to plant growth, such as the ESP and salinity (EC_e) status of irrigated soils, and the assessment of a soil's 'available' P and K status by soil tests (Section 11.2). When classification is based on one or two properties, classes may be separated by subdividing the range of variation in the individual property at the minima in the frequency distribution (Fig. 14.2a), or according to the user's requirements (Fig. 14.2b).

General-purpose classification

Because this kind of classification is intended for many uses, foreseen and unforeseen, the classes are defined on as many properties as possible. Provided that the properties chosen are relevant to the use and management of soil, the classification should be useful; if not, the classification may be academically elegant, but of little practical value. But where the important soil properties are those requiring tedious and expensive laboratory measurement, it is expedient to choose certain diagnostic properties for the separation of classes. These properties are not necessarily relevant to soil management, but are easily assessed in the field, and are assumed (or known) to be correlated with other important properties. For example, the colour of the A horizon may be correlated with organic matter content, and mottling of the subsoil is indicative of periodic waterlogging. Where such a natural grouping or 'clustering' of soil properties exists, the classes created should be more meaningful to the general-purpose user than those defined by arbitrary criteria.

Box 14.1 The reliability of soil classification.

For a given categorical level of classification (Section 14.4), it is found that each class created has a similar range of variation so that a pooled within-class variance (σ_w^2) can be estimated by taking the weighted average of the class variances (the weighting factors are derived from the number of degrees of freedom associated with each class variance). The effect of classification is then measured by the index

$$\tau = 1 - \sigma_w^2/\sigma_t^2, \qquad (B14.1.1)$$

where σ_t^2 is the total variance in the soil population. Clearly, when $\sigma_w^2 \cong \sigma_t^2$ the spread of values within a class is comparable with the total spread of values, and the proportion of the total variance accounted for by the classification (τ) approaches 0 – the classification is therefore of little value. It is easiest to calculate τ for a special-purpose classification based on measurements of a single soil property.

Typically, about half the variance in a physical or mechanical property of a soil, but only about one-tenth of the variance in many chemical properties, can be attributed to differences between classes. The within-class variance can be decreased by increasing the number of observations from which the class mean and variance are calculated, but it is uncommon for σ_w^2 to be reduced much below 25% of the total variance unless a great deal of effort is expended. If the effort and hence cost of classification become too great, the alternative of measuring soil properties directly, when required, may be better than attempting prediction from a classification. As indicated below, it is now also possible to measure some soil properties (or their surrogate variables) at high densities of observation by proximal or remote sensing and store the spatially referenced values in a GIS.

Quality of classifications

Implicit in the decision to classify soils is the assumption that the potential user will be able to make better generalizations about soil properties in a heterogeneous landscape than if there were no classification. In other words, better prediction of soil-dependent behaviour such as regional drainage, salinity risk or crop productivity can be made using the class properties, rather than measurements made on the soil at a few individual sites. For this to be true, a substantial proportion of the variation in the properties of a population of soils must occur *between* classes rather than remain *within* classes. The accepted mathematical measure of variation in a population is the variance σ^2. This parameter measures of the spread of individual values about their mean. This and other statistical terms used in the quantitative analysis of soil variability are explained by Webster and Oliver (2001) to whom the reader should refer for a more detailed discussion. Assessment of the quality of soil classification using variances is discussed in Box 14.1.

14.3 Soil survey methods

What is soil survey?

Soil survey primarily involves collecting information about soils in the field. This information is usually augmented by laboratory measurements on samples from the field. Soil information is required so that informed decisions can be made on land use and the management best suited to different land uses. The effort put into survey and the methods adopted are determined by the size of the area of interest in relation to the human and financial resources available, the intensity of the proposed land use, and the exclusiveness of any classes to be created. Traditionally, the output of soil survey has been a classification, and a map showing the distribution of soil classes at a scale commensurate with the density of sampling on the ground. Alternatively, the survey data can be put into a GIS from which special- or general-purpose soil maps may be created, provided that there is a model for predicting soil properties at unsampled sites and the creation of map classes (see **Soil survey output**).

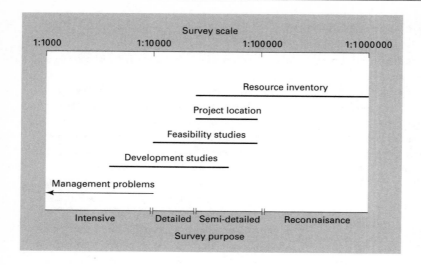

Fig. 14.3 Scale of soil survey in relation to purpose (after Young, 1973).

Sampling to assess the full range of soil variability, and describing the soil at known locations within a target area, are necessary first stages of classification. Soil variability influences the density of sampling required to achieve a certain precision in the classification (measured by τ in Equation B14.1.1). In traditional mapping, variability occurring within blocks < 100 m² (i.e. < 0.01 ha) cannot be mapped at an acceptable scale*, so that this restriction sets an upper limit to the density of sampling for this kind of mapping. However, proximally or remotely-sensed data can obtained at distances as small as 1 m and, together with other spatially referenced soil properties, used to produce digital soil maps. In the context of digital maps, 'scale' is better replaced by the concepts of 'resolution' and 'spacing'; for example, a scale of 1 : 5000 corresponds to a block size of 5×5 m and a resolution of 10×10 m, which is described as a fine resolution.

Purpose of soil survey

The relation between the type and purpose of the survey and the map scale is summarized in Fig. 14.3. There are two main types of survey.

Reconnaissance and semi-detailed surveys. These surveys range in scale from 1 : 2,000,000 to 1 : 25,000 and have the broad objective of providing a resource inventory for large areas (provinces, states, countries). The preliminary work relies heavily on the interpretation of remotely sensed data (Box 14.2) to delineate differences in vegetation, land form and management that may be the outward expression of soil differences. Once boundaries separating notional soil classes have been marked on the satellite images or air photographs, surveyors go into the field to check their significance, relocating them if necessary – the process of 'ground-truthing'. They also describe the main features of the soils and environment within each unit. Areas suitable for various forms of land use can be identified, and guidance given as to the most suitable type of survey to be performed at the next (more detailed) stage.

Detailed and intensive surveys. These range in scale from 1 : 25,000 to 1 : 1000 and are carried out for one of three reasons:
• To assess the technical and economic feasibility of a proposed project in a given area;
• to assess the preliminary work necessary for the development of a specified project; or
• within a developed area, to provide data for the solution of particular management problems.

---

* The practical limit is a scale of 1 : 1000, i.e. 1 cm on the map represents 10 m on the ground. This is a large scale, whereas 1 : 100,0000 is a small scale.

Box 14.2 Remote sensing for soil and land survey.

Devices used for remote sensing in soil and land survey detect electromagnetic radiation (EMR). The sensing may be:
- *Active*, when the sensing device directs EMR at an object and detects the amount of that energy reflected back. Examples are radar and radio wave emissions (wavelength $> 10^3$ µm) and laser-imaging radar in the UV, visible and near-infrared range (wavelength 10^{-2}–10 µm). At the top end of this wavelength range are the electromagnetic induction instruments, flown in an aircraft and used for detecting soil profile salinity (*cf* Box 13.3);
- *passive*, when the sensing device detects EMR originating from another source, such as the sun, or naturally occurring radioactive elements in the soil and rocks. Of the latter, the most useful are the gamma-ray emitters, such as ^{40}K, ^{238}U and ^{232}Th, which have very short wavelengths and hence high energies.

Remote sensing relies on detecting differences in the reflected or emitted radiation from different areas on the land surface over a range of wavelengths. It can be done from a variety of platforms, but mainly aircraft or spacecraft, and the data are recorded either photographically or in digital form. A complete remote sensing exercise usually involves integrating more than one 'level' of imagery with on-site observations, including ground-truthing.

Data must be in digital form for computer processing, so air photographs must be converted to a digital format before undergoing image analysis and data processing. Most satellite and airborne-scanner imagery is initially recorded in digital form. In digital form, remotely sensed data provide a natural input to GIS where they can be put into layers along with other spatial data (cadastral boundaries property, roads, rivers, buildings, etc.), and linked to attribute data that describes the properties of a spatial feature in the GIS. For soil, attribute data could include pH, texture or drainage class, up to the predominant (pre-determined) soil series in a defined area. Further details on remote sensing applications are given in Harrison and Jupp (1989).

The last-mentioned surveys are usually special-purpose; for example, to produce a map of the salinity risk in an irrigation area. In the more intensive surveys, such as for high-value orchards or vineyards, soil data may be collected by proximal sensing, as well as by field observation and from the analysis of soil samples. For example, an EM38 instrument pulled at close spacings over the soil surface can provide a high density of measurements of bulk soil EC (Box 13.3).

Survey methods

Any soil survey should begin with reconnaissance and general fact-finding in the area. The use of remotely sensed data has been mentioned. In addition, the surveyor needs to study the geology, climate, hydrology and vegetation of the area, as well as consult with local land users and agricultural historians. When the preliminaries are complete, the surveyor goes into the field to sample the soil and create a data set that will provide the basis for mapping. The surveyor may intend to create a special-purpose classification, or may work within the framework of an existing general-purpose classification. The sampling procedure is determined primarily by the scale of mapping and may be one of the following.

Grid survey

This procedure is preferred for special-purpose surveys that are mapped at large scales ($> 1:25,000$). The sample sites are located on a grid, and the density of sampling should be adjusted according to the area surveyed so that the number of observations per cm^2 on the final map is independent of the scale. Commonly, 4–5 observations per ha are recommended, which at a scale of $1:10,000$ produces 4–5 observations per cm^2 of the map. Boundaries are drawn to join points of equal value of a specific property, or where general-purpose classes are to be defined, boundaries are drawn between points at which dissimilar soils occur (Fig. 14.4). Grid survey is expensive, the cost increasing roughly in proportion to the square of the map scale.

One advantage of grid survey is that the observation points can be accurately located, which

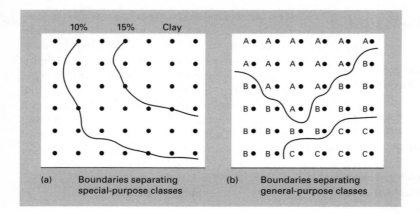

Fig. 14.4 (a) and (b) Soil boundaries drawn from grid surveys (after Beckett, 1976).

(a)   Boundaries separating special-purpose classes

(b)   Boundaries separating general-purpose classes

facilitates the entry of data into a GIS (raster format). Grid surveys (or regular sampling along a transect) with a large number of observation points (> 100 pairs) also permit the analysis of any spatially dependent variation in soil properties. Where such variation is identified, the techniques of 'spatial statistics' or geostatistics can be used to obtain best estimates of soil properties within localized areas (Box 14.3).

Free survey

In free survey, the surveyor chooses sampling points on the assumption that changes in surface features (e.g. soil colour, relief, vegetation or land use) are indicative of soil differences. The density of sampling can be varied as the surveyor concentrates on confirming the inferred boundaries and checking soil uniformity within each boundary (Fig. 14.5a). At small scales, often the inferred boundaries must be accepted and the limited field effort directed to discovering which soils lie within the boundaries and the elucidation of any recurrent patterns, such as catenas (Fig. 14.5b).

Soil survey output

Classification

A list of classes is the classification. The surveyor's aim is to place any soil in the area in an appropriate class. With special-purpose classifications, the class limits are rigidly defined according to the values of one or two properties. For example,

one classification of soils for irrigation purposes sets the following class limits (Section 13.3):

- saline soils $EC_e > 4$ dS/m, ESP < 15;
- saline-sodic soils $EC_e > 4$ dS/m, ESP > 15;
- sodic soils $EC_e < 4$ dS/m, ESP > 15.

In Australia, other special-purpose classifications (also called technical classifications) have been created for mine soils, saline soils, viticultural soils and coastal acid sulphate soils (Fitzpatrick *et al.*, 2003).

For a general-purpose classification, each class is usually built around a central concept covering a limited range in the values of several properties. Data for individual profiles are subsumed into a definitive description of each class, and one or more properties chosen as diagnostic or key criteria to enable an unknown soil to be allocated to a class. Traditionally, class concepts are derived from the soil-climate-landscape model in the surveyor's mind, which is developed from a qualitative assessment of soil morphology and the influential soil forming factors (Jenny's (1941) *clorpt* function – Section 5.1). A certain amount of overlap in property-factor values is permissible at the class limits. Recently, however, soil scientists have argued for a quantitative approach to predicting the existence of soil classes, which may be hard (sharply defined limits) or fuzzy (class membership based on fuzzy logic applied to key input variables) (McBratney *et al.*, 2003).

With modern spatially referenced databases, the distinction between special- and general-purpose classifications becomes less meaningful. Such databases allow the selection and grouping of data

Box 14.3 Geostatistics and its use in soil survey.

As stated in Section 14.1, soil is continuously variable in space. Thus, the value of a soil property at any position in the landscape is a function of that position. The function that determines the property value at a point may be extremely complex and effectively unknowable, but intuitively one expects the values of a property at close positions to be more similar than values at positions farther apart. This is the basis of regionalized variable theory, which provides statistical tools for estimating with known precision the value of a soil property at unsampled sites. For intensive studies of specific soil properties in small areas, the theory is a powerful adjunct to traditional classification methods.

The estimation procedure relies on the spatial dependence in a soil property (s) being quantified through the relationship

$$E[\{s(x) - s(x + h)\}^2] = 2\gamma(h), \qquad (B14.3.1)$$

where $s(x)$ is the property's value at point x and E[ ] is the expected value of the squared differences between the $s(x)$ values and values of s at a distance h away in a specific direction, $s(x + h)$. $\gamma(h)$ is the semi-variance, and a one-dimensional variogram is obtained when $\gamma(h)$ is plotted against the lag h in a chosen direction. This graph can have several forms of which that shown in Fig. B14.3.1 is an example. Note that the semi-variance (or the variance per site when the sites are considered in pairs) increases as h increases up to a local maximum (the sill variance). The lag at this point is called the range and defines the limit of spatial dependence. In theory, the variance at $h = 0$ is zero, but in practice it is often found there is a finite variance, called the nugget variance, when the variogram is extrapolated to $h = 0$. The nugget variance measures variation occurring over

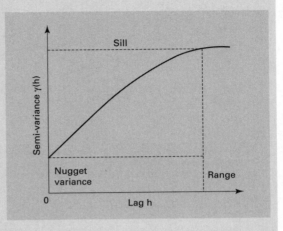

Fig. B14.3.1 Elements of a one-dimensional variogram for a soil property (after Webster, 1985).

distances less than the shortest sampling interval (the short-range variation discussed in Section 14.2).

A mathematical function fitted to the experimental variogram is used to derive weighted averages of the property s within the local area – a technique called kriging. If most or all of the variance is nugget, kriging is of no value – the variance is not spatially dependent and can be calculated from classical statistics as σ^2 (Box 14.1). In this case, the soil property s shows discrete rather than continuous variation. Generally, the value of s at a point can be considered as the sum of a global mean, a class-dependent variation from that mean and a spatially correlated residual (Heuvelink and Webster, 2001). With this conceptual model, prediction of s at unsampled points is essentially kriging with an external drift, which can be used as a basis for classification (McBratney et al., 2003).

Webster and Oliver (2001) give further details of the application of geostatistics to the spatial analysis of soil variability.

relevant to a particular purpose, and the flexible weighting of properties in accordance with that purpose. Hence the boundaries of special-purpose groupings may often cut across the divisions between the taxonomic classes of a general-purpose (often national) classification (Dudal, 2003).

Soil mapping

All soil survey should produce a map to show the geographical distribution of soil classes in an area. A class may be given a locality name and referred to as a soil series. Soils of the same series that differ

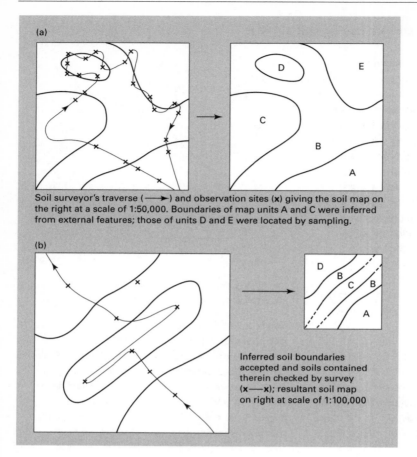

(a)

Soil surveyor's traverse (——▶) and observation sites (**x**) giving the soil map on the right at a scale of 1:50,000. Boundaries of map units A and C were inferred from external features; those of units D and E were located by sampling.

(b)

Inferred soil boundaries accepted and soils contained therein checked by survey (**x——x**); resultant soil map on right at scale of 1:100,000

Fig. 14.5 (a) and (b) Sampling pattern and soil boundaries drawn from free survey (after Beckett, 1976).

only in the texture of the A horizon are separated as types. Because all the occurrences of any one soil series need not be contiguous, their distribution in the landscape is frequently complex. Depending on the map scale, this complexity may necessitate some simplification for mapping. This is achieved by a mapping unit, consisting of one or more series, which is the smallest area of the map that can be delineated by a single boundary at the scale used. At large map scales, the objective is to use simple mapping units, each representing a narrowly defined class (i.e. a series), and to attain a purity > 80% within that unit. But as the map scale decreases (the threshold lying between 1 : 30,000 and 1 : 75,000), or even at larger scales where a complex mosaic of classes occurs in a small area, compound mapping units must be used. The legend to the map lists the classes in each unit, their

proportions and if possible their pattern of occurrence. In addition to the legend, most soil maps provide a memoir containing a detailed description of each soil series, with general comments about the geology and vegetation of the area and the management of soils grouped in each mapping unit.

At the smallest scales (< 1 : 500,000) complex units appear as:
• *Soil associations*, in which the classes are recognized and located first and then grouped for ease of presentation; or
• *land systems*, which are first differentiated by their surface expression (e.g. from air photos) and the included classes identified by subsequent sampling.
A 'nested' arrangement of mapping units, showing land systems and their included map units, is illustrated in Fig. 14.6.

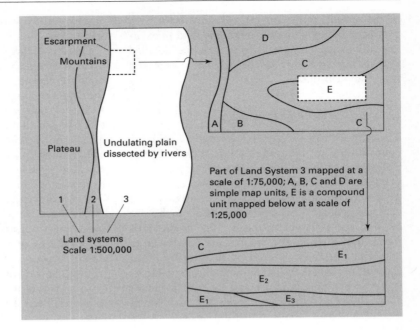

Fig. 14.6 'Nested' arrangement of soil mapping units.

Alternatively, based on empirical quantitative relationships between soil properties and other spatially referenced factors (the *clorpt* factors), a model for predicting soil properties or classes can be constructed. Through a soil-environmental factor database in a GIS, the model can be applied to an area in the landscape (from a local catchment to a large continental area) and a digital soil map produced from scratch. McBratney *et al.* (2003) outline the procedure, and Fig. 14.7 is an example

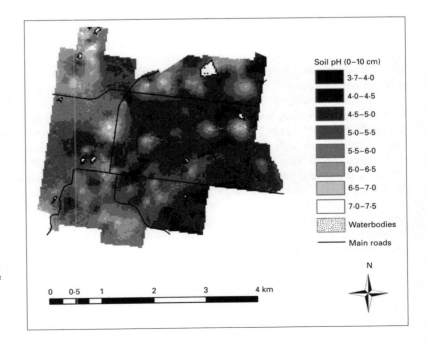

Fig. 14.7 A digital soil map of topsoil pH for a catchment in the Hunter Valley, Australia, based on field observations and a predictive model for soil pH. Courtesy of the Australian Centre for Precision Agriculture, The University of Sydney.

of the output. Other digital soil maps have been produced by digitizing the analogue data in existing soil maps, but these maps lack the detailed information provided by the original survey and topographical data.

Land evaluation

The output from soil survey can make a valuable contribution to land evaluation, which is the assessment of the suitability of land for particular purposes (also referred to as land use capability). This exercise takes account of the influence of climate, geology, soil and vegetation on the use of land for different purposes. The framework for such evaluation at a small scale (i.e. for large areas) has been established by FAO (1976) and is useful for broad-brush planning purposes. However, for planning land use at a large scale (small areas), most government agencies have developed their own land evaluation methods which can be quite different from the generic FAO scheme, but still rely heavily on soil information.

14.4 Soil information systems

The role of classification

Divergent approaches

As an output from soil survey, classification provides a framework for making local generalizations about the soil, based on properties of the soil series identified. The upwards expansion of a classification to allow progressively broader generalizations at higher levels or categories has preoccupied many individual soil scientists and corporate bodies concerned with soil survey, resulting in national and even global classifications of a hierarchical structure*. Such classifications, which are general-purpose, are normally developed from the bottom up by combining small, relatively homogeneous classes to form more inclusive groups; but identification of an unknown soil always proceeds stepwise from the highest

---

* A structure in which individuals are collected into small groups, the small groups belonging to larger groups, and so on.

category downwards, usually by means of a taxonomic key.

Opinions differ widely on the criteria to be used to establish the different categories and to separate classes within categories. Traditionally, the definitive properties are chosen on the basis of:
• The inferred *genesis* of a soil, on the assumption that the soil is in steady-state with its environment and reflects the influence of prevailing soil-forming factors; and/or
• a soil's *morphology*, which is the outward expression of the processes that shaped the soil. The success of the resulting classification depends on the taxonomist's skill and experience in choosing definitive properties that are correlated with a large number of other soil properties. Less traditional taxonomists have employed numerical methods to sort soil profiles into classes (Box 14.4).

Soil classification in the USA, where most effort has been devoted to soil survey over a very large area (> 10,000 series defined), has followed a broadly 'traditionalist' approach which places primary emphasis on soil properties that can be observed or measured, and preferably those that are quantitative, such as clay content, rather than qualitative, such as colour or grade of structure. In establishing classes, the classification aspires to take no account of the imperfectly understood relationships between pedogenesis and soil-forming factors. Nevertheless, one of its underlying premises is that 'the properties selected should be those that either affect soil genesis or result from soil genesis' (Soil Survey Staff, 1960). To make the classification as comprehensive as possible, the taxonomists consulted extensively with pedologists in other countries as the classification progressed through several 'approximations', culminating in the publication of Soil Taxonomy (Soil Survey Staff, 1975, 1999). The outcome of this and other attempts at national or supra-national soil classification is summarized in Table 14.1.

Conceptual classes

No matter how divergent current classifications are in principle, language and format, two main levels of generalization stand out:
• A lower level encompassing the surveyor's series, which are narrow classes in harmony with the landscape; and

Box 14.4 A numerical approach to soil classification.

Numerical classification uses objective, similarity-grouping techniques. The profiles are characterized in terms of those soil properties that are observable or measurable; that is, inferred genesis is not used. Some properties, such as soil water content, are not used because they are transient and influenced by external factors not germane to the objective of the classification. These properties may need to be specified, however, because of their effect on definitive properties, such as the effect of water content on soil colour and grade of structure. Generally, each property is given equal weighting when soil profiles are compared, although some taxonomists consider that the value of a property in a surface horizon should be given greater weight than its value in the subsoil. Some disadvantages of numerical classifications are that:

• Although the classes contain maximum information, they frequently do not correspond to natural soil groupings in the field and so are difficult to map; and

• numerical classifications are unstable because the inclusion of newly described profiles in the similarity-grouping process may completely alter the classification.

Further details are given by Webster (1977).

• a higher level of broad classes with somewhat overlapping boundaries, each of which embodies a central, usually pedogenic, concept.

Classes based on a central concept are valuable in teaching soil science because each conveys a measure of knowledge about the soil, distilled from the experience of scientists in many areas. For this reason the USDA classification of Baldwin *et al.* (1938), revised by Thorp and Smith (1949), was presented in Table 5.4. The suborder descriptions and Great Soil Group names denote the influence of one or more of the soil-forming factors on pedogenesis. The same can also be said of Soil Taxonomy at the higher categorical levels (from the 'order' down), so this widely adopted classification may be used to illustrate the meaning of conceptual classes. The relationship between the ten orders of Soil Taxonomy and the Great Soil Groups of the older Baldwin *et al.* classification is summarized in Box 14.5.

Diagnostic horizons

Below the order level, classes in Soil Taxonomy are defined progressively more narrowly according to measured soil properties and the microclimate in which a soil forms. For example, the suborders are separated on differences associated with hydromorphism (Section 9.5), vegetation (as it influences podzolization and lessivage – Section 9.2), parent material, time and intensity of weathering

Table 14.1 Structures of a selection of national and supra-national soil classifications.

Classification	Categories (the number of actual or potential classes in each category is shown in parentheses)
Soil Taxonomy (Soil Survey Staff, 1999)	Orders (10) → Suborders (47) → Great Groups (185) → Subgroups (970) → Families (4500) → Series (> 10,000)
World Reference Base for Soil Resources (FAO, 1998)	Reference Soil Groups (30) → Soil Units (> 200) → Subunits
Soil Classification for England and Wales (Avery, 1980)	Major Groups (10) → Groups (41) → Subgroups (109) → Series
Canadian System of Soil Classification (Canada Soil Survey Committee, 1978)	Orders (8) → Great Groups (22) → Subgroups (165) → Families (c. 900) → Series (c. 3000)
Australian Soil Classification (Isbell, 2002)	Order (14) → Suborder → Great Group → Subgroup → Family
Factual Key for Australian Soils (Northcote, 1979)	Divisions (3) → Subdivisions (11) → Sections (54) → Classes (271) → Principal Profile Forms (855)
New Zealand Soil Classification (Hewitt, 1998)	Orders (15) → Groups (76) → Subgroups (253)

Box 14.5 Soil Taxonomy orders and some of the included Great Soil Groups. After Butler, 1980.

Order	Brief description	Included Great Soil Groups
Entisols	Recently formed soils, no diagnostic horizons	Azonal soils and some Low Humic Gley soils
Vertisols	With swell-shrink clays, high base status	Grumusols
Inceptisols	Slightly developed soils without contrasting horizons	Acid Brown soils, some Brown Forest soils. Low Humic Gley and Humic Gley soils
Aridosols	Soils of arid regions	Sierozem, Solonchak, some Brown and Reddish Brown soils and Solonetz
Mollisols	Soils with mull humus	Chestnut, Chernozem, Prairie, Rendzinas, some Brown Forest, Solonetz and Humic Gley soils
Spodosols	Soils with iron and humus B horizons	Podzols, Brown Podzolic soils and Groundwater Podzols
Alfisols	Soils with a clay B horizon and > 35% base saturation	Grey-brown Podzolic, Grey Wooded soils, Degraded Chernozem and associated Planosols
Ultisols	Soils with a clay B horizon and < 35% base saturation	Red-yellow Podzolic, Reddish-brown Lateritic soils and associated Planosols
Oxisols	Sesquioxide-rich, highly weathered soils	Lateritic soils, Latosols
Histosols	Organic hydromorphic soils	Bog soils (Peats)

(resulting in ferrallitization – Section 9.4) and so on.

Separation of classes in the next inferior category – the Great Group – is intended to divide the soil continuum into more or less equally spaced groupings according to the variation in soil genesis. Much use is made of diagnostic horizons, which are connotative of an underlying pedogenic process (just as the symbol Ea in the Soil Classification for England and Wales connotes a bleached sandy horizon from which sesquioxides have been eluviated). However, a diagnostic horizon need not coincide with a pedological horizon, such as A1, B2 or Bt: it may include parts of more than one pedological horizon or soil layer. For example, a mollic epipedon – one of the six diagnostic horizons recognized at the surface – may include all of an A horizon and the upper part of a B horizon when soil fauna have uniformly incorporated organic matter to a considerable depth. More weight is given in Soil Taxonomy to subsurface horizons, of which there are ten, because they are less prone to disturbance through ploughing or truncation by erosion. Thus, the Great Groups are separated on the presence or absence of diagnostic horizons and their arrangement in the profile.

Other soil classifications

Soil Taxonomy has been widely deployed outside the USA and purports to be a supra-national classification. It is not of great value for teaching general soil science because the terminology at any level below order lacks intuitive meaning and is intimidating for the non-taxonomist. Many countries have persevered with developing and refining their own national classifications: for example the Australian Soil Classification in Australia (Isbell, 2002) and the New Zealand Soil Classification in New Zealand (Hewitt, 1998). The former scheme is general-purpose and multicategoric (Table 14.1), and designed to incorporate as many as possible of the desirable features of schemes it replaces, such as the Great Soil Group classification (Stace et al., 1968) and the Factual Key (Northcote,

Table 14.2 Orders of the Australian Soil Classification and approximate included Great Soil Groups. After Isbell, 2002.

Order	Connotation	Included Great Soil Groups
Anthroposols	Human influence	None
Calcarosols	Calcareous throughout	Solonized brown soils, grey-brown and red calcareous soils
Chromosols	Often brightly coloured	Non-calcic brown soils, some red-brown earths and podzolic soils
Dermosols	Often with clay skins on peds	Prairie soils, chocolate soils, some red and yellow podzolic soils
Ferrosols	High iron content	Krasnozems, euchrozems, chocolate soils
Hydrosols	Wet soils	Humic gleys, gleyed podzolic soils, solonchaks and some alluvial soils
Kandosols	–	Red, yellow and grey earths, calcareous red earths
Kurosols	Pertaining to clay increase	Many podzolic soils and soloths
Organosols	Dominantly organic materials	Neutral to alkaline and acid peats
Podosols	Bleaching of A horizon due to Fe translocation	Podzols, humus podzols, peaty podzols
Rudosols	Rudimentary soil development	Lithosols, alluvial soils, calcareous and siliceous sands, some solonchaks
Sodosols	Influenced by sodium	Solodized solonetz and solodic soils, some soloths and red-brown earths, desert loams
Tenosols	Weak soil development	Lithosols, siliceous sands, alpine humus soils and some alluvial soils
Vertosols	Shrink–swell clay	Black earths, grey, brown and red clays

1979), and schemes with which it competes (e.g. Soil Taxonomy and the World Reference Base for Soil Resources). Some of the general principles underlying the development of the Australian Soil Classification are that it:

• Is based on diagnostic properties, horizons or materials;
• assumes concepts of pedogenesis in the development of horizons in a vertical sequence in the soil profile;
• is based on Australian data (mainly from the field though supported by laboratory measurements), with the selected properties to be relevant as far as possible to land use and soil management; and
• has simple but unambiguous nomenclature.

The 14 orders of this classification, their connotation and included Great Soil Groups from Stace *et al.* (1968) are given in Table 14.2.

Other countries, particularly developing ones where soil survey data are sparse, have adopted the World Reference Base for Soil Resources (WRB) (FAO, 1998) (formerly the World Soil Classification). This classification arose as the list of classes in the legend to the Soil Map of the World (FAO–Unesco, 1974), which has subsequently been revised several times. The classification is multi-categorical, with the highest level being 30 Reference Soil Groups that convey useful generalizations about the genesis of a soil in relation to the interactive effects of the main soil-forming factors (Table 14.3). However, the Reference Groups are so broad and the total variation within each group so large that they have little value in predicting soil suitability for land use. Their merit is that they constitute a gallery of conceptual classes that convey some general understanding to a worldwide clientele. The second categorical level is the critical one, where Soil Units are separated on the basis of diagnostic horizons that connote the underlying pedogenesis, in a way similar to Soil Taxonomy. Some of the descriptive terminology has been adopted from Soil Taxonomy in simplified form. But a number of traditional Great Soil Group names have also been retained, as well as new names coined that do not suffer in translation nor have different meanings in different countries.

Originally mapped at a scale of 1 : 5,000,000 as soil associations designated by the dominant soil unit, the units are now mapped at

Table 14.3 Reference Soil Groups and their connotations from the World Reference Base for Soil Resources. After FAO, 1988, 1998.

Acrisols	Strongly acid and base saturation < 50%
Alisols	High aluminium content
Albeluvisols	Like a Luvisol but with an irregular A–B horizon boundary
Andosols	Derived from volcanic material
Anthrosols	Influenced by human activity
Arenosols	Weakly developed, coarse-textured soils
Calcisols	Accumulation of $CaCO_3$
Cambisols	Change in colour, structure or consistence from A to B horizon
Chernozems	Soils with a dark chernic or mollic horizon
Crysols	Evidence of freezing
Durisols	Hardened or cemented horizon
Ferralsols	Soils high in sesquioxides
Fluvisols	Formed on alluvial deposits
Gleysols	Show gley features due to waterlogging
Gypsisols	Accumulation of $CaSO_4.2H_2O$ (gypsum)
Histosols	High in fresh or partly decomposed organic matter (> 40 cm deep)
Kastanozems	Brown mollic horizon and $CaCO_3$ accumulation
Leptosols	Shallow soils (< 25 cm deep)
Lixisols	Clay accumulation (duplex) and base saturation > 50%
Luvisols	Clay accumulation through lessivage
Nitosols	Clay accumulation with shiny ped surfaces
Phaeozems	Soils with a mollic horizon
Planosols	Soils with an abrupt texture change
Plinthosols	Hardened sesquioxide horizon within 50 cm
Podzols	Soils with a strongly bleached horizon (spodic)
Regosols	Poorly differentiated rocky soils
Solonchaks	Soils affected by salts
Solonetz	Soils affected by sodium
Umbrisols	Soils with a dark brown umbric horizon
Vertisols	Swell-shrink clay > 35%

1 : 1,000,000 with subunits differentiated on textural class (coarse, medium and fine) and slope class (level to gently undulating, rolling to hilly, and steeply dissected to mountainous). At a lower level, subunits (phases) are now recognized, which are intergrades between first and second level units or have characteristics additional to those used in defining the soil units.

The WRB is a valuable inventory of global soil resources (see **Modern soil information systems**) and the legend of the Soil Map of the World has been used for correlation between different national classifications.

Modern soil information systems

In this age, no soil classification is of much value *unless* it is supported by a comprehensive electronic data base (Section 14.1). Only in this case can major deficiencies in the communication of soil information to, and use by non-specialist users be addressed. Several national and international agencies for soil and land resource information have foreseen this need and acted to satisfy it. For example, the USDA Natural Resources Conservation Service (NRCS) recognized that many of the concepts and tools being used in soil survey and classification pre-date present advanced computer hardware and software technology. These tools and concepts were developed mainly because it was not possible for humans to handle in an orderly way the huge amounts of data derived from soil and land surveys. This constraint is no longer relevant so the NRCS has developed a National Soil Information System (NASIS) consisting of an attribute database linked to a Soil Survey Geographic Database that can produce digitized soil maps (http://soildatamart.nrcs.usda.gov). NASIS is intended to provide the following capabilities:
• Preservation of individual site data (field and laboratory);
• the relation of these individual data and the spatial variability of soil properties to individual, delineated areas on soil maps;
• provision of two types of map units – primary units, which are composed of the most detailed, delineated spatial entities representing the soil scientists' best knowledge, and user-defined

units, which are any unique grouping of primary units defined by a user to meet specific needs; and

• interpretation of individual components of map units as well as what were formerly included soil classes within the units.

In Australia, coordination in the collection, collation, interpretation and dissemination of soil information was hindered by the existence of several State and Federal agencies and the incompatibility of the databases used. This problem was addressed through the creation of the Australian Collaborative Land Evaluation Program (ACLEP), which has collaborated with State and Federal agencies to agree data exchange protocols for use in describing soil profiles and collating spatial and attribute data from various sources, including information in the Atlas of Australian Soils (scale 1 : 2,000,000). In 2001, a national database of soil information suitable for use at a continental or large regional scale – the Australian Soil Resources Information System (ASRIS) – was established (www.clw.csiro.au/aclep/asc/asc.htm). This database, which is being expanded and updated, focuses on the agricultural zone of the country.

At the international level, the International Union of Soil Sciences and agencies such as FAO and the United Nations Environment Program (UNEP) have cooperated to set up a World Soils and Terrain Database (SOTER). The SOTER database comprises digitized spatial data for geology, topography, soils, vegetation and climate manipulated through GIS software and linked to non-spatial attribute data stored in a Relational Database Management System. Overall, SOTER has the following characteristics:

• It is a comprehensive framework for the storage and retrieval of soil and terrain data that can be used at different scales;

• contains sufficient data to allow information to be extracted at a scale of 1 : 1,000,000 in the form of maps and tables;

• is compatible with global databases of other environmental resources; and

• through resource maps, interpretative maps and tabular information, enables international, national and regional specialists to develop, manage and conserve environmental resources (World Soil Resources Reports, 1993).

The SOTER database, which is fully compatible with the WRB, has been tested in various parts of the world and is undergoing continual expansion and revision. A digital soil map of the world and derived soil properties are available on CD-ROM (FAO, 2003).

14.5 Summary

Soil information is collected by the survey of soils in the field and analysis of samples in the laboratory. Field information is collected in many ways including:
• Remote sensing from spacecraft and aircraft;
• geological and vegetation maps;
• proximal sensing, visual observations and measurements made by a soil surveyor; and
• the experience of soil users.

The exact combination of methods used depends on the purpose of the survey, which will determine the intensity of sampling (or spacing) and eventually the scale of any soil map (or resolution in the case of digital mapping). Depending on the map scale, which may vary from 1 : 5,000,000 for continental reconnaissance surveys to 1 : 1000 for intensive surveys of local soil management problems, surveying proceeds on a grid (for large scales) or free survey pattern (for small scales). Geostatistical techniques are useful for intensive surveys on small areas (large map scale) where there is quantifiable spatial dependence in the variation of soil properties.

Traditionally, soil survey output has been used in soil classification, which has been a prerequisite for the orderly presentation of accumulated knowledge to soil scientists and other users. Classification involves grouping soils into classes according to the degree of similarity between individual soil profiles. However, soil classification is made difficult and its results contentious because there is no definable soil entity: the soil mantle varies continuously so that subdivision of the variation in one or more properties to create classes is often arbitrary. Grouping has been attempted using numerical methods, but the results can be difficult to present in a soil map. For a classification to be useful for predictive purposes, the variation remaining within classes should be substantially less than the variation between classes for a range of soil properties. Classes are defined in terms of one or two soil properties relevant to a specific purpose (as in a technical or special-purpose classification), or on many properties considered relevant for various soil uses (as in a general-purpose classification). Diagnostic or key properties, which are not necessarily relevant to use but are easily assessed in the field, are often chosen to define classes on the assumption (or knowledge) they are correlated with more important properties.

Classes recognized by a surveyor (soil series) are usually aggregated to form progressively less homogeneous (and hence more inclusive) classes that facilitate national and international exchange of soil information. Because of the wide choice of criteria by which to separate classes at any one level and establish different levels or categories of generalization, many classifications have been proposed. The USDA's Soil Taxonomy is a classification using only those soil properties that can be observed or measured, preferably quantitatively, in establishing criteria for class separation. The advantage of this approach, particularly at the lower, more exclusive category levels, is that it avoids reliance on imperfectly understood relationships between soil properties, pedogenesis and the soil-forming factors. However, soil scientists are now developing quantitative empirical models to predict the variation in soil properties with soil forming factors in a landscape. A classification can then be developed where the class boundaries are sharp or fuzzy, depending on the logic applied. Coupled with a *digital elevation model (DEM)*, the results are displayed as a digital soil map based on a GIS. Thus far, this approach has been most successful for single properties and hence the development of special-purpose classifications, but progress is being made on multifactor models that could lead to a general-purpose classification.

Lay users of soil maps tend to want information for specific purposes. Non-taxonomic soil scientists seek broader generalizations about soil differences that are embodied in the higher level conceptual classes. For this purpose Soil Taxonomy's Orders and Suborders may be sufficient, or the WRB, which comprises more than 200 Soil Units grouped into 30 Reference Soil Groups.

Advances in information technology now permit a more comprehensive approach to the storage of soil information from a range of sources, its collation and codification to develop classifications, and its retrieval in map or tabular form. Thus, many national and international agencies charged with making soil and land resource inventories are developing modern soil information systems that ideally provide for the preservation of individual site data (field and laboratory observations) in a relational database, and the display of the spatially referenced original or interpreted data in digital soil maps. Examples are the international SOTER database developed by FAO in collaboration with national agencies, the NASIS database in the USA, and the ASRIS database in Australia.

References

Avery B. W. (1980) *Soil Classification for England and Wales*. Soil Survey Technical Monograph No. 14. Soil Survey of England and Wales, Harpenden, UK.

Baldwin H., Kellogg C. W. & Thorp J. (1938) Soil classification, in *Soils and Man*. Yearbook of Agriculture, Washington DC.

Beckett P. H. T. (1976) Soil survey. *Agricultural Progress* **51**, 33–49.

Butler B. E. (1980) *Soil Classification for Soil Survey*. Clarendon Press, Oxford.

Canada Soil Survey Committee (1978) *The Canadian System of Soil Classification*. Publication 1646. Research Branch, Canada Department of Agriculture, Ottawa.

Dudal R. (2003) How good is our soil classification? in *Soil Classification a Global Desk Reference* (Eds H. Eswaran, T. Rice, R. Ahrens & B. A. Stewart). CRC Press, Boca Raton, pp. 11–18.

FAO (1976) *A Framework for Land Evaluation*. FAO Soils Bulletin 32. FAO, Rome.

FAO (1998) *World Reference Base for Soil Resources*. World Resources Report No. 84. FAO, Rome.

FAO (2003) *Digital Soil Map of the World and Derived Soil Properties*. Version 3.6. FAO/Unesco, Rome.

FAO–Unesco (1974) *Soil Map of the World* 1 : 5,000,000 Volume 1, Legend. Unesco, Paris.

FAO–Unesco (1988) *Soil Map of the World, Revised Legend*. World Soil Resources Report No. 60. FAO, Rome.

Fitzpatrick R. W., Powell B., McKenzie N. J., Maschmedt D. J., Schokneckt N. & Jacquier D. W. (2003) Demands on soil classification in Australia, in *Soil Classification a Global Desk Reference* (Eds

H. Eswaran, T. Rice, R. Ahrens & B. A. Stewart). CRC Press, Boca Raton, pp. 77–100.

Harrison B. A. & Jupp D. L. B. (1989) *Introduction to Remotely Sensed Data*. CSIRO Publications, Melbourne.

Heuvelink G. B. M. & Webster R. (2001) Modelling soil variation: past, present and future. *Geoderma* **100**, 269–301.

Hewitt A. E. (1998) *New Zealand Soil Classification*, 2nd edn. Manaaki Whenua Press, Lincoln, New Zealand.

Hodgson J. M. (1978) *Soil Sampling and Soil Description*. Clarendon Press, Oxford.

Isbell R. F. (2002) *The Australian Soil Classification*, revised edn. Australian Soil and Land Survey Handbooks Series Volume 4. CSIRO Publishing, Melbourne.

Jenny H. (1941) *Factors of Soil Formation*. McGraw-Hill, New York.

McBratney A. B., Mendonca Santos M. L. & Minasny B. (2003) On digital soil mapping. *Geoderma* **117**, 3–52.

McDonald R. C., Isbell R. F., Speight J. G., Walker J. & Hopkins M. S. (1998) *Australian Soil and Land Survey Field Handbook*, 2nd edn, reprinted. Australian Collaborative Land Evaluation Program, Canberra.

Northcote K. H. (1979) *A Factual Key for the Recognition of Australian Soils*, 4th edn. Rellim, Glenside, South Australia.

Soil Survey Division Staff (1993) *Soil Survey Manual*, 3rd edn. United States Department of Agriculture, National Soil Survey Center. USDA-NRCS Soil Survey Division Data National STATSGO Database, Washington DC.

Soil Survey Staff (1960) *Soil Classification: a Comprehensive System 7th Approximation*. United States Department of Agriculture, Washington, DC.

Soil Survey Staff (1975) *Soil Taxonomy. A Basic System of Soil Classification for Making and Interpreting Soil Surveys*. Handbook No. 436. United States Department of Agriculture, Washington DC.

Soil Survey Staff (1999) *Soil Taxonomy. A Basic Classification for Making and Interpreting Soil Surveys*, 2nd edn. Handbook No. 436. United States Department of Agriculture Natural Resources Conservation Service, Washington DC.

Stace H. C. T., Hubble G. D., Brewer R., Northcote K. H., Sleeman J. R., Mulcahy M. J. & Hallsworth E. G. (1968) *A Handbook of Australian Soils*. Rellim, Glenside, South Australia.

Thorp J. & Smith G. D. (1949) Higher categories of soil classification. *Soil Science* **67**, 117–26.

Webster R. (1977) *Quantitative and Numerical Methods in Soil Classification and Survey*. Oxford University Press, Oxford.

Webster R. (1985) Quantitative spatial analysis of soil in the field. *Advances in Agronomy* **3**, 1–70.

Webster R. & Oliver M. A. (2001) *Geostatistics for Environmental Scientists*. Wiley & Sons, London.

World Soil Resources Reports (1993) *Global and National Soils and Terrain Digital Databases (SOTER)*. Land and Water Development Division, FAO, Rome.

Young A. (1973) Soil survey procedures in land development planning. *Geographical Journal* **139**, 53–64.

Further reading

Beckett P. H. T. & Webster R. (1971) Soil variability – a review. *Soils and Fertilizers* **34**, 1–15.

Ernstrom D. J. & Lytle D. (1993) Enhanced soil information systems from advances in computer technology. *Geoderma* **60**, 327–41.

Petersen G. W., Bell J. C., McSweeney K., Nielsen G. A. & Robert P. C. (1995) Geographic Information Systems in agronomy. *Advances in Agronomy* **55**, 67–111.

Example questions and problems

1 (a) Give an example of an instrument used for proximal sensing of a soil property at a high density of observations.
  (b) Give an example of an 'active' remote sensing technique.
  (c) How are remotely-sensed soil data checked during soil survey?

2 (a) What is the traditional output of a soil survey?
  (b) A reconnaissance survey has an average of one soil observation every 1 km. What is the largest scale at which the data collected can reasonably be mapped?
  (c) If the soil for an irrigation scheme covering 1000 ha is to be mapped at a scale of 1 : 2000, how many soil observations should be made per ha?
  (d) Is 1 : 2,000,000 a large or small scale?
  (e) What is the statistical measure of variation in a soil property?

3 (a) What comprises a soil classification?
  (b) A special-purpose classification has been carried out on soils to be used for rice growing. The pooled within-class variance is 60% of the total variance in the soil properties sampled. State with a reason whether you think the classification is reliable.

  (c) What is the statistical method for determining whether there is spatially dependent variation in a soil property?
  (d) If there is spatial dependence in soil property variation, what technique is used to interpolate values at unsampled sites?

4 What is the prime purpose of land evaluation as carried out by the FAO method?

5 (a) Name two widely used international soil classifications.
  (b) What is the highest categorical level in Soil Taxonomy?
  (c) In the Australian Soil Classification, into which three orders do soils with duplex profiles fall? (Hint – see Section 9.2.)
  (d) In the Australian Soil Classification, in which class would soils with uniform-texture profiles dominated by sesquioxides normally fall?
  (e) What is the high-level class in Soil Taxonomy equivalent to the class identified in 5(d)?

6 (a) Complete the following: a modern soil information system comprises .... and .... data collated from various sources.
  (b) What generic computer software is used to display spatial soil data?

Chapter 15

Soil Quality and Sustainable Land Management

'The fabric of human life is woven on earthen
looms – it everywhere smells of the clay'
(from Mitchell *et al.*, 1950 –
Saskatoon Soil Survey Report No. 13)

15.1 What is soil quality?

Options for defining soil quality

In Chapter 11, we see that land productivity,
usually measured by the yield of crop or animal
product per hectare, is a function of a number
of factors – soil, climate, pests, disease, genetic
potential of the crop/animal, and management.
Fertility is probably the most important single
soil factor affecting productivity. However, in
recent years groups in society outside the agricultural sector have demanded a more comprehensive
descriptor of the 'condition' of a soil than is
provided by soil fertility. This reflects an overall
concern in society about the state of the environment in which we live, of which soil is increasingly
recognized as a vital part, along with air and
water. Whereas the desirable qualities of air and
water for human and animal life can be adequately
defined and monitored, the same is not true of
soil for several reasons:

• Soil is continuously variable in space and,
being a complex, dynamic biological system, is
variable in time. The rate of change of individual
soil properties is itself very variable.

• In most cases, a soil's properties do not impinge
directly on human and animal well-being. The
effect of soil is experienced indirectly through
the diversity and health of the natural ecosystems
it supports, the quality and quantity of crops
(and animals) grown, the water (and sometimes
sediment) that runs off and drains from a soil,
and the gases absorbed or released by a soil.

• The quality of a soil depends on the soil's use
and its fitness for that use or function. This
premise underlies the assessment of land use
capability (called land evaluation – Section 14.3)
in which land is classed according to its suitability
for various uses. Agriculture has traditionally
been a key use, and soil quality is a major factor
determining the land capability class (ranging from
the best class – no limitations, to the worst
class – severe limitations). The land use capability
approach has been adapted to determine the suitability of soils for a range of crops, usually based
on a correlation between crop yield and a set of
biophysical criteria (e.g. climate, soil physical
properties, soil acidity and organic matter content)
to derive a soil quality index. However, such an
index is specific to the crop or crops being considered and not necessarily applicable to a wider
range of land uses; nor does it indicate the
long-term sustainability of this particular land use
(Section 15.2).

• The community at large has concerns about
soil suitability not only for agriculture but also
as a reserve of biological diversity (biodiversity),
for nurturing natural ecosystems of recreational
and conservation value, as a foundation for stable
landscapes, and as a component of waste management systems.

Thus, although we ideally aspire to defining soil quality in objective terms, in reality the concept is diffuse and must reflect a range of social, cultural and economic factors as well as biophysical properties. Doran and Parkin (1994) offered a broad definition of soil quality as 'the capacity of a soil to function within ecosystem boundaries to sustain biological productivity, maintain environmental quality, and promote plant and animal health'. However, Bezdicek *et al.* (1996) pointed out that the relationship between crop productivity and soil quality was probably the best understood of Doran and Parkin's quality components. Even with respect to crop productivity and soil quality, there are many opinions as to which one or other soil property is a key factor for quality. Further, there is a lack of standardization of methods for measuring the relevant soil properties, particularly the biological ones.

For these reasons, soil quality is often assessed by the management practices that are designed to improve, or at least maintain, a soil's performance for specific purposes, e.g. minimizing erosion, supporting a variety of crops, storing good quality water underground, or minimizing the movement of heavy metals applied in industrial wastes. Soil quality is inferred from:

• The degree to which a management practice protects the soil from degradation (caused by erosion for example); and
• the extent of adoption of a 'best management practice' (BMP), or other practices, within an area of interest (a farm, catchment or region).

Production and environmental impact

Until such questions are resolved as:
• Is there a limited number of key soil properties that can serve as appropriate quality indicators for a range of land uses; and
• is it feasible to measure these indicators over time (monitoring) with proper spatial referencing at a range of scales;

the 'inverse' approach is likely to be preferred whereby landholders, acting on their own knowledge or informed by scientific research, infer soil quality from the land management practices in force. There is, however, an important proviso. The research designed to develop and test BMPs must consider a particular system (e.g. high rainfall pastures) in the context of the whole environment. Too often such research has followed a 'reductionist' approach, focusing on a narrow production problem, without considering the ecological processes that sustain biological activity overall, and the flows of energy, nutrient and water between the production system and the wider environment. The widespread introduction of an exotic legume (subterranean clover) into pastures for beef cattle and sheep in southern Australia is an example of this point. Pastures were fertilized with superphosphate to encourage legumes, which fixed atmospheric N_2 and improved the growth and quality of the pasture. With higher stocking rates, native perennial grasses were grazed out and more N cycled through the grazing animals. Mineral N accumulated in the soils at the end of the dry summers and substantial NO_3^- leaching occurred during autumn and early winter, contributing to soil acidification. The already acidic A horizons of duplex soils in the region slowly became more acid, a condition which adversely affected productivity and the range of species that could be grown (see 'Accelerated soil acidification' – Section 11.3). Watertables have also risen in the higher rainfall areas (> 600 mm/yr) because of greater drainage below the root zone of the short-lived annual grasses compared with perennial grasses and native trees.

Recognition of these shortcomings has led to the idea that research on land and water management should focus on the 'ecosystems services' provided by a natural landscape (including the parts influenced by humans). This approach requires that all the services (e.g. clean water and air, energy, food production, pharmaceuticals, recreation and conservation attributes) provided by an ecosystem are identified and studied so that the processes generating these services can be recognized and fostered (Daily, 2000). More is said about this approach in Section 15.2.

Contamination and soil quality

The definition of soil quality becomes more specific when contaminated land is considered (Box 10.6). Soil contamination by organic and inorganic chemicals occurs in urban and rural areas from various causes such as:

- Mining activities – exploration, extraction and processing;
- industrial activities, including defence, and the disposal of industrial waste;
- urban living, including motor transport, and the concentration of urban waste through storm water runoff, sewage effluent and sewage sludge (Box 11.1); and
- agriculture, forestry and horticulture, through over-use of fertilizers and pesticides.

Contaminants affect human health through several pathways: (a) directly, through ingestion of soil, inhalation of gases or dust, and skin contact, and (b) indirectly through food consumed and water that is drunk or used for washing. To determine whether a soil is contaminated (and its quality adversely affected) through human activities, the concentration of potential contaminants must be compared with their 'background' or normal concentrations in soil. For some anthropogenic compounds (e.g. DDT), the background is zero and any detectable concentration indicates contamination. For most inorganic elements, such as the heavy metals, assessing contamination is more difficult because of their variable concentration in soils naturally, due to differences in parent material and rates of weathering (Chapter 5). In this case, background concentrations must be calculated after excluding high concentrations due to geothermal anomalies, localized pollution due to waste disposal, deposition along roadways and over-use of agricultural chemicals and fertilizers.

In recent years, most countries have become very aware of soil contamination, particularly in urban areas, and have developed strategies for managing and, if necessary, ameliorating the problem through remediation. In Australia and New Zealand, for example, guidelines for the assessment and management of contaminated sites have been drawn up (ANZECC/NHMRC, 1992). These guidelines give background concentrations for a range of inorganic and organic substances, and indicate threshold concentrations above which further investigation of a site is required to determine the extent of contamination. A selection of soil quality criteria for contaminated soils is given in Table 15.1. Many of the values were originally derived from European and North American experience, but are continually being revised as new information for Australasia becomes available.

Table 15.1 Soil quality guidelines (concentrations in mg/kg soil) for environmental purposes. After Barzi et al., 1996.

Substance	Background concentration	Environmental investigation threshold
Heavy metals		
Cd	0.04–2	3
Cr	0.05–110	50
Ni	2–400	60
Pb	2–200	300
Monocyclic aromatic hydrocarbons (MAH)		
Benzene	0.05–1	1
Polycyclic aromatic hydrocarbons (PAH)	0.95–5	5
Chlorinated hydrocarbons		
PCBs (total)	0.02–0.1	1
DDT	< 0.001–1	0.2

15.2 Concepts of sustainability

Land use and land management

Although the terms 'land use' and 'land management' tend to be used interchangeably in the literature on sustainability, it is important to recognize the distinction in their meaning.

The use to which land is put may be urban, industrial or rural. Within each major category, such as rural land use, there are subcategories such as agriculture, forestry (native woodlands and plantations), conservation and recreation. Some of these subcategories can apply to urban or industrial land use. Agriculture may be further subdivided into field crops, pasture, horticulture and so on, and in some cases, such as with sugar cane grown on specially assigned land in Queensland, the particular crop grown can be specified. Land management, on the other hand, refers to the practices adopted for a particular land use, such as the method of cultivation (Section 11.4), erosion control measures such as contour banks (Section 11.5), as well as fertilizer practice and pest control (Chapter 12). Management should encompass farm planning whereby the various uses of land – cropping, grazing, shelter belts and wood lots, and irrigation where feasible – are assigned to the most suitable soils and landscape units, and well-informed decisions are made about financial investment in the farm and its enterprises. Land management also includes such activities as surface and subsurface drainage, groundwater pumping (for watertable control), and decisions on fence positions, paddock sizes (especially for livestock) and the use of animal wastes.

Clearly, a land use that is totally inappropriate for a particular soil will not be viable, irrespective of whether management is good or bad. Similarly, even a land use that is appropriate for a particular soil can become non-viable if the management practices are bad. These distinctions are important in defining the meaning of sustainability.

Ecological, economic and social sustainability

Sustainable agriculture

The concept of 'sustainability' became a widely accepted environmental and political credo during the 1980s. However, with its increasing use the word began to lose precision in its meaning (Viederman, 1993). Implied in its use are concepts of resource quality and availability, economic viability, environmental impact, social justice, intergenerational equity and biodiversity. There are also spatial and temporal dimensions to sustainability, emphasizing the dynamic nature of the concept (Smyth and Dumanski, 1993).

In discussing sustainability, some authors have focused solely on agriculture. For example, Reganold et al. (1990) considered that many formerly productive and labour-efficient farms in the USA were experiencing declining productivity, deteriorating environmental quality, reduced profitability and possible threats to human and animal health. This change in fortune was attributed to a number of causes including:

• Over-dependence on external inputs such as fossil fuels, fertilizers and concentrated feed for animals;
• heavy use of pesticides;
• depletion of groundwater reserves through irrigation;

• cropping and grazing management that induced soil structural decline and accelerated soil erosion; and
• loss of soil biological diversity, particularly beneficial organisms.

As an alternative to 'conventional' agriculture, which depends on high inputs (mostly sourced from outside the farm) and is often environmentally unfriendly, Reganold et al. (1990) advocated 'non-conventional' agriculture – sometimes called organic, alternative, regenerative, ecological or low-input – in which a farmer works in harmony with the biological and ecological processes of the soil-plant system, rather than relying on a large input of chemicals and expenditure of energy. Some of the basic principles of alternative or organic agriculture are described in Box 15.1.

A key question in the debate on the relative merits of non-conventional and conventional agriculture is – can the former be as profitable as the latter has proven to be? The answer is crucial to persuading large numbers of profit-minded farmers to change their practices, because they argue it is 'difficult to be green when you are in the red!' The results of studies carried out to answer this question have been mixed. Cochrane (2003) concluded that although yields from non-conventional farms are lower than from conventional farms, the difference is frequently offset by lower production costs that lead to equal or greater net returns from non-conventional farms. It is also argued that the quality of 'organic' produce is better than that from conventional farms because of smaller pesticide residues and greater concentrations of health-promoting antioxidants.

Notwithstanding the conclusion that non-conventional agriculture can be equally or more profitable than conventional agriculture, the generally acknowledged lower yields (and often greater damage due to pest and disease attack) from the former beg the question – what would happen to world food security if China, India and the USA (the three largest food producing countries) were to switch to organic or low-input agriculture on a massive scale? The success of the Green Revolution in dramatically increasing food production, mainly in Asian countries, through high-yielding varieties, fertilizers and increased irrigation is described in Box 11.2. But as discussed in Box 11.2, this success has imposed severe

Box 15.1 Alternative or organic agriculture.

Organic agriculture is sometimes described as 'biological farming' as opposed to the 'chemical farming' of conventional agriculture. It is more labour-intensive than conventional agriculture, which is capital-intensive. The basic principles of organic agriculture are as follows:

• *Crop rotation* – involves a planned succession of crops on each farm field to achieve weed, disease and insect control, efficient nutrient cycling and good soil structure. By appropriate sequencing of crops and management of residues, the populations of natural predators of pest organisms can increase, allowing pesticide use to be scaled back. This concept is the basis of integrated pest management (IPM).

• *Use of organic manures and crop residues* – by maximizing the return of manures and residues to a soil, nutrient cycling is improved and soil organic matter maintained or increased. The latter effect is beneficial for soil structure, increases water infiltration and storage, and improves soil tilth (Section 2.2). Good tilth promotes seed germination and seedling emergence. Organic matter encourages soil micro-organisms and soil fauna, especially earthworms.

• *Use of green manures* – the benefits of green manures are outlined in Section 11.1. Green manures are especially important in organic farming not only for the return of residues, but also for the incorporation of legumes that contribute N_2 fixed from the air (Section 10.2). In addition to N inputs from legumes, P usually needs to be supplied from external sources, preferably as unprocessed reactive phosphate rock, and S as elemental S (Sections 12.3 and 12.4).

• *Biodiversity* – through a diversity of crops and livestock, farm income can be buffered against market, climatic and biological risks, in contrast to crop monoculture (e.g. cotton) or single enterprise animal farming (e.g. sheep for wool) where income is particularly vulnerable to fluctuations in commodity prices, extremes of weather and the incidence of pests and disease.

stresses on the resource base (soil, plant and water) and the steady increase in food production per head shown in Fig. 11.2 cannot be expected to continue indefinitely. This problem is more acute in some regions than others, especially in Central Africa, Central America and parts of South and East Asia where the population growth rate is high and food production is not keeping pace. With continued growth in the world population, and most of this growth occurring in poorer developing countries, a more sustainable agriculture based on the widespread adoption of low-input, organic agriculture is unlikely to be feasible.

Others such as Rovira (1990) believe it would be a tragedy for agriculture if the term 'sustainable' became synonymous in people's minds with organic farming only. These scientists argue that conventional agriculture, based on mineral fertilizers and the judicious use of chemicals, is sustainable and in many cases improving soil fertility, e.g. in European Union countries. This generalization is examined in more detail in Section 15.3 with reference to Australian agriculture.

Sustainable societies

Although agriculture is a very important land use (and soil is essential to agriculture), agricultural sustainability is only one component of the larger issue of sustainable landscapes, which in turn are a subset of sustainable societies and the sustainable use of global resources. Worldwide concern over the last-mentioned issue led the United Nations (UN) General Assembly in 1983 to ask the World Commission on Environment and Development (WCED) to 'propose long-term strategies to achieve sustainable development, combining global economic and social progress with respect for natural systems and environmental quality'.

In proposing strategies for sustainable development, the WCED Report (1987) illustrates an inherent paradox in the concept of sustainability, namely:

• People in the many diverse countries of the world have basic *needs* and a legitimate expectancy that those needs will be met; but

• technology and society impose *limitations* on the ability of the environment to meet those needs. Thus, continued growth in the world population and the rise in expectations of improved living standards will inevitably increase the demands placed on the Earth's natural resources. Techno-logical change, coupled with economic and social adjustments, can help to meet these demands, but this inevitably places further strains on the Earth's resources and quality of the environment. The response to this paradox is illustrated by differing definitions of what constitutes a 'sustainable society', as the following examples show:

1 The goal of sustainability must be to revive growth, but to make economic growth less energy-intensive. The essential needs of an expanding population in the developing world must be met, but the population must reach a sustainable and stabilized level. The resource base must be conserved and enhanced, technology must be re-orientated, and environmental and economic concerns must be merged in decision-making (Lebel and Kane, 1990);

2 a sustainable society must keep population within the capacity of the resources needed to produce an acceptable quality of life; pollution must not exceed the capacity of the ecosystem to absorb its impacts; and consumption per capita should be minimized by maximizing the recycling of nutrients, minerals and other materials (Roberts, 1992);

3 a sustainable society, with agriculture at its core, must address the human, cultural and ecological dimensions, and include wise man-agement (conservation) of natural resources and restoration of ecological systems (Miller and Wali, 1995).

Future landscapes – an Australian approach

In 1993, following the international trend, the Australian Government adopted a National Strategy for Ecologically Sustainable Development (NSESD) that set the objectives for natural resource management policy in the 1990s. How-ever, while academics and policy-makers debated the definition of 'sustainability', achievement of significant change in the actual management of the Australian landscape was slow and patchy. Programs such as the national Sustainable Grazing Systems (SGS) programme, involving both researchers and producers of red meat and wool, were successful in persuading a majority of par-ticipants (8000 out of nearly 10,000 producers in

the > 600 mm rainfall zone of southern Australia) to make changes that improved the profitability and sustainability of their enterprises (Allan *et al.*, 2003). Other 'on-ground' programmes for improv-ing land and water management, funded through Landcare and the Natural Heritage Trust, have had a small and variable impact relative to the overall investment.

During the 1990s, people wanting to see real change became frustrated with the rhetoric of 'sustainability' and turned to more radical strat-egies for improving the management of Australian landscapes. In 1996, CSIRO and Land and Water Australia launched a new programme called 'Re-designing Agriculture for Australian Landscapes' which asked the basic question 'Can we design agricultural farming systems which mimic natural systems?' (Clarke, 2000). The conceptual change from a focus on land management (primarily agricultural) to landscape management (a holistic approach to the ecosystem of which agriculture is a part) was consistent with the ecosystems ser-vice concept mentioned in Section 15.1. However, the vision of 'future landscapes' outlined so far by Hajkowicz *et al.* (2003) appears to be more agro-centric than holistic-ecosystem in approach, given the five major factors identified as prompt-ing the need for landscape change (Table 15.2). Hopefully, the concept will evolve so that the whole ecosystem is considered, including urban and peri-urban areas where most of the Australian population lives and adverse impacts on the environment are widespread.

Is true sustainability achievable?

When land use is driven by the profit motive there will inevitably be competition between segments of society and conflict in objectives, because each act of environmental modification can have unforeseen consequences that affect some other part of the ecosystem. In such circumstances, a truly sustainable society is not achievable because society relies on as-yet undiscovered scientific/technological advances to solve the problems created by the previous advance (Falvey, 2003). The history of land use in southern Australia since the mid-19th century, in the Murray-Darling Basin (MDB) in particular, is a classic example of this phenomenon.

Table 15.2 Major factors prompting the need for landscape change in Australia. After Hajkowisc *et al.*, 2003.

Causal factor	Summary evidence
Loss in soil productivity	Problems such as salinity, acidity, acid sulphate soils, sodic soils, soil compaction, erosion and soil contamination have worsened during the 20th century and some are likely to become more serious.
Deterioration in water quality	Increasing salinity, nutrient loads and turbidity in rivers, eutrophication and algal blooms, mainly caused by current and historic land management practices.
Biodiversity loss and risk	Agriculture is the major cause of 78 species extinctions and is placing a further 105 species under present or future threat.
Farm income decline	Farm incomes in some rural regions and for some industries have not kept pace with other parts of the country. Increasing emphasis on irrigation (covering only 0.5% of the agricultural area but generating 50% of the net economic return) has exacerbated problems of water supply, salinity and eutrophication.
Rural population decline	The proportion of Australians living in rural areas has dropped from 37% in 1921 to 15% in 1986, with people moving to urban areas and important services being withdrawn from rural areas.

Some 100–200 years ago, settlers were encouraged by governments to clear land, plant crops and export the products. As well as land clearing, irrigation was especially encouraged in northern Victoria and southern New South Wales. However, as early as the 1920s increases in dryland salinization (Box 13.1) were observed, which were subsequently attributed to changes in the regional hydrology due to the large-scale clearance of native forests and scrub. Irrigation also contributed to rising watertables where it was practised. Recognizing this problem of dryland salinity, government agencies and environmental groups in the 1980s and 1990s advocated extensive tree planting, and large-scale modelling by CSIRO suggested that to reverse the trends in salinity, one-third to half of the MDB should be re-planted to trees. However, investors found that with the benefits of tax concessions obtained for plantation forestry, profits could be made provided the trees were planted on fertile soils under high rainfall and reasonably close to a timber or paper mill. So large areas of good agricultural land are being taken out of crops and pasture and put under plantations; but scientists are now advising that this change in land use, especially in the upper parts of important catchments in the MDB, will substantially decrease water yields and exacerbate the problem of insufficient water to

meet the demands of irrigators, towns and the environment (Vertessy *et al.*, 2003). Other examples of how the unforeseen consequences of actions driven by the profit motive have caused repercussions elsewhere in an ecosystem can be found in other parts of Australia and the world.

15.3 Sustainable land management

Sustainable production and land degradation

The physical resource for producing food and fibre, primarily soil and water, has a major influence on the sustainability of an ecosystem or society. There are many examples at local, national and global levels to show the adverse impact of human activity on physical resources. The main causes of degradation are soil erosion, salinization, weed infestation, severe nutrient decline and accelerated soil acidification. The impact of 'slash-and-burn' agriculture in tropical rainforests (c. 500 million ha) on soil erosion is illustrated in Table 11.1. On a global scale, about half the 1500 million hectares of arable land are considered to be degraded, with the attributable causes shown in Table 15.3.

In Australia, a Land and Water Australia report concluded that maintenance of agricultural

Table 15.3 Causes, extent, reversibility and off-site effects of land degradation. After Greenland et al., 1998.

Cause	Area (million ha)	Reversibility	Off-site effects
Water erosion	700	D	I
Wind erosion	280	D	I
Loss of nutrients	135	S	P
Acidification	10*	S	N
Salinization	80	D	P
Pollution	20	D	P
Physical damage	60	D	P

* Less than the area estimated for Australia alone, so figures in this table must be treated with caution. D = difficult, S = simple (provided the necessary inputs are available), I = important, P = possibly important, N = negligible.

productivity and competitiveness had been achieved at considerable cost to the condition of the land, and may have disguised the extent of damage taking place: 'the seriousness of the situation is reinforced by the fact that the effects of past decisions may still take many decades to be fully felt' (Anon., 1995). Subsequently, the National Collaborative Project on Indicators for Sustainable Agriculture, NCPISA (NCPISA, 1998) produced an updated report that concluded *inter alia*:
• While agricultural production was rising in some industries, there were concerns about the sustainability of the resource base;
• although improved management practices such as conservation farming had contributed to a decrease in wind erosion, soil sodicity and acidity were expanding and stream water quality was declining; and
• investment in 'sustainable' practices such as the use of soil conditioners (including lime and gypsum) and appropriate irrigation management was inadequate, contributing to adverse on-site and off-site impacts on soil and water resources. An example of land degradation due to rising saline groundwater is shown in Fig. 13.2, and examples of water and wind erosion are shown in Fig. 15.1a and b.

Compounding the problem of degradation of existing physical resources is the loss of land for agriculture through competition from other uses, e.g. urban and industrial. This problem applies particularly to the densely populated eastern sea-board of China, for example, where rapid industrialization and urban development are occurring on prime land previously used for wheat, maize and rice production. Globally, the reserves of good quality land that can be converted to agriculture are relatively small and the FAO estimates that for every new hectare made arable, one hectare of 'old' land is lost through some form of degradation (Anon., 1993). Certainly, the area harvested per person for grain production in the world decreased from *c.* 0.23 to 0.13 ha from the mid-1950s to the 1990s as a consequence not only of increased yields per ha, but also of land lost by degradation and a shift to other crops.

Thus, the squeeze between the demands of increasingly affluent and expanding populations and our ability to produce from natural resources with an acceptably small environmental impact, has meant that sustainable land management (SLM) is not only a challenge for science and technology but also a political issue. As outlined under 'Future landscapes' above, successive Australian governments have responded to this challenge through the NSESD, which has spawned the National Landcare Program, the Natural Heritage Trust and the National Action Plan (primarily focused on salinity). A prerequisite of any national or international programme of ecologically sustainable development (ESD) is a reliable system for monitoring progress towards improved sustainability over a realistic time period. To this end, international scientists developed a Framework for Sustainable Land Management (FESLM) (Dumanski *et al.*, 1991). FESLM provides a logical pathway to assess the probability of sustainability by connecting the form of land use with a range of environmental, economic and social factors that together determine whether a particular land use is sustainable. Key aspects of FESLM and SLM in general that relate to the soil are:
• Assessment of the soil resource – its properties and their spatial distribution; and
• monitoring the change in the soil resource and/or the effect of management practices on the condition of that resource.
These two aspects of SLM are discussed below.

(a)

(b)

Fig. 15.1 (a) Severe land degradation due to water erosion in the Central Tablelands of New South Wales.
(b) Soil erosion by wind on sandy soils of the Mallee, northwest Victoria. Wind-blown soil accumulates along fences
and tree shelter belts.

Evaluating sustainable land management

Resource assessment

The purpose and methods of collecting soil information by soil survey are discussed in Section 14.2. Remote sensing of soil and land properties and their input to a GIS are briefly described in Box 14.2. These procedures form the core of land resource assessment (LRA) which traditionally has interpreted soil, geology and climatic information in terms of the suitability of land for generalized uses (Land Use Capability). There is still scope for this type of LRA for broad-scale planning purposes, but increasingly more quantitative functions are needed to relate land resource attributes to the yield of specific crops, yield stability, options for alternative land use and environmental impacts, especially on soil and water quality. For example, specific quantitative functions are a prerequisite for 'precision farming' or site-specific soil management (Box 11.4). In the quest for SLM, increasing emphasis is being placed on LRA methods that marry estimates of productivity for specific enterprises (and combinations of enterprises in a landscape) with the probability and magnitude of their environmental impacts. Advanced techniques are now being used to collect data for LRA and in modelling processes at a landscape scale, as outlined in Box 15.2.

Box 15.2 Topographic modelling for LRA.

Topographic data are basic to most land resource databases and can be used to create a digital elevation model (DEM) that functions in a GIS environment. The source data for DEMs are published topographic maps, stereo pairs of aerial photos, or stereo pairs of satellite images, all supplied in digital format. The DEM provides a three-dimensional image of a landscape from which other physical information can be deduced, e.g. slope, aspect, catchment boundaries, the direction of water flow and location of soil wetness zones. Other layers of information – meteorological data, vegetation type, soil properties and land use – can be 'draped' over the DEM and the combined information used to develop a process-based model of soil variation or of ecosystem behaviour in terms of energy, water and nutrient fluxes.

Fig. B15.2.1 is an example of a digital soil map of a small catchment based on soil profile properties and a DEM. ANSWERS-2000 (Bouraoui and Dillaha, 1996) and SWAT (Neitsch et al., 2001) are examples of spatially referenced, process-based models that are used to predict water and nutrient fluxes at a catchment scale. Such GIS coupled process-based models can incorporate economic modules and be adapted for use as Decision Support Systems by land managers.

Order
- Vertosol
- Tenosol
- Rudosol
- Kurosol
- Kandosol
- Hydrosol
- Ferrosol
- Dermosol
- Chromosol
- Anthroposol

Fig. B15.2.1 A map of soil classes for a catchment in the Hunter Valley, Australia, based on a DEM and a biophysical model to predict soil distribution. Courtesy of the Australian Centre for Precision Agriculture, The University of Sydney (see also Plate B15.2.1).

Sustainability indicators

Ecosystems are naturally complex and the complexity of their function is generally made greater by human intervention. In determining whether a particular agro-ecosystem or land use is sustainable or not, we need indicators of the behaviour of the system that are diagnostic of its condition. Ideally, an indicator should be measurable objectively and directly, although surrogate variables may be used to represent the true indicator. To monitor a system effectively, the indicator value must be measured over time. Spatial variability can affect the reliability of the measurements, and there is a need to set realistic threshold and benchmark values for an indicator. These terms are defined as follows:

• A *threshold value* is one below (or sometimes above) which the system changes significantly and is no longer sustainable;
• a *benchmark value* is one which identifies an optimum condition for that particular system.

Unless the unsustainability of a system is very obvious, a single indicator is unlikely to provide a reliable assessment of whether the system is sustainable or not.

Fig. 15.2 shows some examples of biophysical indicators, based on studies of grazed pastures in southeastern Australia (White *et al.*, 2000). The development history of this land is typical of that of much of southeastern Australia, as outlined under **Production and environmental impact** in Section 15.1. The main threats to resource sustainability are:

1 Recharge to the regional groundwater has increased due to an increase in deep drainage relative to that under the original forest vegetation. As the groundwater has risen, the area of land affected by dryland salinity has increased;

2 pH in soil A horizons has decreased by approximately 1 unit (from *c.* 5 to 4 in 0.01 M CaCl$_2$) over 30–50 years, depending on the level of N inputs and losses from the system by NO$_3^-$ leaching and product removal. The profitability of the sheep and beef grazing enterprises is low, and farmers can not afford the cost of applying lime to counteract the accelerated soil acidification;

3 as a consequence of decreased soil pH and poor grazing management, desirable deep-rooted perennial pasture plants, such as phalaris and lucerne

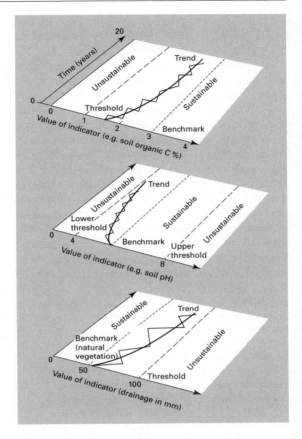

Fig. 15.2 A 'stack' of soil sustainability indicators.

that can decrease deep drainage and produce more productive pastures, have become more difficult to grow.

Fig. 15.2 indicates a 'stack' of potential indicators, each of which has a measurable value and is monitored through time (at least annually). The thresholds and benchmarks must be determined by field experimentation. Although these are well-known for the soil pH indicator, they are not well-known for drainage below the root zone (> 1.5 m for grasses), which affects recharge to groundwater. As Fig. 15.2 shows, there can be large variability from year to year due to climatic variability. In establishing thresholds, benchmarks and trends, measurements should be made over several years (10 or more) to minimize the confusing effects of large annual variations. Such knowledge can then be applied on a farm-by-farm basis, although a composite index of sustainability might need to be developed that takes account of several indicators, and weights them according to their influence on the system's condition.

But what of the larger scale and the sustainability of land use in a large catchment or whole region? An example is given in a recent report (NCPISA, 1998) where the Australian continent was divided into 11 agro-ecological zones and a set of five indicators and associated attributes for assessing the sustainability of agriculture in these zones were identified (Table 15.4). Note that these indicators cover a much broader range of sustainability drivers (biophysical, environmental and socio-economic) than an earlier set focused primarily on the soil resource (Hamblin, 1992). Nevertheless, abstraction of data for the attributes for each indicator at the agro-ecological zone level inevitably means that only broad generalizations

about sustainability trends can be made. The further development of indicator–attribute relationships and improved methods of gathering relevant data are major challenges for future research in SLM.

15.4 Summary

Soil quality is more difficult to define than air or water quality because of the spatial variability of soil and the fact that it is used for a variety of purposes. However, the essential ecosystem functions related to soil quality are the sustaining of biological productivity, maintenance of environmental quality, and promotion of plant and animal health. Quantifying the relationships between one or more of these functions and soil properties is an on-going task and, except in the case of soil contamination where strict guidelines are in force, soil quality is usually inferred from the degree and extent to which management practices protect the soil from degradation. Establishing best management practices (BMPs) requires that the soil-plant system be considered as an integral component of the wider environment, rather than focusing narrowly on production.

Sustainability is an imprecise concept and implies aspects of resource quality and availability, economic viability, environmental impact, social justice and biodiversity. Its spatial and temporal dimensions must be recognized. Conventional agriculture, which relies heavily on external inputs of fertilizers, pesticides and energy, may be less sustainable than non-conventional agriculture (organic or alternative agriculture) relying primarily on nutrient recycling, maintenance of biodiversity and integrated pest management. However, on a global scale, the reduction in yields associated with low-input, alternative agriculture poses a serious impediment to satisfying the legitimate needs of an expanding population, within the environmental-impact constraints imposed by society and technology.

More recently, the sustainability debate has focused on the concept of ecosystem services, whereby ecosystems are evaluated according to the value of services provided, such as clean water and air, energy, food and pharmaceutical products, recreation and conservation. In this concept, the

Table 15.4 Indicators and their attributes selected for assessing the sustainability of Australian agriculture. After NCPISA, 1998.

Indicator	Attributes
Long-term real net farm income	Real net farm income
	Total factor productivity
	Farmers' terms of trade
	Average real net farm income
	Debt servicing ratio
Natural resource condition	Nutrient balance: P and K
	Soil condition: acidity and sodicity
	Rangeland condition and trend
	Agricultural plant species diversity
	Water utilization by vegetation
Off-site environmental impacts	Chemical residues in products
	Salinity in streams
	Dust storm index
	Impact of agriculture on native vegetation
Managerial skills	Level of farmer education
	Extent of participation in training and Landcare groups
	Implementation of sustainable practices
Socio-economic impacts	Age structure of the agricultural workforce
	Access to key services

function of agriculture can be compared with other less well understood but no less important functions provided by an ecosystem. Properly functioning ecosystems are essential for a sustainable society, but one must question whether any society in which land use is driven by the profit motive can be truly sustainable because each act of environmental modification has unforeseen effects, often adverse, on other parts of the ecosystem.

There is much evidence that human activities have degraded natural resources on a national and global scale. Governments and international agencies have responded with strategies to promote ecologically sustainable development (ESD). A prerequisite of any ESD programme is a reliable system of monitoring progress towards achieving more sustainable land management (SLM). Key aspects of SLM are assessment of the soil (land) resource – identifying the relevant properties and their spatial expression, and monitoring changes in the resource or the effect of management on the condition of the resource.

Land resource assessment (LRA) depends heavily on soil survey information and spatially referenced models of landscape form and behaviour. These models dynamically link digital elevation models (DEM) of the landscape with process models of ecosystem function in a GIS. For resource assessment, indicators are measurable variables that are diagnostic of the underlying condition of the resource. Threshold values of the indicators showing when a system becomes unsustainable and benchmarks for good condition must be identified, and the system's performance monitored over time using these values. Surrogate variables are often used as sustainability indicators, especially at the national or regional level, but the relationship between these surrogates and the more fundamental biophysical indicators should be well defined. Incorporating such relationships into Decision Support Systems for SLM requires that social and economic factors, which often determine whether or not individual landholders will change their practices, are taken into account.

References

Allan C. J., Mason W. K., Reeve I. J. & Hooper S. (2003) Evaluation of the impact of SGS on livestock producers and their practices. *Australian Journal of Experimental Agriculture* 43, 1031–40.

Anon. (1993) FAO sounds soil-loss siren. *Science* 261, 423.

Anon. (1995) *Sustaining the Agricultural Resource Base.* Office of the Chief Scientist, Commonwealth of Australia, Canberra.

ANZECC/NHMRC (1992) *Australian and New Zealand Guidelines for the Assessment and Management of Contaminated Sites.* Australian and New Zealand Environment and Conservation Council and National Health and Medical Research Council, Canberra.

Barzi F., Naidu R. & McLaughlin M. J. (1996) Contaminants and the Australian soil environment, in *Contaminants and the Soil Environment in the Australasia-Pacific Region* (Eds R. Naidu *et al.*). Kluwer Academic Publishers, Dordrecht, pp. 451–84.

Bezdicek D. F., Papendick R. I. & Lal R. (1996) Introduction: importance of soil quality to health and sustainable land management, in *Methods for Assessing Soil Quality* (Eds J. W. Doran & A. J. Jones). Special Publication No. 49. Soil Science Society of America, Madison WI, pp. 1–8.

Bouraoui F. & Dillaha T. A. (1996) ANSWERS-2000: runoff and sediment transport model. *Journal of Environmental Engineering* 122, 493–502.

Clarke D. (2000) *Redesigning agriculture for Australian landscapes research and development program: phase two program strategy 2000–2002.* Land and Water Resources Research and Development Corporation, Canberra.

Cochrane W. W. (2003) *The Curse of American Agricultural Abundance. A Sustainable Solution.* University of Nebraska Press, Lincoln, Nebraska.

Daily G. C. (2000) Management objectives for the protection of ecosystem services. *Environmental Science and Policy* 3, 333–9.

Doran J. W. & Parkin T. B. (1994) Defining and assessing soil quality, in *Defining Soil Quality for a Sustainable Environment* (Eds J. W. Doran, D. C. Coleman, D. F. Bezdecik & B. A. Stewart). Special Publication No. 35. Soil Science Society of America, Madison WI, pp. 3–21.

Dumanski J., Eswaran H. & Latham M. (1991) A proposal for an international framework for evaluating sustainable land management, in *Evaluation for Sustainable Land Management in the Developing World* (Eds J. Dumanski *et al.*) Volume 2, Technical Papers. International Board for Soil Research and Management, Bangkok, pp. 24–45.

Falvey L. (2003) Agri-history and sustainable agriculture: a consideration of technology and ancient wisdom. *Asian Agri-History* 4, 279–94.

Greenland D. J., Gregory P. J. & Nye P. H. (1998) Land resources and constraints to crop production, in *Feeding a World Population of More than Eight Billion People* (Eds J. C. Waterlow, D. G. Armstrong, L. Fowden & R. Riley). Oxford University Press, New York, pp. 39–55.

Hajkowicz S., Hatton T., McColl J., Meyer W. & Young M. (2003) *Futures. Exploring Future Landscapes: a Conceptual Framework for Planned Change*. Land and Water Australia, Canberra.

Hamblin A. (Ed.) (1992) *Environmental Indicators for Sustainable Agriculture*. Report of a National Workshop. Bureau of Rural Resources, Canberra.

Lebel G. G. & Kane H. (1990) *Sustainable Development. A Guide to our Common Future*. Oxford University Press, Oxford.

Miller F. P. & Wali M. K. (1995) Soils, land use and sustainable agriculture: a review. *Canadian Journal of Soil Science* 75, 413–22.

National Collaborative Project on Indicators for Sustainable Agriculture (NCPISA)(1998) *Sustainable Agriculture. Assessing Australia's Recent Performance*. SCARM Technical Report 70. CSIRO Publishing, Melbourne.

Neitsch S. L., Arnold J. G., Kinery J. R. & Williams J. R. (2001) *Soil Water Assessment Tool User's Manual Version 2000*.

Reganold J. P., Papendick R. I. & Parr J. F. (1990) Sustainable agriculture. *Scientific American* 262 (6), 112–20.

Roberts B. (1992) An international perspective on sustainable land management, in *Sustainable Land Management* (Ed. P. Henriques). Hawke's Bay Regional Council, Napier, New Zealand, pp. 10–17.

Rovira A. (1990) Biological – organic – biodynamic or conventional – which is it? *Soil News* No. 82. Australian Society of Soil Science Inc, pp. 1–2.

Smyth A. J. & Dumanski J. (1993) *FESLM: an International Framework for Evaluating Sustainable Land Management*. World Soil Resources Report 73. FAO, Rome.

Vertessy R. A., Zhang L. & Dawes W. R. (2003) Plantations, river flows and river salinity. *Australian Forestry* 66, 51–61.

Viederman S. (1993) A dream of sustainability. *Renewable Resources Journal* 11, 14–15.

White R. E., Helyar K. R., Ridley A. M., Chen D., Heng L. K., Evans J., Fisher R., Hirth J., Mele P. M., Morrison G. R., Cresswell H. P., Paydar Z., Dunin F. X., Dove H. & Simpson R. J. (2000) Soil factors influencing the sustainability and productivity of perennial and annual pastures in the high rainfall zone of southeastern Australia. *Australian Journal of Experimental Agriculture* 40, 267–83.

World Commission on Environment and Development (WCED) (1987) *Our Common Future*. Oxford University Press, Oxford.

Further reading

Barr N. & Cary J. (1992) *Greening a Brown Land. The Australian Search for Sustainable Land Use*. Macmillan, Melbourne.

Roberts B. (1995) *The Quest for Sustainable Agriculture and Land Use*. University of New South Wales Press, Sydney.

Sposito G. & Zabel A. (Eds) (2003) The assessment of soil quality. *Geoderma* 114, Nos. 3–4.

Example questions and problems

1 (a) What is the single most important factor affecting soil productivity?
  (b) Distinguish between a benchmark and threshold value of indicators for soil quality.
  (c) According to ANZECC guidelines (1992), what is the threshold concentration for soil Pb that warrants environmental investigation?

2 (a) Name three of the dominant land uses in rural areas.
  (b) Name a common management practice for control of (i) water erosion on arable land,

     (ii) wind erosion on arable land, and
     (iii) waterlogging in orchards.

3 (a) Name three on-ground practices used in alternative or organic agriculture.
  (b) Other than organic sources, name the forms acceptable for organic agriculture of (i) phosphorus, and (ii) sulphur.
  (c) What is the approach to pest management in organic agriculture?

4 (a) Give one of the natural resource factors prompting the need for landscape change in Australia.

(b) With reference to the Murray-Darling Basin in Australia, what is one of the main land degradation processes currently occurring (over large areas)?

(c) On a global scale, what is the main cause of land degradation?

5 (a) What is (i) ESD, and (ii) a DEM?

(b) What combination of modelling tools offers the best option for the development of BMPs for catchments?

6 Consider a legume-based pasture fixing an average of 100 kg N/ha/yr and a grass pasture on the same soil type fertilized with urea at the rate of 100 g N/ha/yr. Of the N mineralized to NO_3^-, the equivalent of 10 and 20% of the input N is lost by leaching under the legume and N-fertilized pasture, respectively. Removal of N in animal products, and loss of NO_3^- by denitrification (2% of N input), are the same for both pastures. (Assume a reduction of NO_3^- to N_2 gas – Section 8.4.)

(a) Calculate the net H^+ gain or loss (kmol/ha/yr) for these two pastures.

(b) With respect to pH change, which system is the more sustainable?

Answers to questions and problems

Chapter 1

1. The residues of plants growing in the soil.
2. (a) O, A, B and C horizons, sequentially from the surface litter layer down
   (b) 'eluvial' means material has been washed out; 'illuvial' means material has been washed into the horizon.
3. Climate, parent material, relief (topography), organisms and time.
4. (i) Transect 1 mean = 1.97%, CV = 32%; transect 2 mean = 1.92%, CV = 33%;
   (ii) variation in soil drainage, pasture growth and return of cattle dung.
5. (a) 0.475, called the porosity;
   (b) (i) 1.355 t (ii) 6.775 g;
   (c) 2032 Mg;
   (d) (i) 1.615 Mg (ii) 2422.5 Mg.

Chapter 2

1. (a) (i) A plane of atoms consists of a single thickness of covalently bonded atoms (ii) a crystal layer is made up of two or more atomic sheets (covalently bonded), stacked in the Z direction;
   (b) (i) 0.72 nm (ii) the crystals do not swell in water because of strong hydrogen bonding between layers;
   (c) no. The CEC of kaolinite is smaller because there is much less isomorphous substitution in kaolinite crystals.
2. (a) (i) A variable charge due to H^+ association or dissociation at surface OH groups (ii) a permanent negative charge due to Al^{3+} substitution for Si^{4+} in the tetrahedral sheet;
   (b) variable or pH-dependent charge.
3. (a) (i) 4.63 min. (ii) 7.73 h;
   (b) (i) 40.1% (ii) 19.9%.
4. Upper limit = 0.113 m^2/g and lower limit = 1.13 m^2/g.
5. (a) CEC = 100 cmols charge/kg;
   (b) 0.5 moles Ca^{2+};
   (c) 5×10^{-4} cmols charge/m^2.
6. (a) 8.5 cmols charge (+)/kg;
   (b) 4.25 cmols charge (+)/kg;
   (c) net CEC would increase (variable positive charge decreases and negative charge increases).

Chapter 3

1. (a) Because plant residues and animal dung fall directly on the soil surface;
   (b) turnover is the flux of C through the soil organic C per unit volume of soil;
   (c) 26.7 yr.
2. (a) Obligate anaerobes;
   (b) (i) saprophytes (ii) parasites (iii) mycorrhizas;
   (c) (i) earthworms (ii) termites.
3. 130 cmol charge (+)/kg.
4. 533 mg/kg.
5. (a) 90;
   (b) 0.11 (11%);
   (c) from the soil mineral N;
   (d) the growth yield.
6. (a) 0.905 t C/ha;
   (b) 40 kg C/ha;

(c) 4 kg N/ha;

(d) N potentially mineralized = 4.95 kg/ha, but 11.25 kg N/ha is required by the micro-organisms for growth – therefore no net mineralization of N is expected.

Chapter 4

1. Through the secretion of gums and mucilage that act as cementing agents and through the binding action of root hairs and fine roots.

2. (a) Positively charged amino groups attracted to negatively charged clay surfaces; clay–polyvalent cation–organic cation interactions ('cation bridges'), and high molar mass organic polymers interacting with mineral particles through H-bonding and van der Waals' forces;

   (b) sesquioxides (Fe and Al oxides), $CaCO_3$ and amorphous SiO_2.

3. (a)

Core sample	Mass of o.d. soil (g)	Soil volume (cm³)	Bulk density	Porosity
A	380	331.8	1.145	0.568
B	419	331.8	1.263	0.524
C	390	331.8	1.175	0.556
D	432	331.8	1.302	0.509

Mean bulk density = 1.221 ± 0.037 Mg/m³; mean porosity = 0.539 ± 0.014 m³/m³;

   (b) 1832 t;
   (c) 0.539 (54%).

4. (a) 0.33 g/g (33%);
   (b) 0.469 m³/m³.

5. 336 mm.

6. (a) A horizon = 15.7% and B horizon = 6%;
   (b) 67.4 mm.

7. (a) 240 Mg;
   (b) 10 yr;
   (c) no, because of the high bulk density and compaction.

Chapter 5

1. (a) Toposequence;
   (b) lithosequence;
   (c) catena.

2. (a) Zonal concept;
   (b) the 1938 USDA soil classification based on Great Soil Groups;
   (c) when a soil is young, e.g. on recent alluvium or volcanic material.

3. Intrusive rocks result from magma forced upwards into existing rock strata and solidifying; extrusive rocks form by solidification of surface lava flows.

4. Olivine has no covalently linked silica tetrahedral; in quartz all the O atoms of each silica tetrahedron are bonded to O atoms in adjacent tetrahedra.

5. (a) Eroded material deposited by a river;
   (b) fragmented rock material that is pulled down slopes by gravity;
   (c) fragmented rock material deposited by retreating ice sheets or glaciers.

6. (a) A deep wind-blown deposit, usually of periglacial origin;
   (b) soil slippage downslope over a frozen subsoil;
   (c) the bouncing of wind-blown particles over a soil surface (also occurs in stream beds under strongly flowing water).

7. (a) Steady state occurs when the rate of soil loss equals the rate of soil formation;
   (b) 2996 yr.

Chapter 6

1. (a) Matric potential ψ_m and gravitational potential ψ_g;
   (b) −30,179 kPa.

2. (a)

Depth (cm)	10	30	50	70	90
Tensiometer potential (kPa)	−60	−34	−16	−5	−8
Gravitational potential (kPa)	−1	−3	−5	−7	−9
Matric potential (kPa)	−59	−31	−11	+2	+1

   (b) 70 cm;
   (c) upwards to the surface and downwards as drainage;
   (d) evaporation of soil water.

3. (a) 6 mm;
   (b) 2 mm/h;
   (c) no.

4. (a) 140 mm/m depth;
   (b) 84 mm;
   (c) 42 mm.

5.

Time (week)	1	2	3	4	5	6	7
Rainfall (mm)	0	5	0	8	0	2	0
E_{to} (mm)	28	24	35	38	42	40	48
C_c	0.25	0.40	0.52	0.65	0.78	0.90	1.02
Cumulative SWD (mm)	−7	−11.6	−29.8	−46.5	−32.8	−66.8	−49
Irrigate? (y/n)	n	n	n	y	n	y	y

6. (a) 1;
   (b) 50 mm/h;
   (c) 30 mm/h;
   (d) 90 mm.
7. (a) 5 mm;
   (b) 10 mm;
   (c) 0.04 m^3/m^3.

Chapter 7

1. (a) (i) 1 : 1 (ii) 2 : 1;
   (b) no, because there is very little isomorphous substitution in kaolinite;
   (c) (i) K^+ fits into the hexagonal holes between adjacent crystal layers (ii) it is replaced by other cations;
   (d) NH_4^+.
2. (a) Permanent charge occurs within the crystal due to isomorphous substitution, whereas variable charge arises due to the association or dissociation of H^+ ions at surface O and OH groups;
   (b) the charge changes from positive to net zero to negative;
   (c) the tendency to deflocculate is increased.
3. (a)

Cation	Leachate concn (mg/L)	mmol of cation in 100 mL of leachate	mmol of cation charge per g of clay	Exchangeable cation (cmol charge (+)/ kg clay)
Ca^{2+}	1600	4	0.8	80
Mg^{2+}	192	0.8	0.16	16
K^+	156	0.4	0.04	4
Na^+	69	0.3	0.03	3

   (b) 103 cmol charge(+)/kg.
4. (a) $Al(H_2O)_6^{3+} + H_2O \leftrightarrow AlOH(H_2O)_5^{2+} + H_3O^+$;

   (b) at pH < 4 neutralization of H^+, pH 4–5.5 neutralization of Al^{3+}, and pH 5.5–7.5 neutralization of hydroxyaluminium ions;
   (c) 150 mL.
5. (a) 1 : 5 ratio;
   (b) $pH_w > pH_{Ca}$ by 0.6–0.8 units;
   (c) 0.03.
6. (a) 3.2 mmol P;
   (b) 19.84 cmols charge/kg.
7. (a) (i) 1.5 nm (ii) 1.9 nm;
   (b) Ca^{2+} − clay + $2Na^+ \leftrightarrow Na_2^+$ − clay + Ca^{2+};
   (c) it would increase to between 1.9 and 4 nm.
8. (a) Deflocculate, due to the expansion of diffuse double layers around the crystals;
   (b) flocculate because, following exchange of Ca^{2+} for adsorbed Na^+, the diffuse double layers should contract.

Chapter 8

1. (a) Diffusion;
   (b) (i) 0.1–1.5% (ii) 0.1–1.5 kPa;
   (c) CO_2.
2. (a) 1.21 L/m^2/day;
   (b) 1.2.
3. (a) Air-filled porosity and tortuosity;
   (b) provides a network of large continuous pores between aggregates;
   (c) 0.1 m^3 air/m^3 soil.
4. (a) 0.018 cm^2/s;
   (b) 0.0023 mL/mL.
5. 1.49 kPa.
6. (a) Autotrophs;
   (b) NH_4^+;
   (c) nitrapyrin and acetylene.
7. (a) Facultative anaerobes;
   (b) $2NO_3^- + 12H^+ + 10e^- \leftrightarrow N_2 + 6H_2O$;
   (c) pH should increase;
   (d) NO_3^-; MnO_2; $Fe(OH)_3$ and SO_4^{2-}.

Chapter 9

1. (a) Soil horizons form by pedogenesis in a single parent material, soil layers represent different parent materials;

(b) an horizon formed at the soil surface, recognized by a change in structure, darkening by organic matter or eluviation of material;

(c) > 40 cm deep in a soil profile ≥ 80 cm, or > 30 cm if on bedrock.

2. Any four of leaching, cheluviation, lessivage, bioturbation, oxidation-reduction (mottling and gleying), rain wash, creep and wind deposition.

3. (a) (i) $FeS_2 + 7/2O_2 + H_2O \leftrightarrow Fe^{2+} + 2SO_4^{2-} + 2H^+$ and $Fe^{2+} + 1/4O_2 + 5/2H_2O \leftrightarrow Fe(OH)_3 + 2H^+$; (ii) pH decreases;

(b) 15.6 kg.

4. (a) A sharp increase in texture between the A and B horizons;

(b) (i) Sodosol, Chromosol or Kurosol (ii) Alfisol or Ultisol.

5. (a) 35%;

(b) montmorillonite (smectite).

6. (a) Vegetation growing in a cool to cold climate with P > E for most of the year;

(b) (i) an ironstone layer formed under seasonal waterlogging in a Groundwater laterite (ii) it hardens.

7. (a) From the sea;

(b) $Na^+ - clay + H_2O \leftrightarrow H^+ - clay + NaOH$;

(c) 10–10.5.

Chapter 10

1. (a) Organic, inorganic and biomass stores;

(b) fertilizers, weathering rock and the atmosphere;

(c) in rain and by dry deposition;

(d) 1.25 kmols H^+/ha.

2. (a) P that is immediately available to plants (easily desorbed P and inorganic P in solution);

(b) soil inorganic P becoming less soluble with time;

(c) ectomycorrhizas and arbuscular mycorrhizas;

(d) a cylinder of low nutrient concentration created around an absorbing root;

(e) diffusion.

3. (a) 0.546 t C/ha;

(b) 36 kg N/ha;

(c) 29 kg N/ha.

4. (a) 100 kg/ha;

(b) two equal applications, one drilled with the seed, the other broadcast at early tillering.

5. (a) $(NH_2)_2CO + 2H_2O \rightarrow (NH_4)_2CO_3$ and $(NH_4)_2CO_3 + 2H_2O \rightarrow 2NH_4OH + H_2CO_3$;

(b) NH_4^+ and NO_3^-;

(c) NO_3^-;

(d) 40 mm.

6. (a) 17.25 kg NH_4-N/ha;

(b) 17.25 kg NO_3-N/ha;

(c) no net H^+ or OH^- produced.

7. (a) Mn^{2+};

(b) Cu^{2+};

(c) $H_3BO_3^-$;

(d) MoO_4^{2-}.

Chapter 11

1. (a) Farmyard manure (and other animal manures), compost, biosolids (sewage sludge);

(b) green crop that is easily decomposed and has a low C : N ratio;

(c) a large increase in per capita food production in developing countries of South and East Asia from 1950 to 1990.

2. (a) Plant analysis (tissue testing) and interpreting visual symptoms;

(b) the element concentration in a plant below which an increase in supply increases yield;

(c) the yield when profit per ha is a maximum.

3. (a)

Soil	Mean pH (water)	Mean pH (CaCl₂)
Dermosol	6.2	5.4–5.6
Podosol	5.1–5.3	4.5

(b) (i) Podosol; (ii) Al^{3+} toxicity;

(c) 1600 kg.

4. (a) 22.2 yr;

(b) 3.92 t/ha.

5. (a) The Chromosol is deficient because its Olsen P test < the critical value;

(b) 667 and 133 kg superphosphate for phase A and B, respectively.

6. (a) Yes;
   (b) their burrowing action creates large, continuous and stable pores (biopores);
   (c) 30%.
7. (a) Raindrop splash;
   (b) 0.02–0.06 mm diameter;
   (c) saltation.
8. (a) Yes;
   (b) 0.12 t/ha/yr.

Chapter 12

1. 7.5, 3.9 and 10 kg/ha of N, P and K, respectively.
2. (a) Slow-release (organic) fertilizers and controlled-release fertilizers;
   (b) use a nitrification inhibitor;
   (c) leaching and denitrification.
3. (a) (i) 10 mg NO_3-N/L;
       (ii) 11.3 mg NO_3-N/L;
   (b) 0.05 mg P/L.
4. (a) Monoammonium phosphate (MAP) and diammonium phosphate (DAP);
   (b) rock phosphate;
   (c) blood and bone, or bone meal.
5. (a) KCl;
   (b) kainit;
   (c) sulphate (SO_4^{2-});
   (d) atmospheric deposition.
6. (a) 59 kg;
   (b) RPR;
   (c) no.
7. (a) Adsorption and photolysis (by sunlight);
   (b) volatilization;
   (c) culturing micro-organisms that can use the pesticide as a substrate for growth.
8. (a) Insecticide B;
   (b) the time taken for the pesticide concentration to decrease by 50%;
   (c) 17.7 g/ha.

Chapter 13

1. (a) The electrical conductivity (EC) of a saturation soil extract;
   (b) (i) $EC_{1.5}$ is the EC of an extract from soil mixed with deionized water in the ratio 1 g soil to 5 mL water, and (ii) EC_e is 5–13 times greater than $EC_{1.5}$, depending on soil texture;
   (c) a soil for which $EC_e \geq 4$ dS/m.
2. Because the ESP is 15%, the soil is sodic and therefore requires gypsum.
3. (a) 35 mm;
   (b) 7 days;
   (c) 0.10 m^3/m^3 (10%), which is just adequate.
4. (a) 0.1;
   (b) 400 mm;
   (c) 40 ML;
   (d) 0.2;
   (e) LF is twice LR, which means the efficiency of water use is poor.
5. (a) 10.6 mmols charge$^{1/2}/L^{1/2}$;
   (b) permeability should be satisfactory;
   (c) permeability should decrease.
6. (a) 1.12 kg;
   (b) 1244 kg/ha;
   (c) 900 kg/ha;
   (d) 344 kg/ha;
   (e) 210 mm.
7. (a) 5.445 GL;
   (b) 259.4 GL;
   (c) 21%;
   (d) seepage.

Chapter 14

1. (a) An EM 38 for bulk soil salinity;
   (b) radar and radiowave reflectance, aerial photographs and satellite imagery over a range of wavelengths;
   (c) by ground-truthing.
2. (a) A classification and a map;
   (b) 1 : 100,000;
   (c) 25;
   (d) a small scale;
   (e) the variance.
3. (a) A list of classes;
   (b) no, because the within-class variance is larger than the between-class variance;
   (c) plotting the semi-variance against lag – a variogram;
   (d) kriging.
4. Broad-scale planning for land use according to land capability.

5. (a) Soil Taxonomy and the World Reference Base for Soil Resources;
   (b) Soil Order;
   (c) Sodosols, Kurosols and Chromosols;
   (d) Ferrosol;
   (e) Oxisol.
6. (a) 'attribute' and 'spatial';
   (b) a GIS.

Chapter 15

1. (a) Soil fertility;
   (b) a 'benchmark' value is the target value for good management, a 'threshold' value should not be exceeded (or fallen below);
   (c) 300 mg/kg soil.
2. (a) Agriculture, forestry and conservation;
   (b) (i) contour banks, strip cropping or terraces (ii) conservation tillage (stubble mulching) and windbreaks (iii) surface and subsurface (pipe) drainage.
3. (a) Crop rotation, organic manures, green manure, return of crop residues;
   (b) (i) unprocessed phosphate rock (ii) elemental S;
   (c) integrated pest management.
4. (a) Loss in soil productivity, decline in water quality, biodiversity loss, farm income and rural population decline;
   (b) dryland salinization;
   (c) water and wind erosion.
5. (a) (i) Ecologically sustainable development (ii) digital elevation model;
   (b) spatially referenced process models of energy, water and nutrients incorporated into a Decision Support System.
6. (a) Gain of 0.71 and zero kmol H^+/ha/yr for grass and legume-based pasture, respectively;
   (b) the legume-based pasture.

Index